A Mathematical Introduction to Data Science

Yi Sun • Rod Adams

A Mathematical Introduction to Data Science

 Springer

Yi Sun ⓘ
Department of Computer Science
University of Hertfordshire
Hertfordshire, UK

Rod Adams
Department of Computer Science
University of Hertfordshire
Hertfordshire, UK

ISBN 978-981-96-5638-7   ISBN 978-981-96-5639-4   (eBook)
https://doi.org/10.1007/978-981-96-5639-4

© The Editor(s) (if applicable) and The Author(s), under exclusive license to Springer Nature Singapore Pte Ltd. 2025

This work is subject to copyright. All rights are solely and exclusively licensed by the Publisher, whether the whole or part of the material is concerned, specifically the rights of translation, reprinting, reuse of illustrations, recitation, broadcasting, reproduction on microfilms or in any other physical way, and transmission or information storage and retrieval, electronic adaptation, computer software, or by similar or dissimilar methodology now known or hereafter developed.
The use of general descriptive names, registered names, trademarks, service marks, etc. in this publication does not imply, even in the absence of a specific statement, that such names are exempt from the relevant protective laws and regulations and therefore free for general use.
The publisher, the authors and the editors are safe to assume that the advice and information in this book are believed to be true and accurate at the date of publication. Neither the publisher nor the authors or the editors give a warranty, expressed or implied, with respect to the material contained herein or for any errors or omissions that may have been made. The publisher remains neutral with regard to jurisdictional claims in published maps and institutional affiliations.

This Springer imprint is published by the registered company Springer Nature Singapore Pte Ltd.
The registered company address is: 152 Beach Road, #21-01/04 Gateway East, Singapore 189721, Singapore

If disposing of this product, please recycle the paper.

Preface

This book is written primarily as a text for a one-semester Data Science and Analytics course: Foundations of Data Science. We hope the book will also introduce this area to people who are not students but have some mathematical knowledge and a willingness to learn more. The reader is assumed proficient in handling numbers in various formats, including fractions, decimals, percentages, and surds. They should also have a knowledge of introductory algebra, such as manipulating simple algebraic expressions, solving simple equations, and graphing elementary functions, along with a basic understanding of geometry including angles, trigonometry, and Pythagoras' theorem. This book introduces the reader to the fundamental mathematical and statistical expertise required to understand the principles of many algorithms used in Data Science.

As with all mathematical textbooks, the worked examples are very important, and the exercises for you, the reader, are even more important. You cannot really understand mathematics without seeing and doing examples yourself. By doing examples, you have to keep looking back to find relevant equations, pieces of text, or worked examples that will allow you to complete the example. This is the way you learn mathematics.

A note on numerical answers in this book. You may not get exactly the same answer as we have. We have often used Python to do our calculations and it will probably be working with more decimal places than you might be using, so do not worry if your answers are slightly different. As a rule of thumb to get a result correct to two decimal places you need to work with at least three decimal places. Brief solutions to all exercises are given at the end of the book. Fuller solutions can be found by following this link: sn.pub/5m5zwx.

Chapter 1 presents the general procedures of Data Science, summarises three case studies used throughout the book, and introduces data types.

Chapter 2 provides the knowledge of basic set theory and functions to set up the foundation for later chapters.

Chapter 3 covers the linear algebra knowledge (vectors and matrices) used in the subsequent chapters.

Chapter 4 focuses on two widely used algorithms in Data Science, Principal Component Analysis (PCA) and Singular Value Decomposition, and shows how these two algorithms work.

Chapters 5 and 6 introduce the basic knowledge of calculus (differentiation and integration) and the main optimisation ideas for finding the minimum value of an objective function.

Chapters 7, 8, and 9 reveal principles behind three methods: Principal Component Analysis, Simple Linear Regression, and training simple artificial Neural Networks using knowledge built up in the proceeding chapters.

Chapters 10, 11, and 12 introduce basic knowledge of probability and statistics. These topics underpin lots of scientific disciplines that deal with vast amounts of data, by considering the probability distributions associated with the data and our confidence in our analysis. In particular, it builds the foundations to extend the material on the linear regression algorithm of Chap. 8.

Chapter 13 revisits the linear regression model of Chap. 8 under a probability and statistical framework. Specifically, the chapter presents the method of Maximum Likelihood Estimation.

Chapter 14 discusses some important issues surrounding data analysis which motivates the introduction of two final algorithms that can improve model generalisation, namely Ridge Regression and early stopping.

The overall structure of the book can be divided as follows:

Part 1 Introduction: Chapter 1
Part 2 Mathematical Knowledge (I): Chapters 2–6
Part 3 Algorithms (I): Chapters 7–9
Part 4 Mathematical Knowledge (II): Chapters 10–12
Part 5 Algorithms (II): Chapter 13
Part 6 Conclusion: Chapter 14

By the end of this textbook, you have met many mathematical concepts and techniques in linear algebra, calculus, probability and statistics. Also, several Data Science algorithms, with and without enhancements, have been introduced and illustrated. These include Principal Component Analysis, Singular Value Decomposition, Linear Regression in two and more dimensions, Simple Neural Networks, Maximum Likelihood Estimation, Logistic Regression, and Ridge Regression.

For any comments and questions, please send an email to mathsfds2025@gmail.com.

Hertfordshire, UK Yi Sun
January 2025 Rod Adams

Acknowledgements The book was originally based on the teaching materials we developed for the Foundations of Data Science module in the Department of Computer Science at the University of Hertfordshire. However, we have significantly expanded its content, the details of the techniques and algorithms, and the number of exercises and examples. We want to express our gratitude for the valuable feedback and suggestions from our students, colleagues, and all tutors in the module team. We are especially grateful for the insights provided by our reviewers. We also thank our editor, Nick Zhu, who has been very supportive throughout the process, and the entire Springer Nature publishing team for their excellence, in particular, for their help with final proof reading and editing and putting the book into the correct format for publication.

In addition, over the years, we have absorbed many teaching ideas and useful bits of information derived from a wide spectrum of materials, including traditional textbooks and digital resources available on the internet. It is important to acknowledge that these elements have become so ingrained in our collective expertise that we cannot specifically acknowledge these since we can no longer determine their origin. Although we have conscientiously referenced sources from which we have drawn in the References section and which also serve as the primary repositories for further exploration, we recognise that inadvertent omissions may occur. Therefore, we apologise for any such oversights and express our gratitude for the vast body of knowledge that has contributed to our understanding and teaching practices.

Contents

1 Introduction 1
   1.1   The Procedures of Data Science 2
   1.2   Supervised Learning and Unsupervised Learning 3
        1.2.1   Supervised Learning 3
        1.2.2   Unsupervised Learning 5
   1.3   Case Studies 7
        1.3.1   Case Study 1: Potential Enhancement Ratio Prediction Using Linear Regression 7
        1.3.2   Case Study 2: Data Visualisation Using Principal Component Analysis 10
        1.3.3   Case Study 3: A Simple Two-Layer Neural Network 14
   1.4   Types of Data 17
        1.4.1   Organised (Structured) and Unorganised (Unstructured) Data 17
        1.4.2   Quantitative and Qualitative 18
        1.4.3   The Four Levels of Measurement 19

2 Sets and Functions 23
   2.1   Sets 23
        2.1.1   Sets and Subsets 23
        2.1.2   Venn Diagrams 26
        2.1.3   Basic Set Operations 26
        2.1.4   Sets Written in Comprehension 30
   2.2   Binary Relations 32
        2.2.1   Binary Relations 33
   2.3   Functions 34
        2.3.1   Graph of a Function 36
        2.3.2   Common Functions with One Variable 37
        2.3.3   Properties of a Function 41
        2.3.4   Inverse Functions 43
        2.3.5   Composition of Functions 45
        2.3.6   Functions of Two or More Variables 46

3 Linear Algebra ... 47
3.1 Vectors ... 47
- 3.1.1 Vectors in Physics ... 48
- 3.1.2 Vector Addition ... 49
- 3.1.3 Scalar-Vector Multiplication ... 49

3.2 The Dot Product of Two Vectors ... 50
- 3.2.1 Dot Product: Algebra Definition ... 50
- 3.2.2 Norm ... 51
- 3.2.3 Vector Magnitude and Direction in $\mathcal{R}^2$ ... 52
- 3.2.4 Dot Product: Geometric Definition ... 53
- 3.2.5 Unit Vector ... 55

3.3 Matrices ... 55
- 3.3.1 Matrix Addition ... 56
- 3.3.2 Scalar Multiplication ... 57
- 3.3.3 Matrix Multiplication ... 58
- 3.3.4 Matrices as Linear Transformations ... 62
- 3.3.5 Representations of Simultaneous Equations ... 64
- 3.3.6 Multiplying a Matrix by Itself ... 66
- 3.3.7 Diagonal and Trace ... 67
- 3.3.8 Diagonal Matrices ... 68
- 3.3.9 Determinants ... 68
- 3.3.10 Identity and Inverse Matrices ... 71
- 3.3.11 Matrix Transposition ... 72
- 3.3.12 Case Study 1 (Continued) ... 75
- 3.3.13 Orthogonal Matrix ... 76

3.4 Linear Combination ... 78
- 3.4.1 Vector Spaces ... 78
- 3.4.2 Linear Combinations and Span ... 80

3.5 Linear Dependence and Independence ... 83
- 3.5.1 Linear Dependence and Independence ... 83
- 3.5.2 Basis of a Vector Space ... 87

3.6 Connection to Matrices ... 87
- 3.6.1 Determinants and Singular Matrices ... 87
- 3.6.2 Rank ... 88

4 Matrix Decomposition ... 91
4.1 Eigendecomposition ... 91
- 4.1.1 Computing Eigenvalues and Eigenvectors ... 92
- 4.1.2 Diagonalisation ... 97

4.2 Principal Component Analysis ... 99
- 4.2.1 Mathematics Behind PCA ... 99
- 4.2.2 The Definition of PCA ... 101
- 4.2.3 PCA in Practice ... 102
- 4.2.4 Case Study 2: Continued (1) ... 104

		4.2.5	A Principal Component Analysis on the Sparrow Dataset	106
	4.3	Singular Value Decomposition		109
		4.3.1	Intuitive Interpretations	109
		4.3.2	Properties of the SVD	111
		4.3.3	Find a Singular Value Decomposition of a Matrix	111
		4.3.4	Case Study 2: Continued (2)	112
		4.3.5	An Example of the Interpretation of SVD on a Small Dataset	114
		4.3.6	An Example of Image Compression Using SVD	117
	4.4	The Relationship Between PCA and SVD		120
5	**Calculus**			121
	5.1	Limits of Functions		121
		5.1.1	Left- and Right-Hand Limits	122
		5.1.2	Theorems on Limits	123
		5.1.3	Continuity	127
	5.2	Derivatives		127
		5.2.1	Derivatives of Some Elementary Functions	131
		5.2.2	Rules for Differentiation	131
		5.2.3	The Second Derivative	137
	5.3	Finding Local Maxima and Minima Using Derivatives		138
	5.4	Integrals		142
		5.4.1	First Fundamental Theorem of Calculus	143
		5.4.2	Indefinite Integrals	144
		5.4.3	Second Fundamental Theorem of Calculus	145
		5.4.4	Integrals of Some Elementary Functions	145
		5.4.5	Two Properties of Integrals	147
	5.5	Further Integration Techniques		148
		5.5.1	Integration by Substitution	148
		5.5.2	Integration by Parts	152
6	**Advanced Calculus**			155
	6.1	Partial Derivatives		155
		6.1.1	The First Partial Derivatives	155
		6.1.2	The Second Partial Derivatives	158
		6.1.3	Differentiation of Composite Functions with Two Variables	159
		6.1.4	Gradient	161
		6.1.5	Jacobian Matrix	163
		6.1.6	Hessian Matrix	164
	6.2	Applications of Partial Derivatives		166
		6.2.1	Local Maxima and Minima	166
		6.2.2	Method of Lagrange Multipliers for Maxima and Minima	169
		6.2.3	Gradient Descent Algorithm	173

	6.3	Double Integrals	176
	6.3.1	Integration of Double Integrals Using Polar Coordinates	181
7	**Algorithms 1: Principal Component Analysis**	185
	7.1	Revisit Principal Component Analysis	185
	7.2	Preliminary Knowledge	186
	7.3	Problem Setting	191
	7.4	The Formulation of Principal Component Analysis	192
	7.4.1	The First Principal Component	192
	7.4.2	The Second Principal Component	193
	7.4.3	Data Normalisation	194
	7.5	Case Study 2 from Chap. 1: Continued	204
8	**Algorithms 2: Linear Regression**	207
	8.1	Simple Linear Regression Algorithm	207
	8.2	Least-Squares Estimation	208
	8.2.1	Deriving the Estimates Using the Least-Squares Objective Function	210
	8.3	Linear Regression with Multiple Variables	216
	8.4	Numerical Computation: Case Study 1 from Chap. 1—Continued	221
	8.5	Some Useful Results	224
	8.5.1	Residuals	224
	8.5.2	The Coefficient of Determination	225
9	**Algorithms 3: Neural Networks**	227
	9.1	Training a Neural Network by Gradient Descent	227
	9.2	A Simple One-Layer Neural Network	228
	9.2.1	Linear Activation Function	229
	9.2.2	Logistic Sigmoid Activation Function	232
	9.3	A Simple Two-Layer Neural Network: Case Study 3 from Chap. 1	237
	9.3.1	The Feed-Forward Propagation	237
	9.3.2	The Error Back-Propagation	239
	9.4	The Delta Rule	247
	9.5	Implementation Details	248
	9.5.1	Bias	248
	9.5.2	Stochastic Gradient Descent and Batch	249
	9.6	Deep Neural Networks	249
10	**Probability**	251
	10.1	Preliminary Knowledge: Combinatorial Analysis	252
	10.1.1	Factorial Notation	252
	10.1.2	Binomial Coefficients	252
	10.1.3	Permutation and Combination	253
	10.2	Probability	256

		10.2.1	Axiomatic Probability Theory	256
	10.3	Discrete Random Variables		260
	10.4	Continuous Random Variables		263
	10.5	Mean and Variance of Probability Distributions		268
		10.5.1	Mean	268
		10.5.2	Variance	275
	10.6	Special Univariate Distributions		279
		10.6.1	Discrete Random Variables	279
		10.6.2	Continuous Random Variables	286
11	**Further Probability**			293
	11.1	The Law of Large Numbers and the Central Limit Theorem		293
		11.1.1	The Law of Large Numbers	294
		11.1.2	Central Limit Theorem	295
	11.2	Multiple Random Variables		299
		11.2.1	Joint Probability Distributions: Discrete Random Variables	299
		11.2.2	Joint Probability Distributions: Continuous Random Variables	305
		11.2.3	Multinomial Distribution	314
		11.2.4	Multivariate Normal Distribution	316
	11.3	Conditional Probability and Corresponding Rules		319
		11.3.1	Conditional Probability	319
		11.3.2	Conditional Means and Conditional Variances	324
		11.3.3	Mutual Exclusivity	326
		11.3.4	The Multiplication Rule	327
		11.3.5	Independence	327
		11.3.6	The Law of Total Probability	328
	11.4	Bayes' Theorem		330
12	**Elements of Statistics**			335
	12.1	Descriptive Statistics		335
		12.1.1	Measures of Centre	336
		12.1.2	Measures of Variation	339
		12.1.3	The Range and the Interquartile	342
	12.2	Elementary Sampling Theory		344
		12.2.1	Random Sampling with and Without Replacement	344
		12.2.2	Sampling Distributions of Means	344
		12.2.3	Sampling Distributions of Proportions	347
		12.2.4	Standard Errors	349
		12.2.5	Degrees of Freedom	349
		12.2.6	Two Specific Sampling Distributions	350
	12.3	Inference		354
		12.3.1	Point Estimation	354
		12.3.2	Interval Estimation	355
		12.3.3	Testing Hypothesis	364

13 Algorithms 4: Maximum Likelihood Estimation and Its Application to Regression 379
  13.1 Maximum Likelihood Estimation 379
  13.2 Revisiting Linear Regression 384
    13.2.1 Linear Regression with Maximum Likelihood Estimation 384
    13.2.2 Sampling Distribution of the Linear Regression Estimators 394
  13.3 The Logistic Regression Algorithm 406

14 Data Modelling in Practice 415
  14.1 Data Pre-Processing 415
    14.1.1 Questions to Ask When Pre-Processing the Data 415
    14.1.2 A Simple Feature Selection Method 420
  14.2 Model Selection 422
    14.2.1 Data Splitting 422
    14.2.2 Model Evaluation 422
    14.2.3 Understanding Bias-Variance Trade-Off 428
    14.2.4 Underfitting and Overfitting 433
  14.3 Ridge Regression 434
    14.3.1 The Closed-Form Solution 434
    14.3.2 Bias and Variance of Ridge Regression Coefficients 436
  14.4 Early Stopping 441

Solutions 443

References 471

Index 473

Chapter 1
Introduction

> *The total amount of data created, captured, copied, and consumed globally is forecast to increase rapidly, reaching 149 zettabytes in 2024. Over the next 5 years up to 2028, global data creation is projected to grow to more than 394 zettabytes.*
>
> Statista Research Department [1]

What is a Zettabyte? A zettabyte is a value of 10 to the power of 21, or if we write it down to show all digits of it, we have 1,000,000,000,000,000,000,000 bytes. According to the figure reported by the Department of Economic and Social Affairs, United Nations, the global population was projected to reach 8 billion, 8,000,000,000, on 15 November 2022. Imagine if everyone, including newborn babies, takes an equal number of bytes, then this would mean each of us can have 125 billion bytes, that is, 125 GB (1 GB = 10^9 bytes). If the file size for 1 hour of 4K video is roughly 20 GB, each of us will have a 6.25-hour video to watch.

We live in the age of data. We access data every day: messages we send via our mobile phones, news we hear on the radio, movies we see on TV, and account statements we receive from the bank. These are all data—a collection of information.

Data Science is an interdisciplinary field that uses principles and methods from mathematics, statistics, computer science, and domain knowledge to tackle data. It involves data engineering, which builds up the pipeline to collect and use the data; data analytics, which analyses data to answer questions and draw conclusions; machine learning, which gives computers the ability to learn from data without explicit instructions; and more. Therefore, Data Science is usually used as a broad term that includes collecting and pre-processing the data, understanding the data by extracting useful information, and creating algorithms and predictive models.

1.1 The Procedures of Data Science

Quite often, people working in a problem domain raise questions and ask for help from data scientists. Solutions to questions are not necessarily apparent to data scientists due to a lack of domain knowledge and the complexity of real-world applications. Data scientists must communicate with clients efficiently to understand the essence of the problem and the challenges they face. A good data scientist can quickly grasp what their clients want by listening to them, learning from the background information, and explaining ideas to their clients using a layperson's terms.

Once the two parties agree on the problems they are going to work on, further discussions about the data that can be used to solve the problem are needed, including the volume of the available data and the features (also called attributes of the data) that can characterise the data. Rather than saving data in a CSV file or a relational database, more and more data are stored in cloud data warehouses. Data scientists need to know how to export raw data from data sources, such as web pages, emails, and SQL servers.

After obtaining the raw data, data scientists need to do data exploration in order to understand the relationships among the data better. They may convert the data to a table format where each row represents an observation and each column represents a feature. They then often need to consider whether all those features are significant or related to the task they will be dealing with. They may apply feature selection or feature extraction methods to the data. There are many algorithms to deal with this task involving mathematics knowledge. For example, in Chap. 4, we will introduce the principal component analysis method that can be used for data visualisation. The same method can be employed to perform feature extraction too. When doing data exploration, the data scientist also needs to know at least some basic statistics to understand, for example, what boxplots and histogram plots tell about the data.

Once the data is thoroughly investigated, data scientists are ready to apply existing computational algorithms to the data and/or to create a new algorithm to tackle the problem. This is the stage of modelling. It aims to produce a trained model with the existing data that can either reveal the natural structure in the dataset or make the most accurate predictions for any unseen data. The data scientist also needs to consider what the most suitable performance metrics to use are. The knowledge of optimisation, differentiation, probability, and statistics is very helpful in understanding the principles behind algorithms at this stage.

A further application of some domain knowledge may be used to improve performance. Finally, data scientists need to visualise results using suitable statistical figures and graphs.

We have just gone through the procedures of general Data Science [2], including the following key steps:

- Identifying the questions we want to answer.
- Collecting the data.

- Exploring the data to understand the natural structure and relationship among the data.
- Modelling the data.
- Presenting and explaining the results.

It is essential to understand that the whole procedure works like a spinning wheel rather than flowing in a linearly top-down mode. For example, when modelling the data, if the data scientist realises that more features are needed, they can go back to the second step to collect more data attributes. Alternatively, they may remove some features and redo the modelling.

1.2 Supervised Learning and Unsupervised Learning

This book focuses on the fundamental knowledge of mathematics, probability, and statistics, which are needed for exploring data, modelling data, and presenting results. To start with, we introduce two main techniques for analysing data: supervised learning and unsupervised learning.

1.2.1 Supervised Learning

In supervised learning, we have a training set of data and a test set of data. The training set will be used to train a predictive model. The test set will determine how well the trained model performs on this new unseen data. The training procedure is also called learning. The type of data used here contains a set of input values and a set of output (or target) values. Supervised learning aims to infer a function that maps the relation between the input values and the output values in the training set. The computer can then use this function to estimate the test data's output when given unseen test data, which has its own input values. Depending on the output value types, there are two categories for supervised learning. These are classification and regression. If the target value is discrete or categorical, we say it is a classification problem; if the target value is continuous, we say it is regression.

Classification aims to find decision boundaries to distinguish patterns from different classes given in the input data. The computer then uses these decision boundaries to predict the class label for new unseen data by seeing which side of the decision boundary the data lies in. On the other hand, regression aims to estimate a curve, a line, or a function learned from training examples so that when given unseen data, the computer can substitute the feature values into the function to estimate the target value.

Fig. 1.1 An example of supervised learning—classification

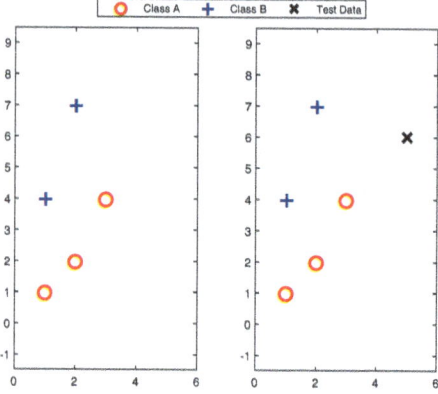

Example 1.1 Figure 1.1 shows an example of supervised learning. The left panel shows the training set, including five data points. There are two classes: Class A, represented by circles, and Class B, represented by plus signs. The learning process is to find a decision boundary between these two classes. In this case, a straight line can be drawn between the two classes of data that will act as a decision boundary. A neural network may be the technique to use to find the best line or decision boundary. Neural networks are covered in Chap. 9. In the right panel, the computer is given a new unseen test point denoted as a cross sign in black. It should be able to estimate the test data's class label using the decision boundary learned from the training set. This is a classification problem.

Example 1.2 Assume we have a heart disease training dataset. It includes 500 patient records with 14 patient attributes and 1 target value. This target value refers to the presence of heart disease in the patient. It is a binary value: 0 means no presence; 1 means there is presence. The 14 attributes may include age, gender, smoker or non-smoker, the measured serum cholesterol, and so on. Thus, we have 500 rows, each representing a patient, and 14 columns, each being a feature, plus a column indicating the corresponding target value. The aim is to find the relation between those 14 attributes and the target from 500 patient records so that when a new patient record with those 14 features is given, the computer can predict whether the patient has heart disease by applying the estimated function. In this example, the target values are discrete, a binary value, so this is a classification problem.

Example 1.3 Suppose an estate agent wants to predict house prices for Hertfordshire. He has 3000 house sale records with 9 features: the income of the householder of the house, the interest rates, the age of the house, the number of bedrooms in the house, the type of house, the population of the local area, the price that the house sold at last time, the postcode of the place, and the garage type. A model is trained to find a relation between these nine features and house prices. This is a regression problem, since house prices are continuous.

The aim is to use the trained model to predict the price for any house not included in those 3000 houses.

1.2.2 Unsupervised Learning

Unsupervised learning is different. Rather than finding a decision boundary or a regression function, it is used to find any natural structure inherent in the dataset. This natural structure may include clusters (data that group together) and outliers among the data (ones that differ considerably from the bulk of the data). It works by using data attribute information without considering data label information or target values, whether or not such information exists.

Example 1.4 Suppose we have 15 data points with two different attributes, as shown in Table 1.1. It is not easy to see if there are any natural clusters among the data, or, if there are any, how many clusters exist, even though this is a small dataset with only two dimensions.

However, the question can be answered easily if we can visualise the data as shown in Fig. 1.2. Data visualisation is a common application of unsupervised learning. It helps us to understand the underlying distribution of the data. In this example, we know there are four clusters in the data.

Table 1.1 A small dataset including 15 data points with 2 attributes

1.1	2
11	1
10	1
10.5	2
9	1
1	1
9	10
1	9
2.1	2.2
8.9	9.2
2	9
1	8
2	1
10	9.5
1.5	10

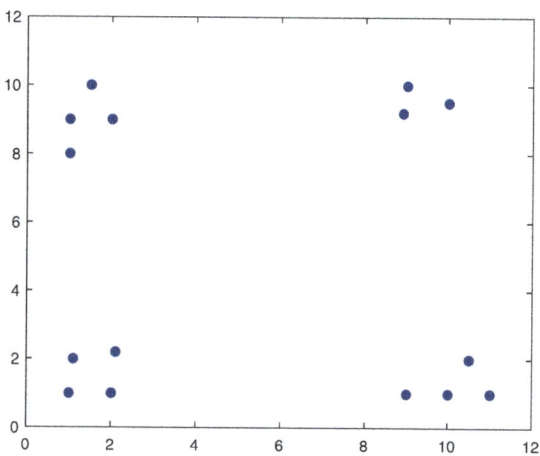

Fig. 1.2 An example of unsupervised learning: the scatter plot of data shown in Table 1.1

Exercise

1.1 Incorporation of certain chemicals into a drug delivery vehicle may lead to the enhancement of drug release and a more rapid clinical response. Such chemicals have been labelled as drug delivery enhancers. Their enhanced ability is measured as an enhancement ratio. You need to develop learning algorithms to address each of the following two problems.

1. You have some chemical compounds. You want to predict the enhancement ratio value for each of these chemicals.

(continued)

2. You have some chemical compounds. You want to decide for each chemical if it is a potential drug delivery enhancer.
   Should you treat these as classification or as regression problems?
   a. Treat both as classification problems.
   b. Treat both as regression problems.
   c. Treat Problem 1 as a classification problem and Problem 2 as a regression problem.
   d. Treat Problem 1 as a regression problem and Problem 2 as a classification problem.

Next, we will introduce three case studies. Each of them applies an algorithm, which will be developed in detail and illustrated using worked examples in Chaps. 7, 8, and 9, respectively. These case studies illustrate some typical problems. The mathematics is introduced so you can see how the mathematics naturally arises when you are characterising and solving the problems. You are not expected to understand all the mathematics yet, but it allows you to see why we need a book like this to deal with Data Science.

1.3 Case Studies

1.3.1 Case Study 1: Potential Enhancement Ratio Prediction Using Linear Regression

Suppose a pharmaceutical researcher wants to study the relationship between the molecular weight and the enhancement ratio of chemicals to identify compounds with potential as transdermal enhancers based on the value of the enhancement ratio.[1]

The researcher has the molecular weight (MW) and enhancement ratio value for three chemicals, as shown in the first three rows in Table 1.2. He wants to apply a computational method to the three chemicals to find the relationship between those two attributes. He then wants to use the learned relationship to estimate the enhancement ratio value of the fourth chemical in Table 1.2. This is a supervised learning problem. We call MW the input to the computational model and

---

[1] Transdermal enhancers are chemicals incorporated into drug delivery vehicles leading to enhancement of drug release through the uppermost layer of the skin, the stratum corneum, thus resulting in a more rapid clinical response. To find potential transdermal enhancers, researchers need to measure the enhancer ratio of each tested chemical by doing experiments in the lab, which are time-consuming and expensive [3].

Table 1.2 A small dataset of compounds with molecular weight values shown against their corresponding enhancement ratio values

Chemicals index	MW	Enhancement ratio
1	295	10
2	305	30
3	300	20
4	301	?

the enhancement ratio the target. The first three chemicals are training examples, including the input and the target. The fourth chemical is test data, for which the researcher knows the input and wants to estimate its target value. The researcher considers estimating a linear line that fits the training examples best. As we know from secondary maths, a linear function passing through the origin is given by

$$y = ax, \qquad (1.1)$$

where x is the input, a is the slope of the line, and y is the output. Readers who have forgotten it may view Sect. 2.3.2.1 of Chap. 2. In this case study, the line needs to be estimated using the first three compounds in Table 1.2, with MW as the input x and enhancement ratio as the output y. Other physio-chemical features may be used together with MW to estimate the enhancement ratio (in practice, there would usually be several more features). That is, the researcher may have more than one physio-chemical feature as the input. Therefore, we rewrite Eq. (1.1) in a more general way as follows:

$$\mathbf{y} = \mathbf{aX}. \qquad (1.2)$$

A careful reader will notice that we have used the bold font in Eq. (1.2). That is because we use bold capital letters to denote a matrix and bold little case letters to denote a vector. The basic knowledge of vectors and matrices is introduced in Chap. 3.

Now the question is how to find $\mathbf{a}$, the gradient or slope, in Eq. (1.2). One way to do it is to use the least-squares regression method, which is a simple but widely used technique in Data Science. The least-squares regression method aims to estimate $\mathbf{a}$ by minimising differences between the estimates of training examples and their actual target values. It is an optimisation problem.

For now, all you need to know is that

$$\mathbf{a} = (\mathbf{X}^T \mathbf{X})^{-1} \mathbf{X}^T \mathbf{y}. \qquad (1.3)$$

Equation (1.3) shows $\mathbf{a}$ is calculated via the matrix and vector multiplication involving the inverse (denoted as $^{-1}$) and transformation (denoted as ^{T}) of a matrix. The explanation of how we get Eq. (1.3) is presented in Chap. 8.

To apply Eq. (1.3), the researcher needs to collect the data into the matrix $\mathbf{X}$ and vector $\mathbf{y}$. Usually, data scientists normalise or re-scale feature values first. This is

1.3 Case Studies

important when there are many features with different ranges, especially those with a large magnitude.

From Table 1.2, we have

$$\text{raw\_X} = \begin{bmatrix} 295 \\ 305 \\ 300 \end{bmatrix} \text{ and } \mathbf{y} = \begin{bmatrix} 10 \\ 30 \\ 20 \end{bmatrix},$$

where the first three molecular weight values in Table 1.2 have been assigned to the input raw_X, and their corresponding enhancement ratio values are assigned to the output **y**. Both raw_X and **y** are called column vectors. If there had been more features than just the molecular weight, raw_X would have had more columns, one for each feature. It would then have been a proper matrix. The mean value of raw_X is 300. Let us subtract the mean value from raw_X and add a new column with $1's$. We obtain the following:

$$\mathbf{X} = \begin{bmatrix} 1 & -5 \\ 1 & 5 \\ 1 & 0 \end{bmatrix}.$$

X is now a matrix with three-row vectors and two-column vectors in it. The reason we add $1's$ into **X** and the idea behind the least-squares regression algorithm will be introduced in Chap. 8.

Once the line is fitted to the data, that is, vector **a** in Eq. (1.2) is estimated, the pharmaceutical researcher can substitute the known MW value of the fourth chemical compound to Eq. (1.2) to obtain its estimated y value for the enhancement ratio. To assess the model performance, he can compare the estimated value with the lab-measured value for this chemical. If he does this, the fourth chemical compound is called test data.

As a practitioner, the pharmaceutical researcher only needs to collect **X** and **y**, substitute them to Eq. (1.3) to estimate **a** first, and then to obtain the estimation of enhancement ratio by substituting **a** and the value of a new input into Eq. (1.2). The knowledge he needs will be taught in Chaps. 2 and 3. If he is lucky, the data scientist may have pre-calculated **a** for him using data that he has supplied to the data scientist. However, as data scientists, we do not stop there. We go further. We want to know how **a** is obtained and how the least-squares technique works, since these will help us understand how to adjust a model when necessary. To do that, knowledge of calculus, shown in Chaps. 5 and 6, is needed.

In addition, rather than obtaining a single estimated enhancement ratio value for a chemical compound, can we determine a range, with a certain confidence level, for the estimate? We may answer this question using probabilistic models. For instance, we can apply the maximum likelihood method covered in Chap. 13 to estimate model parameters and model predictions of a simple linear regression model and derive confidence intervals using statistical principles like standard error calculation. It is built up on the knowledge of probability and statistics, which is introduced in Chaps. 10, 11, and 12.

In this case study, **a** can be obtained by applying Eq. (1.3) for the three chemicals shown in Table 1.2. If the researcher can collect more chemicals or three different chemicals, he may get different **a** values. Which estimated **a** is the most suitable one to use? We will discuss the related issues and model selection in Chap. 14.

Remark 1.1 The maximum likelihood method mentioned in Case Study 1 deals with a regression problem. It can also be applied to classification problems. For a classification task, it can provide the probability that a pattern belongs to each class. The class with the highest probability would be the estimated class for that pattern. ♦

1.3.2 Case Study 2: Data Visualisation Using Principal Component Analysis

Usually, before training a computational model on a dataset, we want to investigate the underlying distribution of the data, the relationships among data attributes, and the correlation between each data attribute and the data targets. This investigation is called data exploration. One of the data exploration methods is data visualisation. For example, a scatter plot of an attribute against another attribute may be used to observe relationships among data points and to detect whether there are clusters in the data.

Let us use the Iris dataset. The dataset contains 3 classes of 50 data items each, where each class refers to a type of Iris plant [4], namely, Setosa, Versicolour, and Virginica, respectively. The dataset includes four features: sepal length, sepal width, petal length, and petal width in centimetres. Figure 1.3 shows scatter plots of one feature against another. Plots along the main diagonal are histogram plots of the data in the corresponding feature, since otherwise, they would just be a comparison of a feature with itself. Each scatter plot shows the correlation between the two features involved. It also displays clustering information: the class Setosa (represented by circles) is separated from the other two classes in all scatter plots, and there is some overlap between classes Versicolour (represented by squares) and Virginica (represented by triangles).

Are there any visualisation methods that consider all features in one single plot panel? The answer is yes, and the classical principal component analysis (PCA) is one of these methods. It is a widely used method for data visualisation and data dimensionality reduction. PCA is an unsupervised learning method. Figure 1.4 shows a PCA plot of the Iris data in the coordinate system constructed by the first two principal components. It is also a scatter plot, and it presents similar clustering information. Looking at the figure, readers who do not know the principal component analysis method may ask the following questions:

1.3 Case Studies

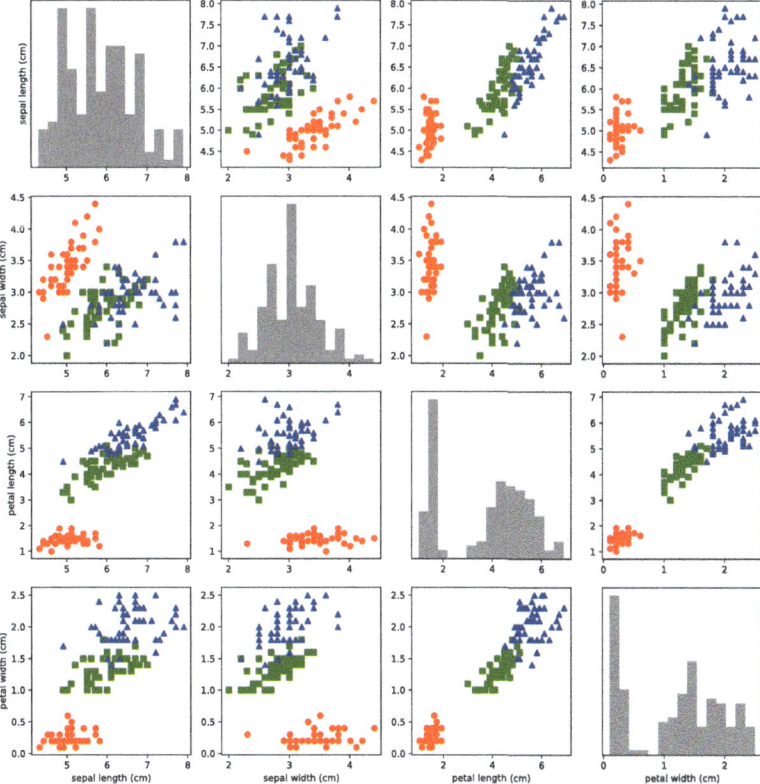

Fig. 1.3 Scatter plots of the Iris dataset, with circles representing the class Setosa, squares representing the class Versicolour, and triangles representing the class Virginica

Fig. 1.4 A PCA visualisation plot of the Iris dataset, where the first two principal components capture about 95.8% of the total variance in the data

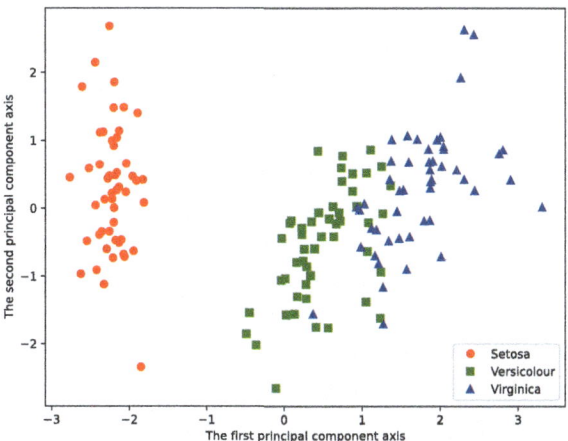

- What are those principal components (PCs)?
- What is the relationship between those PCs and the original four features in the dataset?
- Why is it necessary to report the variance percentage value (shown in the figure caption)?
- How is the variance percentage value calculated?
- How is the position of each data item in the coordinate plane determined?

It is not easy to see what the PCA has done with this dataset. Now let us see another, simpler, example, which we refer to as a toy dataset. It is a much smaller dataset **X**, including just five data points and just two features, and is shown in the following matrix:

$$\mathbf{X} = \begin{bmatrix} 1 & 5 \\ 2 & 2 \\ 3 & 3 \\ 4 & 4 \\ 5 & 1 \end{bmatrix}.$$

The average value of each column vector is the same, 3. We remove the average value of each column. That is, we subtract the mean value from each element in the matrix, and we get the following matrix:

$$\mathbf{newX} = \begin{bmatrix} -2 & 2 \\ -1 & -1 \\ 0 & 0 \\ 1 & 1 \\ 2 & -2 \end{bmatrix}.$$

The left panel of Fig. 1.5 shows the five data points in the $x - y$ Cartesian coordinate system after removing the average value. The first column of **newX** is the vector x_1 which is plotted on the horizontal (or x-axis), and the second column is x_2 and is plotted on the vertical (or y-axis). The right panel of the figure shows projections of those data points in the PCA coordinate system, with the first principal component (PC1) plotted horizontally and the second principal component (PC2) plotted vertically. The PCA projection plot seems to result from the axes in the original Cartesian coordinate system having been rotated, so that the largest distance among the data is displayed along the horizontal axis.

For now, all readers need to know are:

1. The PCA has been performed on the data using the features only, excluding the target value. The class label information (or the target value) shown in the Iris dataset is used only for colouring classes in the plot.
2. The projection, that is, the position of each data along each PC axis, is determined by a linear combination of the original features. As will be shown later in Chap. 7,

1.3 Case Studies

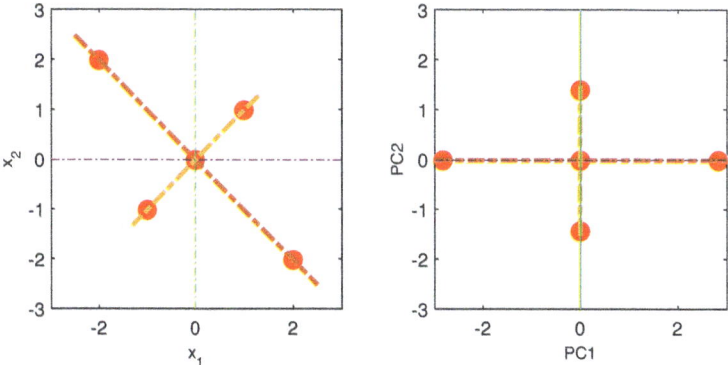

Fig. 1.5 The example of the toy dataset. The left panel shows the scatter plot of the data in the original $x - y$ Cartesian coordinate system; the right panel shows the scatter plot of the data projected onto the PCA space, where PC1 denotes the first principal component axis, and PC2 denotes the second principal component axis

for the Iris dataset, this linear combination is:

$$\text{projection} = c_1 \times \text{sepal.length} + c_2 \times \text{sepal.width} + c_3 \times \text{petal.length}$$
$$+ c_4 \times \text{petal.width},$$

where c_1, c_2, c_3, and c_4 are coefficients that need to be determined when doing the PCA analysis. The four attribute (or feature) values are normalised values, that is, each attribute has had its mean value subtracted and usually has been divided by its standard deviation (this is not necessary for this toy dataset since the two standard deviations are the same).

Remark 1.2 Normalisation is an important pre-processing step when analysing data to make all attributes have the same magnitude. This is useful when doing a distance-based calculation, since it avoids those attributes with large magnitudes dominating the distance. Normalisation may change the range of the data, but it does not change the data's structure and trend. An example can be seen in Fig. 1.6, where the original data with two attributes is shown in the left panel, and the normalised data having a zero mean and unit variance for each attribute is shown in the right panel. ♦

Chapter 4 describes how PCA is carried out after introducing the relevant linear algebra knowledge in Chaps. 3 and 4. Readers should be able to fully understand the idea behind PCA used in Chap. 7 after learning further knowledge regarding the relevant aspects of calculus in Chaps. 5 and 6.

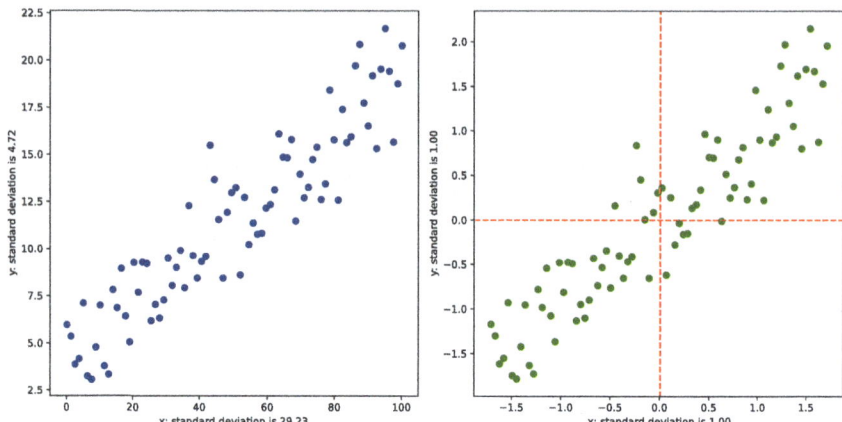

Fig. 1.6 A comparison of the data structure without data normalisation (left panel) and with data normalisation (right panel)

Fig. 1.7 An illustration of a simple two-layer feed-forward neural network

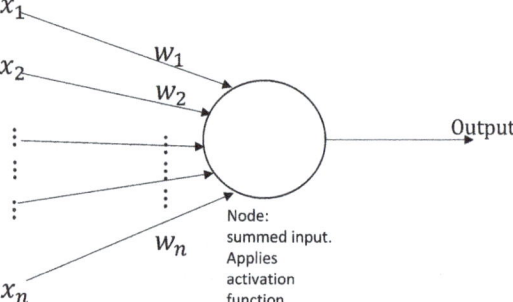

1.3.3 Case Study 3: A Simple Two-Layer Neural Network

Inspired by the biological information processing mechanism of the brain, the neural network (NN) with artificial neurons was first proposed in the 1940s. Since then, many different types of NN have been developed. Especially after 2010, with the growth of computing power, the increased requirements of processing a massive amount of data, and the need to achieve better solutions to optimisation problems, the deep neural network (DNN) has rapidly developed to deal with different types of data such as time series data, text, and images. Even with the development of DNN, the basic building blocks of DNN are still similar to the traditional NN; they all have activation functions and layers.

There are many different sorts of neural networks, each doing a different job and having different complexity and depth. However, the basic element of a neural network is the neuron, or unit, which has n inputs, and usually, each input has a weight w associated with it. This is illustrated in Fig. 1.7. The input of this neuron, or unit, is then the weighted sum of the input values and the weights. From the

1.3 Case Studies

figure, we see that this weighted sum is:

$$x_1w_1 + x_2w_2 + \cdots + x_nw_n = \sum_i x_iw_i = \mathbf{x} \cdot \mathbf{w}.$$

This uses the scalar product of vectors that will be introduced in Chap. 3. Figure 1.7 could be a node that collects the inputs, so the x_i are the input values, or any other unit, where the x_i are the outputs from previous units.

A neural network is usually arranged in layers, going from the input layer, which takes in the input values, to the output layer, which gives the output(s). Any units in layers in between are called hidden units. A network with inputs, a hidden layer, and an output layer is referred to as a two-layer neural network. The input layer is not usually counted as a layer, since it just contains the inputs and does not have adjustable weights. The input values are said to be fed-forward to give the output values of the network.

Having taken a weighted sum of its inputs, each neural unit performs some activation function on its input to transform the input to create an output value. Sample activation functions are the threshold, a linear function, a logistic sigmoid function, and a hyperbolic tangent. The same happens at each layer until output values(s) are produced. The output values are then compared to some target values, and the difference is called the error.

In this book, we are going to concentrate on networks where the weights are trained by gradient descent using some form of propagation of the error back through the network. Hence, we are only interested in the last three of the above activation functions, since they are differentiable, and hence, we can use gradient descent learning to train the network. Figure 1.8 illustrates the different sorts of activation functions introduced here. The concept of differentiable and gradient will be introduced in Chaps. 5 and 6.

The case study in this section illustrates artificial neural networks using a simple example of a two-layer neural network (NN) with only two hidden units in its middle layer. Of course, such simple neural networks have many limitations on what they can represent. However, we use this example to illustrate an activation function in

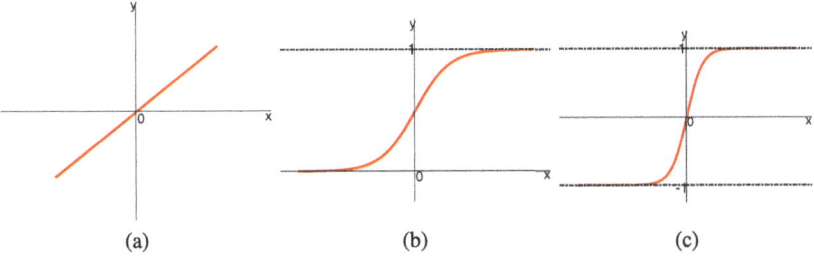

Fig. 1.8 Activation functions: (**a**) linear, (**b**) logistic sigmoid, (**c**) hyperbolic tangent

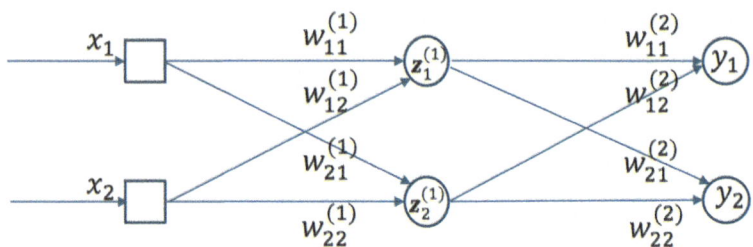

Fig. 1.9 An illustration of a feed-forward simple two-layer neural network

operation and how information is propagated through the layers and how the error is fed back to update the weights.

Figure 1.9 shows the architecture of a two-layer neural network used in this example, where we consider that each input (training) example **x** has only two attributes, or features, x_1 and x_2. Squares in Fig. 1.9 represent two input features, forming the neural network's input layer. Suppose each input example has two targets, denoted as t_1 and t_2. y_1 and y_2 are the outputs or predictions of the neural network for the given **x**, and they form the output layer of the neural network. The two-layer in the name means there are two layers of adaptive weights. The nodes in between two weight layers are called hidden units. Each input in the input layer is connected to hidden units via weights of the first layer. Each hidden unit is a linear combination of the input attributes. Usually, an activation function, which is most often a non-linear function and can transform the total input, is applied to each hidden unit to simulate the complexity of the brain.

We follow the notations used in [5] for weights. That is, we denote each weight as $w_{ji}^{(l)}$, where (l) denotes the lth layer, j the jth hidden unit in the corresponding layer, and i the ith node of the immediate layer to the left. For example, $w_{21}^{(1)}$ denotes the weight going from the first input feature x_1 to hidden unit 2 in the first layer. The training of this neural network aims to adjust weight values to reduce the error, that is, the difference between the targets and the predictions or outputs. We show how to use and train this simple neural network in Chap. 9.

Remark 1.3 The simple two-layer neural network shown in Case Study 3 can be used in a supervised learning task. It can be used for both regression and classification problems. ♦

We have focused on approaches that can be applied to understand the data and make predictions for unseen data. As we can see, these approaches need mathematical and statistical knowledge almost everywhere. In the following chapters, we will equip our readers with the essential skills for data analysis.

However, before we start, let us have a look at data types. You may have noticed that data are in a format as shown in Tables 1.1 or 1.2. However, how do we deal with data in free forms, such as audio signals, email, and survey comments with some numerical scores? In the final section of this chapter, we will briefly discuss data types.

1.4 Types of Data

Looking into the types of data is one of the most important steps you need to take to perform Data Science. It will help you to understand the data and choose the correct class of algorithms that can be used to analyse the data.

1.4.1 Organised (Structured) and Unorganised (Unstructured) Data

When given a dataset, the first question you need to ask yourself is whether it is structured data or unstructured data. Structured data is usually organised using a table method; unstructured data exists as a free entity and does not follow any standard format. Most data analysis algorithms are built with structured data in mind.

Let us have a look at an organised/structured data example. Table 1.3 shows 15 sample rows of the Iris dataset mentioned in Case Study 2 in Sect. 1.3.2 of this chapter. As you can see, the data is sorted into a row and column structure. Each row represents a single observation; each column represents either a feature or class information. This data set has four attributes or features. They are all continuous

Table 1.3 Examples of data items from the Iris dataset, illustrating feature values and corresponding class labels

sepal.length	sepal.width	petal.length	petal.width	Variety
5.1	3.5	1.4	0.2	Setosa
4.9	3	1.4	0.2	Setosa
4.7	3.2	1.3	0.2	Setosa
4.6	3.1	1.5	0.2	Setosa
5	3.6	1.4	0.2	Setosa
⋮				
7	3.2	4.7	1.4	Versicolor
6.4	3.2	4.5	1.5	Versicolor
6.9	3.1	4.9	1.5	Versicolor
5.5	2.3	4	1.3	Versicolor
6.5	2.8	4.6	1.5	Versicolor
⋮				
6.7	3	5.2	2.3	Virginica
6.3	2.5	5	1.9	Virginica
6.5	3	5.2	2	Virginica
6.2	3.4	5.4	2.3	Virginica
5.9	3	5.1	1.8	Virginica

values. The last column, with a head denoted as *variety*, gives the class label information for each plant, indicating which category the plant belongs to.

Examples of unstructured data include genetic sequences and molecular structure graphs. These are unstructured data, since we cannot form features of the sequence using a row-column format without taking a further look. Feedback left for a product review on Amazon and messages on Twitter are also unstructured free text. Images are another type of unstructured data. To read the information saved in an image, we need to open the file with an image viewer.

Exercise

1.2 Is the following data structured or unstructured?

(1) Speech signals,
(2) Emails,
(3) Medical X-ray,
(4) Student ID numbers.

Remark 1.4 Most real-world data are unstructured data. To apply most data analysis algorithms, we must first convert unstructured data to structured data using pre-processing techniques.

For example, consider speech signals. People may decompose each signal into a set of signals with different frequencies, and then values related to the amplitude of frequencies can be used as signal features. How to decompose the signal is not our focus here. However, it is essential to know that converting data from unstructured to structured is a crucial step in data analysis.

As another example, let us consider text data. We have many options to transform the free text into a structured format. We could apply new features that describe the data. For instance, we can define a set of words or phrases first and then count the particular words or the specific phrases appearing in each file. This way, we can use the features defined here to convert them into structured data. Of course, we may also have a topic as a class label for each text file. Then, we can put all of them into one big table: each row representing one text, such as a tweet, columns showing the counts of specific words or phrases, and one column indicating the class label information. ♦

1.4.2 Quantitative and Qualitative

Another classification of data is quantitative and qualitative. Quantitative data can be described using numbers, including discrete and continuous data. Discrete data has limited values, while continuous data can take on any value between two values.

1.4 Types of Data

For example, the number of PC labs in a university is discrete, since it is a whole integer value. In contrast, the average number of hours a student sleeps daily is a continuous value, such as 5 or 7.5 hours, or any value in between.

Qualitative data is also called categorical data and is non-numerical in nature. However, qualitative data may appear as a number, though they cannot be used meaningfully in the computation. For example, Level 4, Level 5, Level 6, and Level 7 denote modules running for the first year, the second year, the third year of undergraduate courses, and the postgraduate course, respectively. It does not make sense if you do an addition between a Level 6 module and a Level 7 module.

Remark 1.5 To tell if a number is quantitative or qualitative, ask yourself whether it still makes sense after adding them together. ♦

Exercise

1.3 Is the following data qualitative or quantitative?

(1) Book title,
(2) Welcome to Year 1,
(3) Maximum daily temperature,
(4) Car registration number.

1.4.3 The Four Levels of Measurement

A more detailed classification of data types is the four levels of measurement. These include the nominal level, the ordinal level, the interval level, and the ratio level. This detailed classification allows us to recognise what mathematical operations can be applied for each level.

1.4.3.1 The Nominal Level

At the nominal level, the data is described by name or category, for example, hair colour, gender, and house types, such as terrace houses, semi-detached houses, and bungalows. At this level, possible mathematical operations to the data include set membership and equality. For example, suppose a colour set has three colours: green, yellow, and red. If a student's hair colour is black, then the colour black is not a member of that colour set. We will introduce the knowledge of sets in Chap. 2.

1.4.3.2 The Ordinal Level

The ordinal level gives us a rank order. For example, the customer rating of a product or a book and the award someone receives after completing a maths competition. Possible mathematical operations that can be applied at this level are ordering and comparison, whilst we cannot do addition and other computations.

> **Exercise**
>
> **1.4** A group of students is asked the following questions. Is the answer collected from each of the following nominal or ordinal?
> (1) Are you an international student?
> (2) What is your gender: male, female, prefer not to say, other?
> (3) How many countries have you visited?
> (4) What is your preferred contact method: email or telephone?
> (5) What is your usual lunchtime: 11 a.m.–12 p.m., 12 p.m.–1 p.m., 1 p.m.–2 p.m., or later than 2 p.m.?

1.4.3.3 The Interval Level

Data measured at the interval level is like the ordinal level, placing numerical values in order. Unlike the ordinal level, however, the interval level has a known and equal distance between each value. For instance, consider the Celsius temperature. The difference between 10 and 30 degrees is a measurable 20 degrees, as is the difference between 40 and 60 degrees. However, the interval level data does not have a natural zero. For example, if we consider the Celsius temperature at zero degrees, then Celsius zero does not mean the absence of temperature.

More complicated mathematical operations are allowed at the interval level. Compared with the ordinal level, we can do addition and subtraction at this level besides ordering and comparison.

1.4.3.4 The Ratio Level

The ratio level allows us to multiply and divide too. Data has a clear definition of zero at this level. For example, students' marks for assessments in numerical values are at the ratio level. We can have zero marks in the final score, and it makes sense to say 90 out of 100 marks is twice as much as 45.

Remark 1.6 Let us consider two continuous number lines: one for the interval level and the other for the ratio level. Data indicate positions along each line. There is no actual zero position along the interval level line. In other words, the zero position

1.4 Types of Data

is arbitrary along the line. For example, zero degrees Celsius and zero degrees Fahrenheit are different temperatures. Zero degrees Celsius is the freezing point of water, while zero degrees Fahrenheit is colder than that, and the freezing point of water in the Fahrenheit scale is 32 degrees Fahrenheit. On the contrary, zero grams and zero pounds mean the same thing along the ratio level line. There are no values less than zero on the ratio level line. ♦

Example 1.5 Calculating the increase or decrease in percentage terms is not useful at the interval level. Suppose the temperature increases from 10 degrees Celsius to 15 degrees Celsius. The increase in percentage terms is $\frac{15-10}{10} = 50\%$. If we convert the temperatures to Fahrenheit, we have $(10 \times \frac{9}{5}) + 32 = 50$ and $(15 \times \frac{9}{5}) + 32 = 59$, respectively. The increase in percentage terms is $\frac{59-50}{50} = 18\%$. It does not make sense to say 15 degrees Celsius is 50% warmer than 10 degrees Celsius, while we get only 18% warmer in Fahrenheit.

Example 1.6 Calculating the increase or decrease in percentage terms is valid at the ratio level. Suppose weight increases from 10 grams to 15 grams. The increase in percentage terms is $\frac{15-10}{10} = 50\%$. If we convert the unit to pounds, we have $10 \times 0.0022 = 0.022$ pounds and $15 \times 0.0022 = 0.033$ pounds, respectively. The increase in percentage terms is $\frac{0.033-0.022}{0.022} = 50\%$. It does make sense to say 15 grams is 50% heavier than 10 grams, since we also get 50% heavier in pounds.

Remark 1.7 Quantitative data includes the interval level and ratio level, while qualitative data includes the nominal level and the ordinal level. ♦

Understanding the data type and its measurement level will help us to select models or statistical procedures to analyse the data. For example, for continuous or ordinal data with a large number of categories, say the number of categories greater than 4, we may use regression models, including ordinary linear regression and neural networks and the Gaussian normal distribution to analyse the data. For nominal or ordinal data (usually with a small number of categories, say 2, 3, or 4), we can use the Chi-square statistical test to examine whether the observed values follow the assumed theoretical distribution, or we can use logistic regression to make predictions on unseen data. Chapter 8 explains how linear regression works; Chap. 9 introduces the principle behind the traditional neural networks. Data following a univariate Gaussian distribution and multivariate Gaussian distributions are described separately in Chaps. 10 and 11. The primary statistical analysis techniques and the Chi-square test are presented in Chap. 12. More linear regression and the logistic regression model will be explained in Chap. 13.

Exercise

1.5 Identify the level of measurement for the following:

(1) Military title: Lieutenant, Captain, Major.
(2) Categorisation of property: Flats, Detached, Semi-detached, Terraced, End-of Terrace, Cottage, Bungalows.
(3) A list of temperatures in degrees Celsius for last week.
(4) Heights of a group of Year 6 students.
(5) Calendar years.
(6) Temperature in Kelvin scale.

Chapter 2
Sets and Functions

Basic set theory and functions are the foundation of later chapters. Although we assume that readers are familiar with the rudiments of basic set theory and basic notions of functions, let us refresh our memory and define notations in this chapter. We focus on functions with one variable in this chapter.

2.1 Sets

In this section, we will introduce set membership, how to find the cardinality of a set, and how to represent sets using a Venn diagram. In addition, we will discuss four basic normal set operations: set union, set intersection, set subtraction, and set complement. Moreover, we will show how to write sets in comprehension and define what a binary relation is.

2.1.1 Sets and Subsets

Definition 2.1 (Sets) A set is a collection of objects. The objects are known as the elements of the set or its members.

For small sets, we can define a set by writing out the names of all the elements of the set, separating them by commas, and enclosing the whole list in curly brackets. For example: {apple, pear, orange, melon}. An empty set with no elements can be represented by { } or ∅. Sets have the following two properties:

- Sets are not ordered. For example, {a, b, c} is a representation of the same set as {c, a, b}.

- There are no repeats. For example, {a, b, c, b} is a representation of the same set as {a, b, c}.

Two sets are equal if they have the same members. The equal sign, =, can be used when two sets are equal. For example, $\{1, 2\} = \{2, 1\}$. The not equal sign, $\neq$, can be used when two sets are not equal.

Definition 2.2 (Cardinality) The Cardinality of a finite set is the number of elements in the set.

It is denoted using the symbol #, or with a vertical bar on each side of the name of the set. For example:

- $\#\{1, 2, 3\} = 3$.
- $|\{3, 3, 7, 2, 1\}| = 4$.

2.1.1.1 Infinite Sets

Sets can be infinite as well as finite. For example, three infinite sets that we will use are:

- $\mathbb{N}$: the set of natural numbers. In this book, it includes 0, that is, it includes all non-negative numbers, $\{0, 1, 2, \ldots\}$.
- $\mathbb{Z}$: the set of integers, $\{\ldots, -3, -2, -1, 0, 1, 2, \ldots\}$.
- $\mathbb{R}$: the set of real numbers, that is, any decimal number.

2.1.1.2 Intervals

A set that contains all the real numbers between two given numbers is called an *interval*. For instance, all the real numbers between 2 and 3, including both numbers, are a *closed interval* and denoted: [2, 3]. If we do not include both the endpoints, then it is an *open interval* and denoted: (2, 3) or]2, 3[. Of course, we can include one endpoint and not the other, giving mixed intervals, that is, [2, 3) and (2, 3].

2.1.1.3 Set Membership

The symbol $\in$ denotes *is a member of* a set. For example,

$$apple \in \{apple, pear, orange, melon\}$$

is a true statement, since *apple* is a member of the given set, while $strawberry \in \{apple, pear, orange, melon\}$ is a false statement, since *strawberry* is not a member of the given set. $\notin$ denotes *is not a member of* a set. So $strawberry \notin \{apple, pear, orange, melon\}$ is a true statement.

Example 2.1

If the real number $x \in [2, 3]$ then $2 \leq x \leq 3$.
If the real number $x \in (2, 3)$ then $2 < x < 3$.
If the real number $x \in (2, 3]$ then $2 < x \leq 3$.
If the real number $x \in [2, 3)$ then $2 \leq x < 3$.

Definition 2.3 (Subsets) The set A is a subset of the set B if and only if either the set A is empty or every element of A is also an element of B. $A \subseteq B$ means *A is a subset of B*.

Definition 2.4 (Proper Subset) The set A is a proper subset of the set B if and only if A is a subset of B but not equal to B. $A \subset B$ means *A is a proper subset of B*.

Note that both $\subset$ and $\subseteq$ can be used with a line through them to denote their opposite. For example, $\{1, 2, 3, 4\} \not\subseteq \{1, 2, 4\}$.

Exercises

2.1 What is the value (True or False) of each of the following statements?

(1) $\{4, 8, -1\} = \{4, 8, 4, -1, -1\}$.
(2) $\{x, y, z, z\} = \{z, y, x\}$.
(3) $\{\} = \{0\}$.
(4) $\#\{x, y, z, z\} = 4$.
(5) $\{\} \subset \{1, 2, 4\}$.
(6) $[1, 2] = [1, 2)$.
(7) $(1, 2) \subset [1, 2]$.

2.2 Write down the value of each of the following.

(1) $\#\{a, b, c, d, e, f\}$.
(2) $\#\{\}$.
(3) $\#\{\{\}\}$.

Definition 2.5 (Power Sets) The power set of a set S is the set containing all possible subsets of S. The cardinality of the power set of a set S is $2^{\#S}$.

Example 2.2 Given $A = \{3, 4, 5\}$ with a cardinality of 3, the power set of A is

$$\{\{\}, \{3\}, \{4\}, \{5\}, \{3, 4\}, \{3, 5\}, \{4, 5\}, \{3, 4, 5\}\},$$

and its cardinality is $2^3 = 8$.

Exercise

2.3 Do the following:

(1) Write down the power set of $\{-1, 1\}$. What is its cardinality?
(2) Write down the power set of $\{0, 1, 2, 3\}$. What is its cardinality?
(3) What is the cardinality of the power set of $\{a, b, c, d, e, f, g, h\}$?

2.1.2 Venn Diagrams

Venn diagrams are a way of describing sets and how they are related to one another in pictures. Each set is represented as a circle within a universe and contains values written inside the circle's boundary. The overlapping part of the two circles shows elements shared by both sets.

Example 2.3 Suppose universe = $\{11, 12, 13, 14, 15, 16, 17, 18, 19, 20\}$, Set A = $\{13, 14, 16, 17, 19\}$, and Set B = $\{12, 13, 15, 16\}$. Figure 2.1 shows the relationship between Set A and Set B.

Another example is the classification of numbers using sets considered in Sect. 2.1.1.1. In Fig. 2.2, $\mathbb{Z}$ and $\mathbb{N}$ are represented as circles within the universe $\mathbb{R}$.

2.1.3 Basic Set Operations

Definition 2.6 (Set Union) Two sets may be joined together to form a new set containing all of the elements in one or the other or both of them. This operation is known as set union and is denoted using the symbol $\cup$.

2.1 Sets

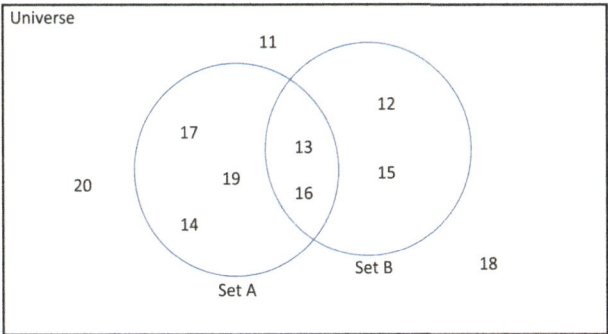

Fig. 2.1 An example of a Venn diagram used to visualise the relationships and intersections among two data sets

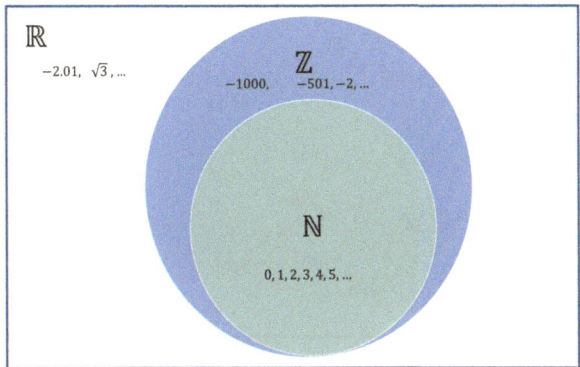

Fig. 2.2 A Venn diagram showing the relationships among $\mathbb{Z}$ (integers), $\mathbb{N}$ (natural numbers), and $\mathbb{R}$ (real numbers)

Example 2.4 The union of two sets A and B in Fig. 2.1 is $A \cup B = \{12, 13, 14, 15, 16, 17, 19\}$, whose elements are highlighted using underscores in Fig. 2.3.

Definition 2.7 (Set Intersection) Two sets may be joined together to form a new set containing only the elements in both of them. This operation is known as set intersection and is denoted using the symbol $\cap$.

Example 2.5 The intersection of two sets A and B in Fig. 2.1 is $A \cap B = \{13, 16\}$, whose elements are highlighted using underscores in Fig. 2.4.

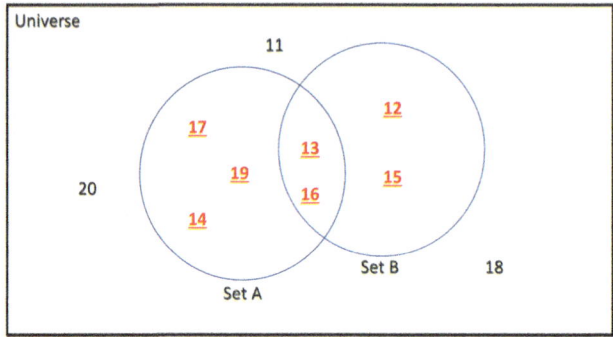

Fig. 2.3 A Venn diagram illustrating the union of sets $A \cup B$, representing all elements that belong to A, B, or both, highlighted using underscores

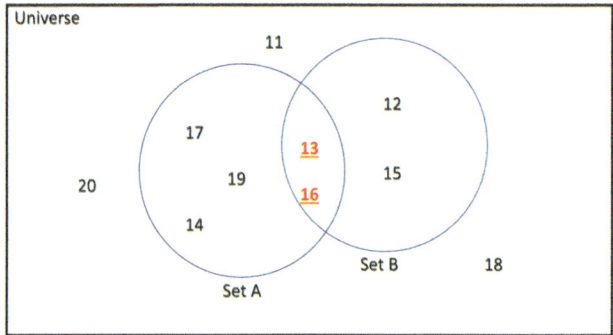

Fig. 2.4 A Venn diagram illustrating the intersection of sets $A \cap B$, representing comments elements that belong to both A and B, highlighted using underscores

Definition 2.8 (Set Subtraction or Difference) Two sets may be joined together to form a new set containing only the elements in the first but not the second. This operation is known as set subtraction and is denoted using the symbol \.

> **Example 2.6** The set formed by subtracting set B from set A in Fig. 2.1 is $A \backslash B = \{14, 17, 19\}$, whose elements are highlighted using underscores in Fig. 2.5.

2.1 Sets

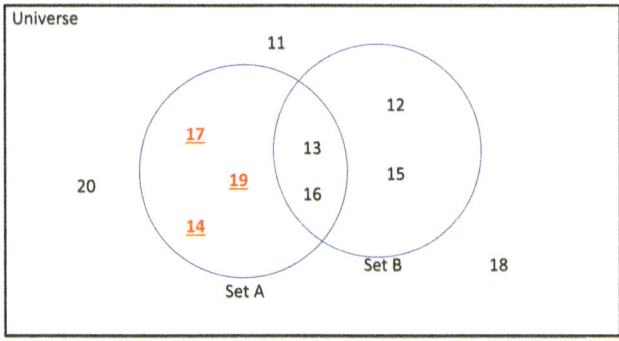

Fig. 2.5 A Venn diagram illustrating the set difference $A \setminus B$, representing elements that belong to A, but not to B, highlighted with underscores

Definition 2.9 (Set Complement) The complement of a set A is the set of all elements in the universe but not in A. $\overline{A}$ denotes the set formed from the complement of set A.

Example 2.7 The complement of set A in Fig. 2.1 is $\overline{A} = \{11, 12, 15, 18, 20\}$, whose elements are highlighted using underscores in Fig. 2.6.

Fig. 2.6 A Venn diagram illustrating the complement of set A, $\overline{A}$, representing elements that do not belong to the set A, highlighted with underscores

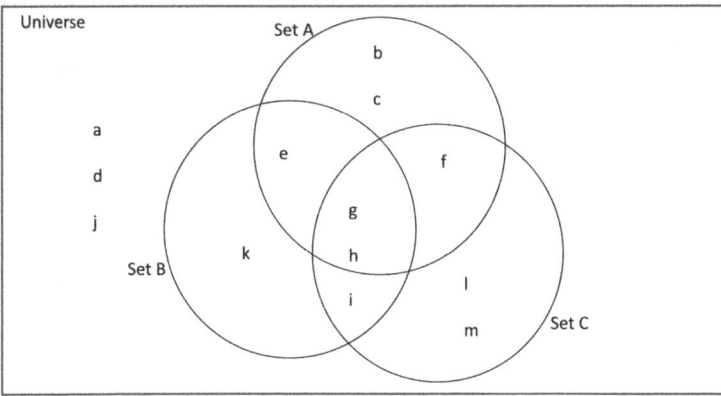

Fig. 2.7 The Venn diagram referenced in Exercise 2.4, representing the sets and their relationships

Exercise

2.4 Given a Venn diagram shown in Fig. 2.7, find the following:

(1) $A \cup B$.
(2) $C \cap B$.
(3) $\overline{A \cup B}$.
(4) $A \backslash (B \cap C)$.
(5) $A \cup B \cup C$.
(6) $A \cap B \cap C$.
(7) $(A \cup B) \backslash C$.
(8) $\overline{(A \cup B) \backslash C}$.

2.1.4 Sets Written in Comprehension

So far, we have written sets in extension. That is to list all of the values in the set separated by commas within a pair of curly brackets. There is another way to write sets. That is to give a typical element and a condition for its inclusion in the set. It is called sets written in comprehension.

2.1 Sets

Example 2.8 Suppose we have $S = \{x \in \mathbb{N} | x < 5\}$, where the curly brackets tell us that this is a set. Inside the brackets, there are two sections. The left of the vertical bar gives us the signature of the values in the set. A signature tells us which universe the value named by x is drawn from. In this example, it is drawn from the universe of values $\mathbb{N}$. The right of the vertical bar states the condition that each member of the set must satisfy. If we write this set in extension, we have $S = \{0, 1, 2, 3, 4\}$. The vertical bar | (sometimes written as a colon:) is usually read as **such that**. So, set S can be described as the set of all numbers x in $\mathbb{N}$ **such that** $x < 5$.

The condition in a set comprehension expression is a truth-valued expression. All values that make this expression true are members of the set. All values that make the expression false are not members of the set (they are in its complement).

2.1.4.1 Using Logic

We can define more complex conditions using operations from *logic*:

- AND is represented by the symbol $\wedge$.
- OR is represented by the symbol $\vee$.
- NOT is represented by the symbol $\neg$.

Example 2.9 Let $Z = \{x \in \mathbb{N} | x > 10 \wedge x < 16\}$. If we write set Z in extension, we have $Z = \{11, 12, 13, 14, 15\}$.

Exercise

2.5 Write each of the following sets in an extension.

(1) $A = \{y \in \mathbb{N} | \neg(y > 10)\}$.
(2) $B = \{x \in \mathbb{N} | (x < 12) \wedge (x \geq 6)\}$.
(3) $C = \{z \in \mathbb{N} | (z < 8) \wedge (z < 5)\}$.
(4) $D = \{y \in \mathbb{N} | (y < 8) \vee (y < 5)\}$.
(5) $E = \{x \in \mathbb{N} | \neg(x \geq 12) \wedge (x > 3)\}$.

2.2 Binary Relations

Definition 2.10 (Cartesian Product Sets) Given two sets A and B, the set that contains all ordered pairs (x, y) such that x belongs to A and y belongs to B is called the Cartesian Product. It is denoted as $A \times B$, which can be expressed as follows:

$$A \times B = \{(x, y) | x \in A \land y \in B\}.$$

A and B may be subsets of different universes. If either A or B is an empty set, then $\emptyset \times B = A \times \emptyset = \emptyset$. If $A \neq B$ and both A and B are not the empty set, then $A \times B$ is not equivalent to $B \times A$ because the inside of each pair is ordered.

Example 2.10 If $A = \{a, b\}$ and $B = \{0, 1, 2\}$, then

$$A \times B = \{(a, 0), (a, 1), (a, 2), (b, 0), (b, 1), (b, 2)\};$$

$$B \times A = \{(0, a), (0, b), (1, a), (1, b), (2, a), (2, b)\}.$$

Remark 2.1 Although the inside of each pair in a Cartesian product set is ordered, it should be remembered that the actual set, like all sets, is not ordered.
So if $A = \{a, b\}$ and $B = \{1\}$, then:

$$A \times B = \{(a, 1), (b, 1)\} = \{(b, 1), (a, 1)\} \neq \{(1, a), (1, b)\} = B \times A.$$

♦

Exercise

2.6 Find the value (True or False) of each of the following:

(1) $(2, 8) \in \{(8, 1), (2, 10), (1, 10), (8, 2)\}$.
(2) $(3, 7) \in \{(1, 3), (3, 7), (7, 3), (1, 7)\}$.
(3) $\{(1, 4), (0, 2), (10, 9)\} = \{(4, 1), (2, 0), (9, 10)\}$.
(4) $\{(1, 4), (0, 2), (10, 9)\} = \{(10, 9), (0, 2), (1, 4)\}$.

2.7 Write each of the following Cartesian products as a single set in extension:

(1) $\{2, 3, 5\} \times \{0, 1\}$.
(2) $\{0, 1\} \times \emptyset$.

2.2 Binary Relations

We can form the Cartesian product of any number $n \in \mathbb{N}$ of sets and whose elements will be n-tuples. For example, we can use it to model customer accounts in the following way:

$$A \times B \times C = \{(a, b, c) | a \in A \land b \in B \land c \in C\},$$

where A represents customer account numbers, B customer names, and C customer addresses.

2.2.1 Binary Relations

Definition 2.11 (Relation) Given two sets A and B, a relation from A to B is a subset of the Cartesian Product $A \times B$.

Any subset of the set $A \times B$ can be considered as a relation. That could be the empty set, the entire set, $A \times B$, or anything in between.

Definition 2.12 (Binary Relation) A binary relation relates values from one universe to the values of another. The from-universe is called the source; the to-universe is called the target.

2.2.1.1 Kinds of Relation

Figure 2.8 shows four types of relation. In each panel, the left rectangle shows the source universe, and the right shows the target universe. The oval in the source includes input values of a relation, called the *domain* of the relation, while the oval in the target includes output values of a relation, called the *range* of the relation.

Panel (a) shows a one-to-one relation, where each value in the domain has only one corresponding value in the range.

Panel (b) presents a many-to-one relation, where two (can be more) values in the domain have the same value in the range.

Panel (c) displays a one-to-many relation, where one value in the domain has two (can be more) different values in the range.

Panel (d) represents a many-to-many relation, where a value in the domain may have more than one output in the range, and a value in the range may have more than one corresponding value in the domain.

Remark 2.2 Note that not all points are necessarily in the ovals in Fig. 2.8, since a relation is any subset of the whole Cartesian product, $source \times target$. Hence, the domain is a subset of the source, and the range is a subset of the target. That is, $domain \subseteq source$ and $range \subseteq target$. Note the illustrated relations in Fig. 2.8 also do not include every possible pair from the domain to the range. Again, this is

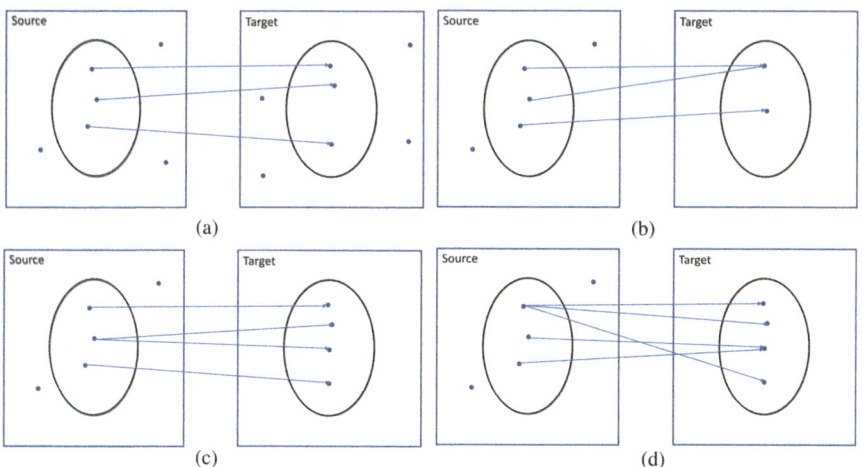

Fig. 2.8 An illustration of four kinds of relations: (**a**) one-to-one, (**b**) many-to-one, (**c**) one-to-many, and (**d**) many-to-many

because the relation is a subset of the whole Cartesian product, so the relation does not have to contain all pairs. ♦

2.3 Functions

Many-to-one and one-to-one relations are very common in computing and have a special status, and they are called functions.

Figure 2.9 presents examples of relations, some of which are functions and others are not. The left column represents the input values (possible domain), while the right column represents the possible output values.

Panels (a) and (b) are two functions, since all the elements in the domain have just one corresponding image, or output, in the target universe and so construct the range of actual values in the target. Values in the right column do not all have a value in the domain with their image in the target universe.

Panel (c) illustrates a case of a relation that is not a function. Every element in the possible domain should have an image in the range by applying the given function. But 3 in green does not have a related value in the range. Panel (d) is another example of a relation that is not a function, since 3 has two images in the range, b and d, indicating a one-to-many relation.

So, in summary, a function has to have an image for every value in the domain, and it has to have just one image. The set of images in the target universe is called the range.

2.3 Functions

Fig. 2.9 An illustration depicting various relations, distinguishing between those that are functions and those that are not

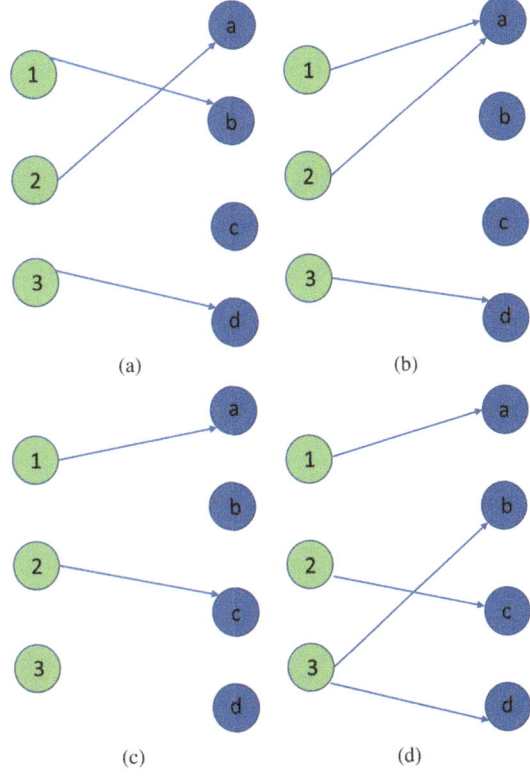

Exercise

2.8 Given the domain as $\{0, 1, 2\}$, are the following relations also functions (assume the target universe includes $\{0, 1, 2, 3, 4, 5, 6, 7, 8\}$):

(1) $\{(0, 1), (2, 2), (1, 3)\}$?
(2) $\{(0, 2), (1, 3)\}$?
(3) $\{(0, 7), (1, 5), (2, 6), (0, 6)\}$?
(4) $\{(1, 4), (0, 4), (2, 4)\}$?

Definition 2.13 (Function) Let x represent the elements of the domain (denoted as D), y represent the elements of the range (denoted as W), and f symbolise the function, then we have $y = f(x)$. It can be written as follows:

$$W = \{y | y = f(x) \wedge x \in D\}.$$

$f(x)$ is also called the image of x with respect to the function f. The domain variable x is called the independent variable, while the range variable y is called the dependent variable.

Example 2.11 Given $W = \{y | y = x^3 \wedge x \in [-3, 3]\}$, we know that $f(x) = x^3$, the domain is $-3 \leq x \leq 3$, and the corresponding range is $-27 \leq x \leq 27$.

2.3.1 Graph of a Function

The plot of pairs $(x, f(x))$ in a coordinate system is the graph of $f(x)$.

Example 2.12 Figure 2.10 is a pictorial representation of the function $\{y | y = x^3 \wedge x \in [-3, 3]\}$.

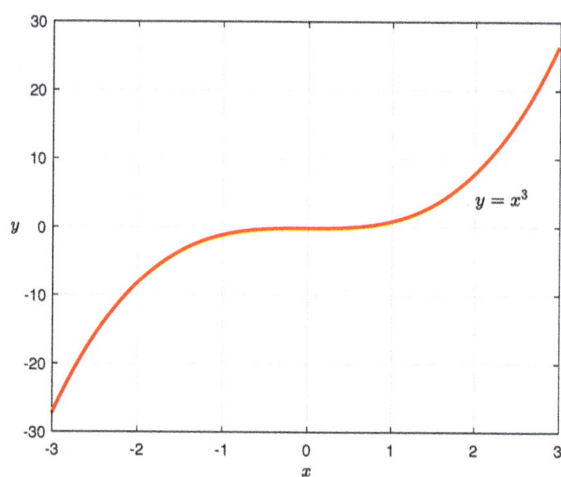

Fig. 2.10 An example of the graph of a function

2.3.2 Common Functions with One Variable

2.3.2.1 Linear Function

A linear function with one independent variable has the following form:

$$y = f(x) = a_0 + a_1 x,$$

where a_0 is the intercept on the vertical axis in the graph (the constant term) and a_1 is the slope of the line in the graph (the coefficient). If $a_1 = 0$, then $y = a_0$, that is, the line is horizontal. If $a_0 = 0$, the line will pass through the origin. If $a_0 \neq 0$ and $a_1 \neq 0$, there are two cases with $a_1 > 0$ and $a_1 < 0$, respectively.

> **Example 2.13** Figure 2.11 shows two linear functions, $y = 5 - 2x$ and $y = 5 + 2x$. Both functions have the same intercept, but one has a negative slope (solid line), and the other has a positive slope (dashed line).

2.3.2.2 Polynomial Function

These are functions built out of non-negative integer powers of the independent variable.

Fig. 2.11 Two linear functions with the same intercept: a dashed line with a positive slope and a solid line with a negative slope

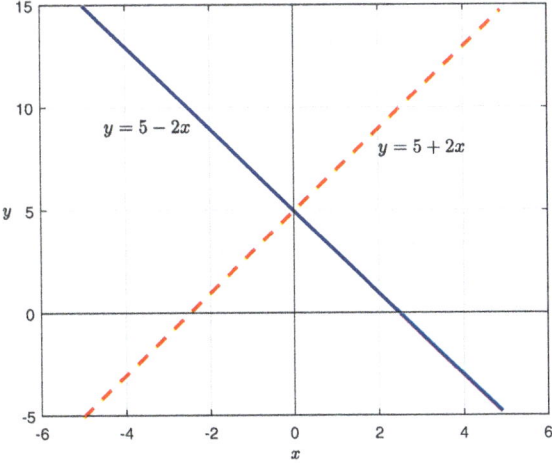

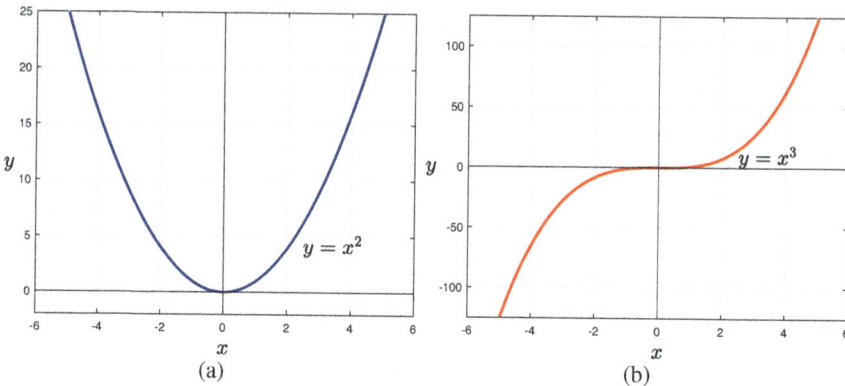

Fig. 2.12 Examples of two simple polynomial functions

Example 2.14 Figure 2.12 illustrates two simple polynomial functions. Panel (a) shows $f(x) = x^2$ and panel (b) depicts $f(x) = x^3$, both defined for $x \in [-5, 5]$.

To have a more complicated polynomial function, we can start with the building blocks, such as $1, x, x^2, x^3$, and so on, and we can multiply these basic functions by numbers and then add a finite number of them together. For example, $f(x) = 7x^2 + 4x^3 - 2$.

2.3.2.3 Exponential Function

The exponential function is defined as $f(x) = a^x$, where the base $a > 0$ and $a \neq 1$. The domain of the function is $(-\infty, \infty)$; the range of the function is $(0, \infty)$. The graph of the function always passes $(0, 1)$ since $a^0 = 1$.

Example 2.15 Figure 2.13 shows two exponential functions with a domain of $[-5, 5]$. Panel (a) displays $f(x) = 2^x$ with a base equal to 2, and panel (b) presents $f(x) = 2^{-x} = (\frac{1}{2})^x$ with a base of $\frac{1}{2}$.

There is a horizontal asymptote at $y = 0$. The curve does not touch the x-axis, no matter what it looks like on the graph. In fact, the graph of $f(x) = 2^x$ is just the reflection of $f(x) = 2^{-x}$ in the y-axis. A common base is $a = e = 2.71828$, using the irrational number e, giving the function $f(x) = e^x$.

2.3 Functions

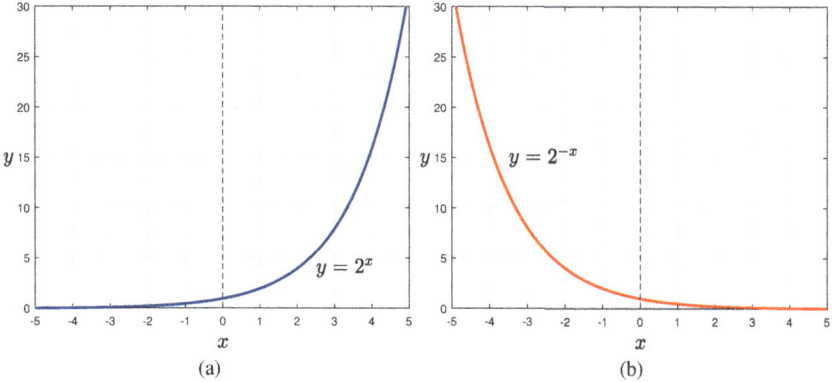

Fig. 2.13 Examples of two simple exponential functions

Exponential functions are important in the Data Science field. When we introduce probability distributions in Chap. 10, we will see that a Gaussian distribution of a random variable is, in fact, a member of the exponential function family. In addition, there are many applications using a type of function that is a combination of exponential functions, for example, $\frac{e^x - e^{-x}}{2}$ and $\frac{e^x + e^{-x}}{2}$. These are special functions: hyperbolic functions. Readers are referred to [6] to find more details.

2.3.2.4 Logarithmic Function

A logarithmic function is denoted as $f(x) = \log_a x$, where a is a constant and $a > 0$, but $a \neq 1$. The domain of a logarithmic function is $(0, \infty)$. The graph of a^x is symmetric to the graph of $\log_a x$ about the line of $y = x$ (see Fig. 2.14). If the base $a = e = 2.71828$, then we denote $\log_a x = \log_e x$ as $\ln x$.

The logarithmic function is also important in the Data Science field. We will discuss a cost function or error function defined in a log probability format in Chap. 13.

2.3.2.5 Trigonometric Functions

The variable x in these functions is generally expressed in radians (π radians = 180°). Figure 2.15 shows $\sin x$ and $\cos x$, respectively, in the domain [−8radians, 8radians]. The relations between $\sin x$ and $\cos x$ can be summarised as:

$$\sin x = \cos(\frac{\pi}{2} - x), \qquad (2.1)$$

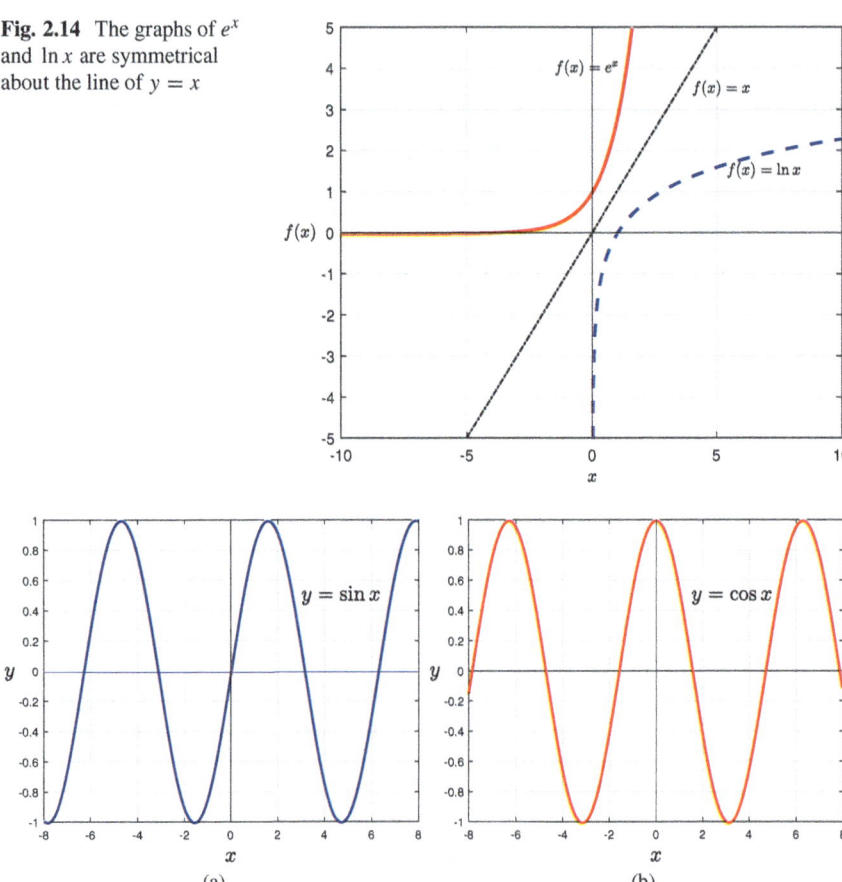

Fig. 2.14 The graphs of e^x and $\ln x$ are symmetrical about the line of $y = x$

Fig. 2.15 Panel (**a**) shows the $\sin x$ function, while panel (**b**) shows the $\cos x$ function

$$\cos x = \sin(\frac{\pi}{2} - x), \qquad (2.2)$$

$$\sin^2 x + \cos^2 x = 1. \qquad (2.3)$$

Equations (2.1), (2.2), and (2.3) are trigonometric identities involving trigonometric functions. There are quite a lot of them, but it is not necessary to remember all the identities. However, it is important to know that these identities are useful when we need to simplify trigonometric functions, and sometimes, they can be used to solve certain types of integrals. We will introduce integrals in Chaps. 5 and 6.

2.3.3 Properties of a Function

Definition 2.14 (Bounded Functions) If there is a constant M such that $f(x) \leq M$ for all x in an interval, then M is called an *upper bound* of the function. On the other hand, if there is a constant M such that $f(x) \geq M$ for all x in an interval, then M is called a *lower bound* of the function. Usually, to indicate that a function is bounded both above and below by M, we write $|f(x)| \leq M$, where M is a non-negative real value and $|f(x)|$ means *the absolute value of $f(x)$*.

Example 2.16 $f(x) = \cos x$ is bounded in the interval $(-\infty, \infty)$ since for all $x \in \mathbb{R}$, $|\cos x| \leq 1$ is valid. Here $M = 1$. Of course, M can be any value not less than 1 in this example.

Example 2.17 Let us consider $f(x) = \frac{1}{x}$. First, suppose $x \in (0, 1)$. We notice that however big a value M takes, we can always find a small value approaching zero for x, so that $\frac{1}{x}$ is greater than M. Therefore, the function is unbounded, since there does not exist an M, so that $|\frac{1}{x}| \leq M$ is valid in the interval $(0, 1)$. Next, suppose $x \in (1, 3)$. Then, the function is bounded. For example, taking $M = 1$, then $|\frac{1}{x}| \leq 1$ is valid to all x in the interval $(1, 3)$.

Example 2.18 Other bounded function examples are the sigmoid functions. One common sigmoid function, defined by $f(x) = \frac{1}{1+e^{-x}} = \frac{e^x}{e^x+1}$, is bounded inside the interval $(0, 1)$ (see Fig. 2.16, where the domain is defined as $[-7.8, 7.8]$).

Definition 2.15 (Monotonic Functions) Given a function $f(x)$ in an interval, for any two points x_1 and x_2 in the interval, if we have:

- $x_1 < x_2$ and $f(x_1) \leq f(x_2)$, then the function is monotonic increasing. If $f(x_1) < f(x_2)$, then the function is called strictly increasing.
- $x_1 < x_2$ and $f(x_1) \geq f(x_2)$, then the function is monotonic decreasing. If $f(x_1) > f(x_2)$, then the function is called strictly decreasing.

Fig. 2.16 A plot of the sigmoid function

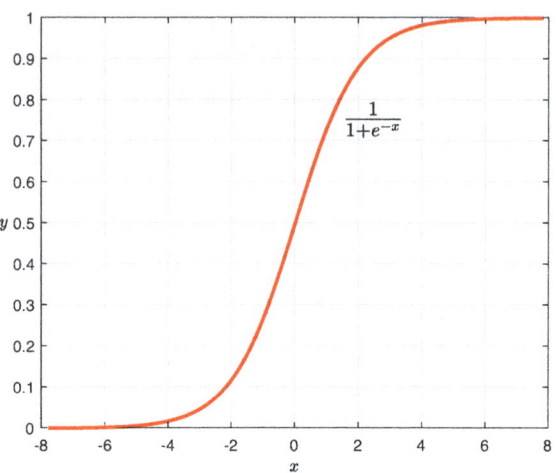

Example 2.19 $f(x) = x^2$ is monotonic increasing in the interval of $[0, \infty)$ and monotonic decreasing in $(-\infty, 0]$. It is not monotonic in the domain of $(-\infty, \infty)$ (see Fig. 2.12a). However, $f(x) = x^3$ is monotonic increasing in the domain of $(-\infty, \infty)$ (see Fig. 2.12b).

Definition 2.16 (Odd and Even Functions) A function $f(x)$ is called odd if $f(-x) = -f(x)$ for all x in the domain. A function $f(x)$ is called even if $f(-x) = f(x)$ for all x in the domain.

An odd function is symmetrical about the origin of the coordinate system. An even function is symmetrical about the y-axis of the coordinate system. A function does not have to be even or odd.

Example 2.20 $f(x) = x^3$ is an odd function, because $f(-x) = (-x)^3 = -x^3 = -f(x)$, and it is symmetrical about the origin (see Fig. 2.12b). $f(x) = x^2$ is an even function, because $f(-x) = (-x)^2 = x^2 = f(x)$, and it is symmetrical about the y-axis (see Fig. 2.12a).

2.3 Functions

Exercise

2.9 Is each of the following functions even, odd, or neither even nor odd? [Hint: some can be answered by looking at the figures given above.]
(1) $x^5 - 2x^3 + 3x$.
(2) $x^3 - 2x + 1$.
(3) e^x.
(4) $\ln x$.
(5) $\sin x$.
(6) $\cos x$.
(7) Sigmoid: $\frac{1}{1+e^{-x}}$.
(8) Hyperbolic Cosine: $\frac{e^x + e^{-x}}{2}$.
(9) Hyperbolic Tangent: $\frac{e^x - e^{-x}}{e^x + e^{-x}}$.

Definition 2.17 (Period of a Function) Given a function $f(x)$, if there exists $l \neq 0$, so that $f(x + l) = f(x)$ is valid for any x value in the domain, then the function is a periodic function and l is called the period of the function.

Example 2.21 Functions $\sin x$ and $\cos x$ are periodic functions with a period of 2π (see Fig. 2.15).

Here are two more trigonometric identities:

$$\sin(x + 2\pi) = \sin x, \tag{2.4}$$

$$\cos(x + 2\pi) = \cos x. \tag{2.5}$$

2.3.4 Inverse Functions

Definition 2.18 (Inverse Functions) Suppose $y = f(x)$. If the relation between the domain and range values is one-to-one, then a new function f^{-1} can be created by interchanging the domain and range of f. f^{-1} is called the inverse function. It can be denoted as $x = f^{-1}(y)$. However, it is usually convenient to rename the domain variable as x and the range variable as y, giving the notation $y = f^{-1}(x)$. The inverse function is symmetric about the line $y = x$ with the original function.

Example 2.22 Suppose $y = x^2$ with $x \in (-\infty, \infty)$ and $y \in [0, \infty)$. There is no inverse function here since, for instance, both 2^2 and $(-2)^2$ are equal to 4, and therefore the function is not one to one. This means the inverse relation is one to many and so not a function.

If we limit the domain to be $x \in [0, \infty)$, the inverse function is $y = \sqrt{x}$. On the other hand, if we limit the domain to be $x \in (-\infty, 0]$, the inverse function is $y = -\sqrt{x}$.

Example 2.23 The logarithmic function $f(x) = \ln x$ is the inverse of the exponential function $f(x) = e^x$ (see Fig. 2.14).

Example 2.24 Inverse trigonometric functions.

By convention, we denote the inverse of $\sin x$ as $arcsin(x)$, and the inverse of $\cos x$ as $arccos(x)$.

Trigonometric functions are periodic and so not one to one, hence to define an inverse function we have to restrict the domain of the original function so that it is one to one. This can be done by restricting the domain to $[-\frac{\pi}{2}, +\frac{\pi}{2}]$ for $\sin x$ and restricting the domain to $[0, \pi]$ for $\cos x$. The domain of x in these two inverse functions for real results is, therefore, $[-1, 1]$.

2.3.4.1 How to Find the Inverse Function

We can apply the following procedure to find the inverse function:

- Step 1—set $y = f(x)$;
- Step 2—make x the subject;
- Step 3—replace y with x to obtain $f^{-1}(x)$.

Example 2.25 Suppose $f(x) = 6x - 3$, then this function is one to one and so $f^{-1}(x)$ exists. Find $f^{-1}(x)$.

Solution

- Step 1—set $y = f(x)$, that is $y = 6x - 3$.

(continued)

Example 2.25 (continued)
- Step 2—make x the subject, then we have $x = \frac{y+3}{6}$.
- Step 3—replace y with x to obtain $f^{-1}(x)$, then we have $f^{-1}(x) = \frac{x+3}{6}$.

Exercise

2.10 Find the inverse function for each of the following functions.
(1) $f(x) = x^3 + 10$;
(2) $f(x) = 3\sin x$;
(3) $f(x) = 4 + \ln(x+1)$;
(4) $f(x) = \frac{3^x}{3^x+1}$.

2.3.5 Composition of Functions

Definition 2.19 (**Composite Functions**) Let f and g be functions. $f \circ g$ (read as f composite g) is called a composite function, denoted as $f \circ g = f(g(x))$, where the range values of $g(x)$ are the domain values of f.

That is, to obtain the composition of functions f and g, $f \circ g$, we need to first apply the function g to x and then apply function f to $g(x)$. Similarly, to find the composition of functions f and g, $g \circ f$, we need to first apply the function f to x and then apply function g to $f(x)$

Example 2.26 Let $f(x) = 5x + 2$ and $g(x) = x^2$. Find $f \circ g$.
Solution

$$f \circ g = f(g(x)) = f(x^2) = 5x^2 + 2.$$

Exercise

2.11 Find $g \circ f$ for the following given functions g and f.
(1) $f(x) = 5x + 2$ and $g(x) = x^2$.
(2) $f(x) = 2x$ and $g(x) = \sin x$.
(3) $f(x) = e^x$ and $g(x) = x^2$.
(4) $f(x) = e^x$ and $g(x) = \ln x$.
(5) $f(x) = \cos x$ and $g(x) = x^3$.

Remark 2.3 The order in the composition of functions is important because, in general, $f \circ g(x)$ is not the same as $g \circ f(x)$.

Remark 2.4 Two functions cannot always be composited to obtain a new function. For example, let $f(x) = \arcsin x$ and $g(x) = 2 + x^2$. The composition of the functions f and g, $f \circ g(x)$, does not exist, because the range value for any x in the domain $(-\infty, \infty)$ of $g(x)$ is a value equal or greater than 2, which cannot be an input to $f(x) = \arcsin x$ (see Example 2.24 in Sect. 2.3.4 of this chapter). ♦

2.3.6 Functions of Two or More Variables

Functions may have more than one independent variable. The domain of a function of two or more variables is a set of n-tuples, while the range is one-dimensional with an interval of numbers.

Example 2.27 Let $f(x, y) = x^3 + 2\sqrt{y}$ with two independent variables x and y, whose domain can be written as follows:

$$\{(x, y) | -\infty < x < \infty \wedge y \geq 0\}.$$

We conclude this chapter here. Readers interested in gaining a deeper fundamental understanding of sets and functions are encouraged to explore classical textbooks, such as Chapter 2 of [7].

Chapter 3
Linear Algebra

In this chapter, we introduce vectors and matrices. We show how to do the basic operations on them and why we need to use both vectors and matrices. In this book, we only consider vectors and matrices that consist of finite real numbers, and not, for instance, complex numbers.

3.1 Vectors

The input in most machine learning applications is an ordered list of numbers; for instance, as indicated in Case Study 3 in Chap. 1, most neural networks consist of one or more layers, and the initial input data goes to the set of first-layer units or neurons. This input could be:

- an ordered list of features of the skin that may determine the ability of the skin to absorb medical drugs (such as Nicotine).
- the list of pixel values from scanning a picture.

In all cases, this list of ordered values is a vector, and knowledge of how to manipulate vectors is essential for a full understanding of the operation of machine learning techniques.

Definition 3.1 (Vector) A vector is an ordered list of numbers and subscripts. Each subscript denotes the position of the value in the list. Such a list of values, denoted as $\mathbf{x} = (x_1, x_2, \cdots, x_d)$, where d is the number of elements in the list, is called a linear array or vector.

Example 3.1 Five students' Maths grades are listed as follows: 82, 90, 65, 78, 46. We can denote all the values in the list using only one symbol, for instance, **x** with different subscripts, that is, x_1, x_2, x_3, x_4, x_5. Sometimes a vector is written vertically. For example,

$$\begin{bmatrix} 2 \\ 1 \\ 5 \end{bmatrix}.$$

Note that we use both a pair of square brackets and round brackets to denote vectors and matrices in this book.

3.1.1 Vectors in Physics

Vectors can be represented by arrows having appropriate lengths and directions and emanating from some given reference point. In Fig. 3.1, the reference point is $(0, 0)$, and the ending point is the vector $\mathbf{w} = (4, 3)$, whose magnitude is denoted as $||\mathbf{w}||$ and θ is the angle from the positive horizontal axis to the vector measured in an anticlockwise direction. In Fig. 3.1, the vector has two elements or components and is referred to as a vector in $\mathcal{R}^2$, where $\mathcal{R}$ is the field of real numbers. It is one of the infinite number of possible vectors in $\mathcal{R}^2$. In general, a vector has d elements over the field of real numbers in $\mathcal{R}^d$. The field of real numbers means we can do addition and scalar multiplication of vectors as with all real numbers.

Fig. 3.1 An illustration of a specified vector (4, 3), where $||\mathbf{w}||$ denotes the length of the vector, and the arrow indicates its direction from the origin

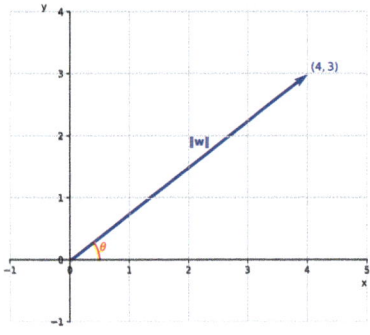

3.1.2 Vector Addition

Considering two vectors in $\mathcal{R}^d$: $\mathbf{x} = (x_1, x_2, \cdots, x_d)$ and $\mathbf{w} = (w_1, w_2, \cdots, w_d)$, their sum is given by

$$\mathbf{x} + \mathbf{w} = (x_1 + w_1, x_2 + w_2, \cdots, x_d + w_d). \tag{3.1}$$

Example 3.2 Suppose we have two vectors: $\mathbf{a} = (2, 4)$ and $\mathbf{b} = (5, 1)$. According to Eq. (3.1), $\mathbf{a} + \mathbf{b} = (2 + 5, 4 + 1) = (7, 5)$ and $\mathbf{a} - \mathbf{b} = (2 - 5, 4 - 1) = (-3, 3)$.

Figure 3.2 shows $\mathbf{a}$ and $\mathbf{b}$ in a plane. The reference point is $(0, 0)$ for both vectors. For $\mathbf{a}$, the ending point is $(2, 4)$. For $\mathbf{b}$, the ending point $(5, 1)$. If we draw a parallelogram with $\mathbf{a}$ and $\mathbf{b}$ as its two sides, then $\mathbf{a} + \mathbf{b}$ actually is the longer diagonal line; $\mathbf{a} - \mathbf{b}$, in fact, can be considered as $\mathbf{a} + (-\mathbf{b})$, that is the shorter diagonal line.

3.1.3 Scalar-Vector Multiplication

The scalar product of a vector with a real number k is given by:

$$k\mathbf{x} = (kx_1, kx_2, \cdots, kx_d), \tag{3.2}$$

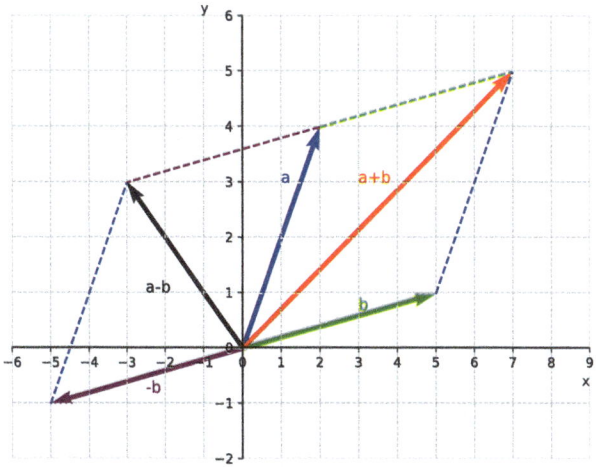

Fig. 3.2 An example of vectors

where k is any real number. When we multiply a vector with a scalar, we actually are compressing or stretching the vector. For instance, if k is 2, then we are doubling the length of the vector. We can also change the direction of the vector to the opposite direction by multiplying by a negative real number.

> **Exercise**
>
> **3.1** Compute the following:
>
> (1) $\begin{bmatrix} 2 \\ 1 \\ 5 \end{bmatrix} + \begin{bmatrix} 3 \\ 6 \\ 1 \end{bmatrix}$, and $\begin{bmatrix} 2 \\ 1 \\ 5 \end{bmatrix} - \begin{bmatrix} 3 \\ 6 \\ 1 \end{bmatrix}$.
>
> (2) $2 \times \begin{bmatrix} 2 \\ 1 \\ 5 \end{bmatrix}$, and $-2 \times \begin{bmatrix} 2 \\ 1 \\ 5 \end{bmatrix}$.

3.2 The Dot Product of Two Vectors

3.2.1 Dot Product: Algebra Definition

Definition 3.2 (Dot Product) Considering arbitrary vectors
$\mathbf{w} = (w_1, w_2, \cdots, w_d)$ and $\mathbf{x} = (x_1, x_2, \cdots, x_d)$ in $\mathcal{R}^d$,
the dot product (also referred to as an inner product) of $\mathbf{w}$ and $\mathbf{x}$ is denoted and defined by

$$\mathbf{w} \cdot \mathbf{x} = <\mathbf{w}, \mathbf{x}> = w_1 \cdot x_1 + w_2 \cdot x_2 + \cdots + w_d \cdot x_d. \tag{3.3}$$

The use of $\mathbf{w}$ and $\mathbf{x}$ is not entirely a coincidence, since it is the type of calculation that occurs in machine learning, such as in a neural network. If $\mathbf{x}$ is the input vector to a neural network, then each element is multiplied by a corresponding weight from a weight vector $\mathbf{w}$, and then all are added together to get a net input. The dot product exactly represents this operation.

> **Example 3.3** Suppose we have two vectors: $\mathbf{a} = (2, 4)$ and $\mathbf{b} = (5, 1)$. According to Eq. (3.3), we have $\mathbf{a} \cdot \mathbf{b} = 2 \times 5 + 4 \times 1 = 14$.

Exercise

3.2 Compute the dot product for the following vectors:

(1) $\mathbf{m} = (-2, -1)$ and $\mathbf{n} = (1, 3)$.
(2) $\mathbf{u} = (-2, -1, 3)$ and $\mathbf{v} = (1, 3, -2)$.
(3) $\mathbf{s} = (-2, 2)$ and $\mathbf{t} = (3, 3)$.
(4) $\mathbf{a} = (4, 3, 5)$ and $\mathbf{b} = (-4, -3, 5)$.

3.2.2 Norm

Definition 3.3 (Norm) The norm or length of a vector $\mathbf{w}$ in $\mathcal{R}^d$, denoted by $\|\mathbf{w}\|$, is defined to be the non-negative square root of $\mathbf{w} \cdot \mathbf{w}$. That is, if $\mathbf{w} = (w_1, w_2, \cdots, w_d)$, then

$$\|\mathbf{w}\| = \sqrt{\mathbf{w} \cdot \mathbf{w}} = \sqrt{w_1^2 + w_2^2 + \cdots + w_d^2}. \quad (3.4)$$

Because a norm is the length of a vector, we can use it to measure the distance between two points.

Example 3.4 Continue Example 3.2.
In Fig. 3.3, the distance between two points P and Q is measured as the norm of $\mathbf{a} - \mathbf{b}$, where $\mathbf{a}$ is $\overrightarrow{OP}$ and $\mathbf{b}$ is $\overrightarrow{OQ}$. That is

$$\|QP\| = \sqrt{(2-5)^2 + (4-1)^2} = 3\sqrt{2}.$$

The distance between O and M is measured as the norm of $\overrightarrow{OM}$. That is

$$\|OM\| = \sqrt{(7-0)^2 + (5-0)^2} = \sqrt{74}.$$

Considering arbitrary vectors $\mathbf{w} = (w_1, w_2, \cdots, w_d)$ and $\mathbf{x} = (x_1, x_2, \cdots, x_d)$ in $\mathcal{R}^d$, in general, the distance between these two vectors is defined as follows:

$$d(\mathbf{w}, \mathbf{x}) = \sqrt{(w_1 - x_1)^2 + (w_2 - x_2)^2 + \cdots + (w_d - x_d)^2}. \quad (3.5)$$

Fig. 3.3 An illustration of distances between two points

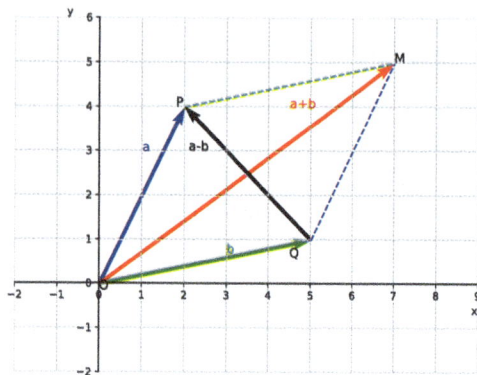

Exercise

3.3 Compute the distance between the following vectors:

(1) $\mathbf{w} = (1, 10, 3, 2)$ and $\mathbf{z} = (5, 4, -1, 0)$.
(2) $\mathbf{a} = (4, 3, 5)$ and $\mathbf{b} = (-4, -3, 5)$.

3.2.3 Vector Magnitude and Direction in $\mathcal{R}^2$

Suppose we have a vector $\mathbf{u} = (x_1, y_1)$ (see Fig. 3.4). Its magnitude is its norm, that is, $\|\mathbf{u}\| = \sqrt{\mathbf{u} \cdot \mathbf{u}} = \sqrt{x_1^2 + y_1^2}$; its direction is given by $\theta = tan^{-1}(\frac{y_1}{x_1})$. Each component of the vector can be obtained as follows: $x_1 = \|\mathbf{u}\| \cos(\theta)$ and $y_1 = \|\mathbf{u}\| \sin(\theta)$, respectively. Note that we define θ so that $0 \leq \theta < 2\pi$.

Fig. 3.4 An illustration of a vector, with the point marking the vector's endpoint

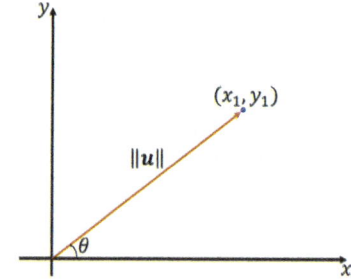

Example 3.5 Suppose we have a vector: $\mathbf{a} = (2, 4)$, then its magnitude is $\|\mathbf{a}\| = \sqrt{\mathbf{a} \cdot \mathbf{a}} = \sqrt{2^2 + 4^2} = 2\sqrt{5}$; and its direction is $\theta = tan^{-1}(\frac{4}{2}) = 1.1071$ radians.

In reverse, the components can be found from the magnitude and direction as $x_1 = 2\sqrt{5}\cos(1.1071) = 2$ and $y_1 = 2\sqrt{5}\sin(1.1071) = 4$.

Exercise

3.4 Suppose we have two vectors: $\mathbf{u} = (2, -3)$ and $\mathbf{v} = (5, 4)$.

(1) Compute the direction of each vector.
(2) Compute the length of each vector.
(3) Compute the distance between $\mathbf{u}$ and $\mathbf{v}$.

3.2.4 Dot Product: Geometric Definition

Definition 3.4 (Dot Product) Considering two vectors $\mathbf{w}$ and $\mathbf{x}$, their dot product can also be defined as follows:

$$\mathbf{w} \cdot \mathbf{x} = \|\mathbf{w}\| \|\mathbf{x}\| \cos \theta. \tag{3.6}$$

where θ is the angle between $\mathbf{w}$ and $\mathbf{x}$.

This definition of dot product can be proved to be equivalent to the previous one, Definition 3.2. A detailed explanation can be viewed in [8].

Remark 3.1 Definition 3.4 can be used to measure the similarity between two vectors in terms of the direction of the vectors.

- When $\cos \theta = 0$, that is, two vectors are at right angles to each other, referred to as being orthogonal to each other, we have $\mathbf{w} \cdot \mathbf{x} = 0$. Intuitively, it says two vectors have zero similarity.
- When $\cos \theta = 1$, that is, two vectors are pointing in the same direction, we have $\mathbf{w} \cdot \mathbf{x} = \|\mathbf{w}\| \|\mathbf{x}\|$. This is the largest value one can get for $\mathbf{w} \cdot \mathbf{x}$.
- When $\cos \theta = -1$, that is, two vectors are opposed to each other, we have $\mathbf{w} \cdot \mathbf{x} = -\|\mathbf{w}\| \|\mathbf{x}\|$. This is the most negative value one can get for $\mathbf{w} \cdot \mathbf{x}$.

♦

Fig. 3.5 An illustration of two vectors in Example 3.6

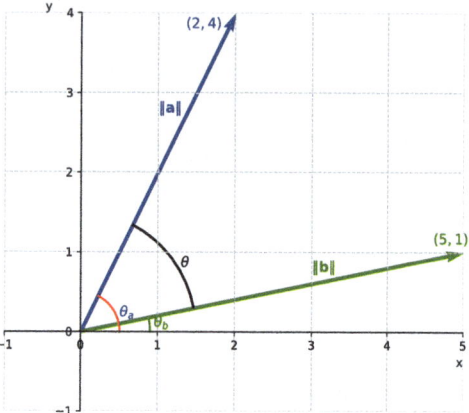

Example 3.6 Calculate the dot product of the two vectors: **a** = (2, 4) and **b** = (5, 1) in Example 3.3 using Eq. (3.6).

Solution $\|\mathbf{a}\| = \sqrt{2^2 + 4^2} = \sqrt{20}$ and $\|\mathbf{b}\| = \sqrt{5^2 + 1^2} = \sqrt{26}$.

Since the direction of **a** is $\theta_\mathbf{a} = tan^{-1}(\frac{4}{2})$ and the direction of **b** is $\theta_\mathbf{b} = tan^{-1}(\frac{1}{5})$, the angle between **a** and **b** is $\theta = \theta_\mathbf{a} - \theta_\mathbf{b} \approx 0.9098$ radians (see Fig. 3.5). The cos of 0.9098 radians is about 0.6139. Thus, we have

$$\|\mathbf{a}\|\,\|\mathbf{b}\|\cos\theta = \sqrt{20} \times \sqrt{26} \times 0.6139 \approx 14.$$

This is the same answer as we got before in Example 3.3.

Exercise

3.5 Compute the dot product for the following vectors using both Eqs. (3.3) and (3.6), and check that they are the same.

(1) **u** = (2, 2) and **v** = (3, 3).
(2) **u** = (2, 2) and **w** = (−2, 2).
(3) **u** = (2, 2) and **s** = (−2, −2).
(4) **u** = (2, 2) and **t** = (0, 5).

3.2.5 Unit Vector

Definition 3.5 (Unit Vector) For any non-zero vector **u** in $\mathcal{R}^d$, the vector $\hat{\mathbf{u}} = \frac{\mathbf{u}}{\|\mathbf{u}\|}$ is a unit vector in the same direction as **u**. The process of finding $\hat{\mathbf{u}}$ from **u** is called normalising **u**.

Example 3.7 For a given vector $\mathbf{u} = (2, 3, 1)$, its unit vector can be calculated as follows:

$$\hat{\mathbf{u}} = \frac{\mathbf{u}}{\|\mathbf{u}\|} = \frac{(2, 3, 1)}{\sqrt{2^2 + 3^2 + 1^2}} = (\frac{2}{\sqrt{14}}, \frac{3}{\sqrt{14}}, \frac{1}{\sqrt{14}}).$$

Exercise

3.6 Calculate the unit vector for each of the following vectors:

(1) $\mathbf{w} = (2, 1)$.
(2) $\mathbf{s} = (3, 1)$.
(3) $\mathbf{t} = (3, 1, -1)$.
(4) $\mathbf{v} = (-1, 2, 4, 1)$.

3.3 Matrices

Definition 3.6 (Matrix) A matrix is a rectangular arrangement of numbers made up of rows and columns. A matrix with m rows and n columns is called an $m \times n$ matrix. Each element in a matrix (**M**) is identified by two indices: the first one indicates the specific row, and the second indicates the column.

A matrix is usually labeled with a (bold) capital letter. For example,

$$\mathbf{M} = \begin{bmatrix} M_{1,1} & M_{1,2} & M_{1,3} \\ M_{2,1} & M_{2,2} & M_{2,3} \end{bmatrix}.$$

M is a rectangular matrix, and $M_{1,2}$ represents the element in the first row and the second column of M. A matrix with the same number of rows and columns is called a square matrix.

Example 3.8 Five students' Maths grades are listed as follows:

82, 90, 65, 78, 46.

Their corresponding English grades are listed as follows:

76, 78, 60, 50, 60.

The matrix looks like

$$\mathbf{M} = \begin{bmatrix} 82 & 76 \\ 90 & 78 \\ 65 & 60 \\ 78 & 50 \\ 46 & 60 \end{bmatrix},$$

where each column corresponds to the specific subject marks and each row represents one student's marks for two subjects.

Example 3.9 In terms of neural networks, if there are multiple input units, then each unit will have a weight vector **w**. So, we will need a compact method to represent this collection of weights, and matrices represent just what is needed. Also, as we shall see in Sect. 9.2 of Chap. 9, the multiplication of a vector and a matrix of these weights is just the operation we need to represent the complete operation of finding the inputs to the first layer of the neural network.

3.3.1 Matrix Addition

The sum of two matrices **M** and **N**, where **M** and **N** must be the same size, is the matrix obtained by adding corresponding elements from **M** and **N**.

Example 3.10

$$\begin{bmatrix} 2 & 4 \\ 0 & 1 \end{bmatrix} + \begin{bmatrix} 1 & 4 \\ 1 & 0 \end{bmatrix} = \begin{bmatrix} 2+1 & 4+4 \\ 0+1 & 1+0 \end{bmatrix} = \begin{bmatrix} 3 & 8 \\ 1 & 1 \end{bmatrix}.$$

3.3.2 Scalar Multiplication

The product of the matrix **M** by a scalar k written $k \cdot \mathbf{M}$ or simply $k\mathbf{M}$ is the matrix obtained by multiplying each element of **M** by k.

Example 3.11

$$5 \times \begin{bmatrix} 1 & 2 \\ 3 & 5 \end{bmatrix} = \begin{bmatrix} 5 \times 1 & 5 \times 2 \\ 5 \times 3 & 5 \times 5 \end{bmatrix} = \begin{bmatrix} 5 & 10 \\ 15 & 25 \end{bmatrix}.$$

Exercise

3.7 Let

$$\mathbf{U} = \begin{bmatrix} -3 & 10 \\ 9 & 0.6 \\ 1 & -5 \end{bmatrix}, \mathbf{V} = \begin{bmatrix} -1 & 2 \\ 1 & 0 \\ 0 & 1 \end{bmatrix}.$$

Find

(1) $\mathbf{U} + \mathbf{V}$.

(2) $2\mathbf{U} - 4\mathbf{V}$.

(3) $-3\mathbf{U} + 2\mathbf{V}$.

Fig. 3.6 An illustration of matrix multiplication: **C = AB**

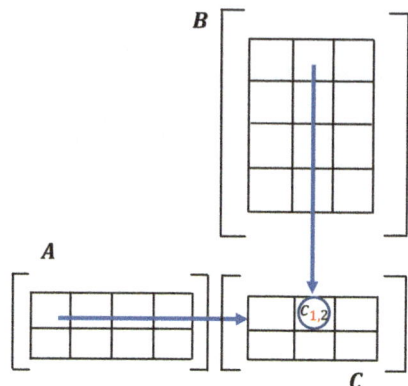

3.3.3 Matrix Multiplication

The product of two matrices **A** and **B** is somewhat complicated. Each element of the resultant matrix (**C** = **AB**) is the dot product of a row from the first matrix **A** and a column from the second matrix **B**. The first index of the element of the resultant matrix tells us which row we need to use from the first matrix **A**, and the second index tells us which column we need to use from the second matrix **B**. Figure 3.6 shows an example of **C** = **AB**, where **A** has two rows and four columns and **B** has four rows and three columns. $c_{1,2}$ is the dot product of the first row of **A** and the second column of **B**.

Example 3.12 Suppose we have two matrices

$$\begin{bmatrix} 2 & 3 & 0 \\ 1 & 4 & 5 \end{bmatrix} \text{ and } \begin{bmatrix} 1 & 2 \\ 7 & 3 \\ 0 & 5 \end{bmatrix},$$

the following shows how we compute the matrix multiplication:

$$\begin{bmatrix} 2 & 3 & 0 \\ 1 & 4 & 5 \end{bmatrix} \begin{bmatrix} 1 & 2 \\ 7 & 3 \\ 0 & 5 \end{bmatrix} = \begin{bmatrix} 2 \times 1 + 3 \times 7 + 0 \times 0 & 2 \times 2 + 3 \times 3 + 0 \times 5 \\ 1 \times 1 + 4 \times 7 + 5 \times 0 & 1 \times 2 + 4 \times 3 + 5 \times 5 \end{bmatrix}$$

$$= \begin{bmatrix} 23 & 13 \\ 29 & 39 \end{bmatrix}.$$

3.3 Matrices

Fig. 3.7 This is an example where the two matrices cannot be multiplied

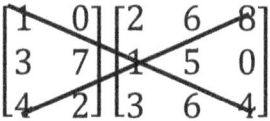

Remark 3.2 Note that to calculate the multiplication of two matrices, the number of columns of the first matrix must equal the number of rows of the second matrix. In Fig. 3.7, the first matrix has two columns, while the second matrix has three rows. Since they are not equal, one cannot calculate the matrix multiplication of these two matrices. ♦

Remark 3.3 If two matrices can be multiplied, the final resultant matrix has the same number of rows as the first matrix and the same number of columns as the second matrix. ♦

Matrices and vectors can also be multiplied together, providing they have appropriate sizes. For instance, a matrix with two columns can be multiplied by a two-component vector written vertically (which can be thought of as a matrix with just one column).

Example 3.13 If you want to multiply the matrix $\begin{bmatrix} 8 & 7 \\ 9 & 6 \end{bmatrix}$ by the vector $\begin{bmatrix} 3 \\ -4 \end{bmatrix}$, we get:

$$\begin{bmatrix} 8 & 7 \\ 9 & 6 \end{bmatrix} \begin{bmatrix} 3 \\ -4 \end{bmatrix} = \begin{bmatrix} 8 \times 3 + 7 \times (-4) \\ 9 \times 3 + 6 \times (-4) \end{bmatrix} = \begin{bmatrix} -4 \\ 3 \end{bmatrix}.$$

Exercise

3.8 Compute the following:

(1) $\begin{bmatrix} 10 & 3 & 2 \\ 1 & 2 & 5 \end{bmatrix} + \begin{bmatrix} 10 & 7 & 0 \\ 16 & 6 & 9 \end{bmatrix}$.

(2) $3 \times \begin{bmatrix} 3 & 2 \\ 0 & 1 \\ 5 & -1 \end{bmatrix}$.

(continued)

(3) $\begin{bmatrix} 9 & -2 \\ 2 & 6 \end{bmatrix} \begin{bmatrix} 2 \\ 1 \end{bmatrix}.$

(4) $\begin{bmatrix} 10 & 3 & 2 \\ 1 & 2 & 5 \end{bmatrix} \begin{bmatrix} 2 & 2 \\ 1 & 8 \\ 2 & 5 \end{bmatrix}.$

3.3.3.1 Properties of Matrix Multiplication

Let $\mathbf{A}$, $\mathbf{B}$, and $\mathbf{C}$ are $n \times n$ matrices.

- Associative property of multiplication

$$(\mathbf{AB})\mathbf{C} = \mathbf{A}(\mathbf{BC}). \tag{3.7}$$

Example 3.14 Suppose we have

$$\mathbf{A} = \begin{bmatrix} 2 & 2 & 0.6 \\ 1 & 0.1 & 8 \\ 3 & 2 & 10 \end{bmatrix}, \mathbf{B} = \begin{bmatrix} 2 & 1 & 7 \\ 5 & 0 & 3 \\ 1 & 6 & -2 \end{bmatrix} \text{ and } \mathbf{C} = \begin{bmatrix} 3 & 2 & -1 \\ 4 & -2 & 10 \\ 7 & 0.5 & 5 \end{bmatrix}.$$

We can compute the following:

$$\mathbf{AB} = \begin{bmatrix} 2 & 2 & 0.6 \\ 1 & 0.1 & 8 \\ 3 & 2 & 10 \end{bmatrix} \begin{bmatrix} 2 & 1 & 7 \\ 5 & 0 & 3 \\ 1 & 6 & -2 \end{bmatrix} = \begin{bmatrix} 14.6 & 5.6 & 18.8 \\ 10.5 & 49 & -8.7 \\ 26 & 63 & 7 \end{bmatrix}.$$

$$(\mathbf{AB})\mathbf{C} = \begin{bmatrix} 14.6 & 5.6 & 18.8 \\ 10.5 & 49 & -8.7 \\ 26 & 63 & 7 \end{bmatrix} \begin{bmatrix} 3 & 2 & -1 \\ 4 & -2 & 10 \\ 7 & 0.5 & 5 \end{bmatrix} = \begin{bmatrix} 197.8 & 27.4 & 135.4 \\ 166.6 & -81.35 & 436 \\ 379 & -70.5 & 639 \end{bmatrix}.$$

Then we compute the following:

$$\mathbf{BC} = \begin{bmatrix} 2 & 1 & 7 \\ 5 & 0 & 3 \\ 1 & 6 & -2 \end{bmatrix} \begin{bmatrix} 3 & 2 & -1 \\ 4 & -2 & 10 \\ 7 & 0.5 & 5 \end{bmatrix} = \begin{bmatrix} 59 & 5.5 & 43 \\ 36 & 11.5 & 10 \\ 13 & -11 & 49 \end{bmatrix}.$$

$$\mathbf{A}(\mathbf{BC}) = \begin{bmatrix} 2 & 2 & 0.6 \\ 1 & 0.1 & 8 \\ 3 & 2 & 10 \end{bmatrix} \begin{bmatrix} 59 & 5.5 & 43 \\ 36 & 11.5 & 10 \\ 13 & -11 & 49 \end{bmatrix} = \begin{bmatrix} 197.8 & 27.4 & 135.4 \\ 166.6 & -81.35 & 436 \\ 379 & -70.5 & 639 \end{bmatrix}.$$

That is, we have $(\mathbf{AB})\mathbf{C} = \mathbf{A}(\mathbf{BC})$.

3.3 Matrices

- Distributive properties

$$C(A + B) = CA + CB. \tag{3.8}$$

Example 3.15 Continue with matrices in Example 3.14. We have

$$A + B = \begin{bmatrix} 2 & 2 & 0.6 \\ 1 & 0.1 & 8 \\ 3 & 2 & 10 \end{bmatrix} + \begin{bmatrix} 2 & 1 & 7 \\ 5 & 0 & 3 \\ 1 & 6 & -2 \end{bmatrix} = \begin{bmatrix} 4 & 3 & 7.6 \\ 6 & 0.1 & 11 \\ 4 & 8 & 8 \end{bmatrix}.$$

$$C(A + B) = \begin{bmatrix} 3 & 2 & -1 \\ 4 & -2 & 10 \\ 7 & 0.5 & 5 \end{bmatrix} \begin{bmatrix} 4 & 3 & 7.6 \\ 6 & 0.1 & 11 \\ 4 & 8 & 8 \end{bmatrix} = \begin{bmatrix} 20 & 1.2 & 36.8 \\ 44 & 91.8 & 88.4 \\ 51 & 61.05 & 98.7 \end{bmatrix}.$$

In addition, we can compute and get the following:

$$CA = \begin{bmatrix} 3 & 2 & -1 \\ 4 & -2 & 10 \\ 7 & 0.5 & 5 \end{bmatrix} \begin{bmatrix} 2 & 2 & 0.6 \\ 1 & 0.1 & 8 \\ 3 & 2 & 10 \end{bmatrix} = \begin{bmatrix} 5 & 4.2 & 7.8 \\ 36 & 27.8 & 86.4 \\ 29.5 & 24.05 & 58.2 \end{bmatrix},$$

$$CB = \begin{bmatrix} 3 & 2 & -1 \\ 4 & -2 & 10 \\ 7 & 0.5 & 5 \end{bmatrix} \begin{bmatrix} 2 & 1 & 7 \\ 5 & 0 & 3 \\ 1 & 6 & -2 \end{bmatrix} = \begin{bmatrix} 15 & -3 & 29 \\ 8 & 64 & 2 \\ 21.5 & 37 & 40.5 \end{bmatrix},$$

and

$$CA + CB = \begin{bmatrix} 20 & 1.2 & 36.8 \\ 44 & 91.8 & 88.4 \\ 51 & 61.05 & 98.7 \end{bmatrix}.$$

That is, we have $C(A + B) = CA + CB$.

$$(A + B)C = AC + BC. \tag{3.9}$$

Example 3.16 Continue with matrices in Example 3.14 again. We have calculated $(\mathbf{A} + \mathbf{B})$ in Example 3.15. So, we can compute

$$(\mathbf{A}+\mathbf{B})\mathbf{C} = \begin{bmatrix} 4 & 3 & 7.6 \\ 6 & 0.1 & 11 \\ 4 & 8 & 8 \end{bmatrix} \begin{bmatrix} 3 & 2 & -1 \\ 4 & -2 & 10 \\ 7 & 0.5 & 5 \end{bmatrix} = \begin{bmatrix} 77.2 & 5.8 & 64 \\ 95.4 & 17.3 & 50 \\ 100 & -4 & 116 \end{bmatrix}.$$

In addition, we can compute and get the following:

$$\mathbf{AC} = \begin{bmatrix} 2 & 2 & 0.6 \\ 1 & 0.1 & 8 \\ 3 & 2 & 10 \end{bmatrix} \begin{bmatrix} 3 & 2 & -1 \\ 4 & -2 & 10 \\ 7 & 0.5 & 5 \end{bmatrix} = \begin{bmatrix} 18.2 & 0.3 & 21 \\ 59.4 & 5.8 & 40 \\ 87 & 7 & 67 \end{bmatrix},$$

$$\mathbf{BC} = \begin{bmatrix} 2 & 1 & 7 \\ 5 & 0 & 3 \\ 1 & 6 & -2 \end{bmatrix} \begin{bmatrix} 3 & 2 & -1 \\ 4 & -2 & 10 \\ 7 & 0.5 & 5 \end{bmatrix} = \begin{bmatrix} 59 & 5.5 & 43 \\ 36 & 11.5 & 10 \\ 13 & -11 & 49 \end{bmatrix},$$

and

$$\mathbf{AC} + \mathbf{BC} = \begin{bmatrix} 77.2 & 5.8 & 64 \\ 95.4 & 17.3 & 50 \\ 100 & -4 & 116 \end{bmatrix}.$$

That is, we have $(\mathbf{A} + \mathbf{B})\mathbf{C} = \mathbf{AC} + \mathbf{BC}$.

3.3.4 Matrices as Linear Transformations

Now you know how to do matrix addition and matrix multiplication. But what is a matrix? Why do you need to use matrices? There are several different very useful properties of matrices. In this book, let us consider two of them. First, let us consider matrices that model linear transformations. We will consider linear transformations in just two dimensions.

Suppose you want to rotate the triangle with vertices $\mathbf{u} = \begin{bmatrix} 1 \\ 3 \end{bmatrix}$, $\mathbf{v} = \begin{bmatrix} 2 \\ 3 \end{bmatrix}$ and $\mathbf{w} = \begin{bmatrix} 1 \\ 1 \end{bmatrix}$ through 90° anticlockwise about the origin as shown in Fig. 3.8.

3.3 Matrices

Fig. 3.8 This is an example of matrices used to represent linear transformations

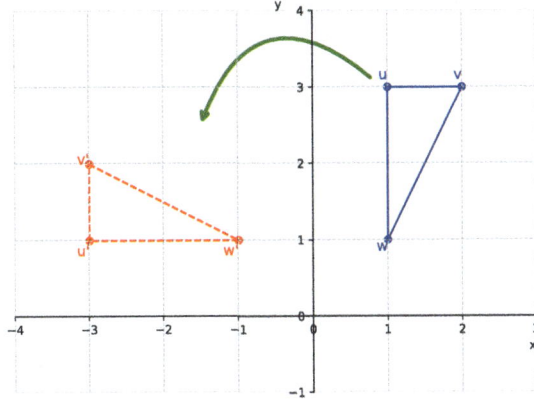

As you can see in Fig. 3.8, $\begin{bmatrix} 1 \\ 3 \end{bmatrix}$ has transformed to $\begin{bmatrix} -3 \\ 1 \end{bmatrix}$, $\begin{bmatrix} 2 \\ 3 \end{bmatrix}$ to $\begin{bmatrix} -3 \\ 2 \end{bmatrix}$ and $\begin{bmatrix} 1 \\ 1 \end{bmatrix}$ to $\begin{bmatrix} -1 \\ 1 \end{bmatrix}$. It looks as if all three points follow the following rule:

$$\begin{bmatrix} x \\ y \end{bmatrix} \text{ is transformed to } \begin{bmatrix} -y \\ x \end{bmatrix}.$$

This can be expressed as a matrix equation:

$$\begin{bmatrix} x' \\ y' \end{bmatrix} = \begin{bmatrix} 0 & -1 \\ 1 & 0 \end{bmatrix} \begin{bmatrix} x \\ y \end{bmatrix},$$

where the matrix $\begin{bmatrix} 0 & -1 \\ 1 & 0 \end{bmatrix}$ represents a 90° anticlockwise rotation about the origin. If we wish to have a more general rotation with any degree (θ) about the origin, we can use the following matrix:

$$\begin{bmatrix} \cos(\theta) & -\sin(\theta) \\ \sin(\theta) & \cos(\theta) \end{bmatrix}.$$

Matrices can also represent other linear transformations, such as reflections.

Fig. 3.9 Vector **u** is rotated by 45° in the anticlockwise direction around the origin

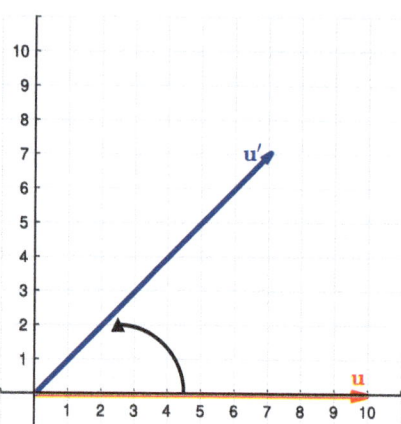

Example 3.17 Suppose there is a vector $\mathbf{u} = \begin{bmatrix} 10 \\ 0 \end{bmatrix}$. You wish to rotate **u** through 45° in the anticlockwise direction around the origin. This is illustrated in Fig. 3.9, where the original vector is along the horizontal axis, and the new vector is located at $\mathbf{u}' = \begin{bmatrix} 7.1 \\ 7.1 \end{bmatrix}$ after the rotation. The new position keeping one decimal place is calculated as follows:

$$\mathbf{u}' = \begin{bmatrix} \cos(45°) & -\sin(45°) \\ \sin(45°) & \cos(45°) \end{bmatrix} \begin{bmatrix} 10 \\ 0 \end{bmatrix} = \begin{bmatrix} 7.1 \\ 7.1 \end{bmatrix}.$$

3.3.5 Representations of Simultaneous Equations

Let us look at the second useful property of matrices by considering the following example. Suppose we have two TV producers (*A* and *B*) that send different proportions of TVs they produce to three warehouses (1, 2, and 3), as shown in Fig. 3.10.

We can model this as a matrix shown as follows:

$$\mathbf{U} = \begin{bmatrix} 0.5 & 0.5 & 0.0 \\ 0.3 & 0.3 & 0.4 \end{bmatrix},$$

3.3 Matrices

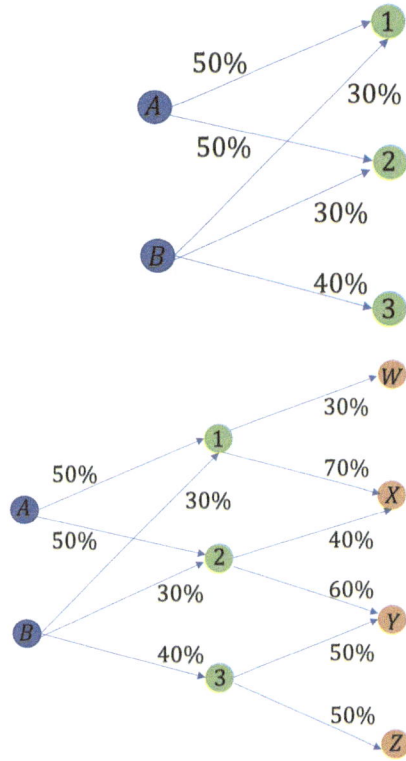

Fig. 3.10 Two TV producers (A and B) that send different proportions of TVs they produce to three warehouses (1, 2, and 3)

Fig. 3.11 The three warehouses then forward TVs in different proportions to four shops (W, X, Y, and Z)

where each row represents one of the two producers, each column represents one of the three warehouses, and each element in the matrix is the proportion of the TVs each producer sends to the specific warehouse.

The three warehouses then forward TVs in different proportions to four shops (W, X, Y, and Z), as shown in Fig. 3.11. Similarly, we can model this as a matrix as well:

$$\mathbf{V} = \begin{bmatrix} 0.3 & 0.7 & 0.0 & 0.0 \\ 0.0 & 0.4 & 0.6 & 0.0 \\ 0.0 & 0.0 & 0.5 & 0.5 \end{bmatrix}.$$

Notice that the number of columns in the first matrix, $\mathbf{U}$, is equal to the number of rows in the second one, $\mathbf{V}$, so that they can be multiplied together. This allows us to calculate the proportions of the TVs that would be sent to the shops in the simplified situation where the two producers send TVs directly to the shops. The following shows the product of these two matrices:

$$\mathbf{UV} = \begin{bmatrix} 0.5 & 0.5 & 0.0 \\ 0.3 & 0.3 & 0.4 \end{bmatrix} \begin{bmatrix} 0.3 & 0.7 & 0.0 & 0.0 \\ 0.0 & 0.4 & 0.6 & 0.0 \\ 0.0 & 0.0 & 0.5 & 0.5 \end{bmatrix} = \begin{bmatrix} 0.15 & 0.55 & 0.30 & 0.00 \\ 0.09 & 0.33 & 0.38 & 0.20 \end{bmatrix}.$$

3.3.6 Multiplying a Matrix by Itself

Definition 3.7 (Square Matrix) A square matrix is a matrix with the same number of rows as columns.

An $n \times n$ square matrix is called a square matrix of order n. For example,

$$\mathbf{A} = \begin{bmatrix} 1 & 5 \\ 2 & 4 \end{bmatrix} \text{ is a square matrix of order 2,}$$

and

$$\mathbf{B} = \begin{bmatrix} 10 & 1 & 2 \\ 8 & 3 & 5 \\ 6 & 0 & 1 \end{bmatrix} \text{ is a square matrix of order 3.}$$

3.3.6.1 Multiplying a Matrix by Itself

Only square matrices can be multiplied by themselves, since the number of columns of the first matrix must equal the number of rows of the second matrix. Let A be a square matrix $\begin{bmatrix} 1 & 5 \\ 2 & 4 \end{bmatrix}$. We have

$$\mathbf{A}^2 = \mathbf{A}\mathbf{A} = \begin{bmatrix} 1 & 5 \\ 2 & 4 \end{bmatrix} \begin{bmatrix} 1 & 5 \\ 2 & 4 \end{bmatrix} = \begin{bmatrix} 11 & 25 \\ 10 & 26 \end{bmatrix},$$

$$\mathbf{A}^3 = \mathbf{A}\mathbf{A}\mathbf{A} = \begin{bmatrix} 1 & 5 \\ 2 & 4 \end{bmatrix} \begin{bmatrix} 1 & 5 \\ 2 & 4 \end{bmatrix} \begin{bmatrix} 1 & 5 \\ 2 & 4 \end{bmatrix} = \begin{bmatrix} 61 & 155 \\ 62 & 154 \end{bmatrix},$$

$$\vdots$$

3.3.7 Diagonal and Trace

Suppose we have

$$A = \begin{bmatrix} a_{11} & \cdots & a_{1n} \\ \vdots & \ddots & \vdots \\ a_{n1} & \cdots & a_{nn} \end{bmatrix}.$$

Definition 3.8 (Diagonal) The diagonal of a matrix consists of the elements with the same subscripts, that is, $a_{11}, a_{22}, \cdots, a_{nn}$.

Example 3.18 Taking the matrix **A** defined in Sect. 3.3.6, its diagonal elements are 1 and 4.

Definition 3.9 (Trace) The trace of a square matrix **A**, denoted as $tr(\mathbf{A})$, is the sum of the diagonal elements, that is, $tr(\mathbf{A}) = a_{11} + a_{22} + \cdots + a_{nn}$.

Example 3.19 Again, considering the matrix **A** in Sect. 3.3.6, its trace, $tr(\mathbf{A})$, is $1 + 4 = 5$.

Exercise

3.9 Given

$$\mathbf{A} = \begin{bmatrix} 1 & 5 & 0 \\ 2 & 4 & 10 \\ -1 & 2 & 3 \end{bmatrix} \text{ and } \mathbf{B} = \begin{bmatrix} 10 & 1 & 2 \\ 8 & 3 & 5 \\ 6 & 0 & 1 \end{bmatrix},$$

compute:

(1) $tr(\mathbf{A})$.
(2) $tr(\mathbf{B})$.

3.3.8 Diagonal Matrices

Definition 3.10 (Diagonal Matrix) A diagonal matrix is a matrix in which the entries outside the main diagonal are all zeros.

Let us have a look at the following two examples.

$$\mathbf{A} = \begin{bmatrix} 1 & 0 & 0 & 0 \\ 0 & 3 & 0 & 0 \\ 0 & 0 & 9 & 0 \end{bmatrix} \text{ and } \mathbf{B} = \begin{bmatrix} 1 & 0 & 0 \\ 0 & 3 & 0 \\ 0 & 0 & 9 \end{bmatrix}.$$

A is called a rectangular diagonal matrix, where the number of rows is not equal to the number of columns. **B** is called a symmetric diagonal matrix, which is a square matrix.

3.3.9 Determinants

Definition 3.11 (Determinant) Each n-square matrix $\mathbf{A} = [a_{ij}]$, where $i = 1, \cdots, n$ and $j = 1, \cdots, n$, is assigned a special scalar called the determinant of $\mathbf{A}$, denoted by $\det(\mathbf{A})$ or $|\mathbf{A}|$ or

$$\begin{vmatrix} a_{11} & \cdots & a_{1n} \\ \vdots & \ddots & \vdots \\ a_{n1} & \cdots & a_{nn} \end{vmatrix}.$$

Note that an $n \times n$ array of scalars enclosed by straight lines called a determinant of order n, is not a matrix but denotes the determinant of $\mathbf{A}$. A general way to calculate the determinant of order n can be learned from [8]. This book covers how to compute determinants of order 1, 2, and 3 only.

3.3.9.1 Determinants of Order 1, 2, and 3

- Determinant of order 1
  The determinant of order 1 is defined as follows:

$$|a_{11}| = a_{11}.$$

That is, the determinant of a 1×1 matrix is that number itself.

3.3 Matrices

- Determinant of order 2

  The determinant of order 2 is defined as the product of elements along the main diagonal minus the product of elements along the reverse diagonal. That is

  $$\begin{vmatrix} a_{11} & a_{12} \\ a_{21} & a_{22} \end{vmatrix} = a_{11}a_{22} - a_{12}a_{21}. \tag{3.10}$$

Example 3.20 Suppose $\mathbf{M} = \begin{bmatrix} 3 & -1 \\ 4 & 11 \end{bmatrix}$. Compute the determinant of $\mathbf{M}$.

Solution

$$\begin{vmatrix} 3 & -1 \\ 4 & 11 \end{vmatrix} = 3 \times 11 - (-1) \times 4 = 37.$$

Exercise

3.10 Compute the determinant of the following matrices.

(1) $\mathbf{N} = \begin{bmatrix} 13 & 1 \\ -4 & 2 \end{bmatrix}$.

(2) $\mathbf{U} = \begin{bmatrix} 1 & 1 \\ 5 & 2 \end{bmatrix}$.

(3) $\mathbf{V} = \begin{bmatrix} 10 & 4 \\ 5 & 2 \end{bmatrix}$.

- Determinant of order 3

  Considering a 3×3 matrix $\mathbf{A} = \begin{bmatrix} a_{11} & a_{12} & a_{13} \\ a_{21} & a_{22} & a_{23} \\ a_{31} & a_{32} & a_{33} \end{bmatrix}$, its determinant is defined as follows:

  $$\det(\mathbf{A}) = |\mathbf{A}| = \begin{vmatrix} a_{11} & a_{12} & a_{13} \\ a_{21} & a_{22} & a_{23} \\ a_{31} & a_{32} & a_{33} \end{vmatrix}$$

  $$= a_{11}a_{22}a_{33} + a_{12}a_{23}a_{31} + a_{13}a_{21}a_{32} - a_{13}a_{22}a_{31}$$
  $$- a_{11}a_{23}a_{32} - a_{12}a_{21}a_{33}. \tag{3.11}$$

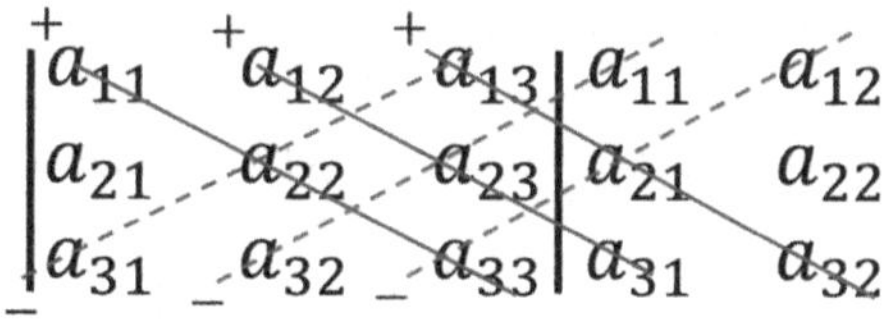

Fig. 3.12 An illustration of the calculation of the determinant of order 3

$$\begin{vmatrix} \overset{+}{a_{11}} & \overset{+}{a_{12}} & \overset{+}{a_{13}} \\ a_{21} & a_{22} & a_{23} \\ a_{31} & a_{32} & a_{33} \end{vmatrix} \begin{matrix} a_{11} & a_{12} \\ a_{21} & a_{22} \\ a_{31} & a_{32} \end{matrix}$$

Looking at Eq. (3.11), we can see six terms on the right-hand side of the equation. Each of these terms is a product of three elements. It may be easier to see which three elements from Fig. 3.12. Here, we copy the matrix first and add the first two columns after the third column. Then, we draw the main diagonal line for every three columns, and similarly, we can draw the reverse diagonal line. The first three terms in the equation are the products along the main diagonals, and the last three are the products along the reverse diagonals. We add up the first three products and subtract the last three products.

Example 3.21 Suppose $\mathbf{X} = \begin{bmatrix} 3 & -1 & 10 \\ 4 & 2 & 0.5 \\ 2.5 & -2 & 6 \end{bmatrix}$. Compute the determinant of $\mathbf{X}$.

Solution

$$\det(\mathbf{A}) = 3 \times 2 \times 6 + (-1) \times 0.5 \times 2.5 + 10 \times 4 \times (-2) - 10 \times 2 \times 2.5$$
$$- 3 \times 0.5 \times (-2) - (-1) \times 4 \times 6$$
$$= -68.25.$$

Exercise

3.11 Compute the determinant of the following matrices.

(1) $\mathbf{W} = \begin{bmatrix} 3 & 1 & 0.5 \\ -4 & 2 & 10 \\ 3 & 0.2 & 6 \end{bmatrix}$.

(2) $\mathbf{X} = \begin{bmatrix} 1 & 1 & 0 \\ 5 & 2 & 3 \\ -4 & 10 & 0.1 \end{bmatrix}$.

3.3.9.2 What Is the Determinant for?

As you will see soon, the determinant can help us find the inverse of a matrix. In addition, it can tell us information about the matrix that is useful in solving systems of linear equations.

3.3.10 Identity and Inverse Matrices

3.3.10.1 Identity Matrices

Definition 3.12 (Identity Matrix) Square matrices with all zeroes except 1s on the main diagonal are identity matrices.

For example,

$$\mathbf{I}_2 = \begin{bmatrix} 1 & 0 \\ 0 & 1 \end{bmatrix}, \mathbf{I}_3 = \begin{bmatrix} 1 & 0 & 0 \\ 0 & 1 & 0 \\ 0 & 0 & 1 \end{bmatrix}, \text{ and } \mathbf{I}_4 = \begin{bmatrix} 1 & 0 & 0 & 0 \\ 0 & 1 & 0 & 0 \\ 0 & 0 & 1 & 0 \\ 0 & 0 & 0 & 1 \end{bmatrix},$$

where we use $\mathbf{I}$ to denote each identity matrix and a value as the subscript denotes the size of the matrix.

Multiplying a matrix $\mathbf{A}$ by an identity matrix $\mathbf{I}$ equals the original matrix $\mathbf{A}$. That is,

$$\mathbf{AI} = \mathbf{IA} = \mathbf{A}.$$

3.3.10.2 Inverse Matrices

Given a matrix $\mathbf{A}$, can we find an inverse matrix $\mathbf{A}^{-1}$ such that $\mathbf{AA}^{-1} = \mathbf{A}^{-1}\mathbf{A} = \mathbf{I}$? Sometimes we can, but only square matrices can have inverses, and not all square matrices do. Finding inverses is complicated, so we shall only consider inverses for 2×2 matrices in this book.

Given a 2×2 matrix $\mathbf{M} = \begin{bmatrix} a & b \\ c & d \end{bmatrix}$, the determinant of this matrix is calculated as follows:

$$\det(\mathbf{M}) = ad - bc.$$

If the determinant is not zero, then the inverse exists, and it can be calculated by swapping the elements on the main diagonal, changing the signs of the elements on the reverse diagonal, and then dividing by the determinant, as shown:

$$\mathbf{M}^{-1} = \frac{1}{\det(\mathbf{M})} \begin{bmatrix} d & -b \\ -c & a \end{bmatrix} = \frac{1}{ad - bc} \begin{bmatrix} d & -b \\ -c & a \end{bmatrix}.$$

Example 3.22 If $\mathbf{A} = \begin{bmatrix} 1 & 5 \\ 2 & 4 \end{bmatrix}$, then its determinant is $4 - 10 = -6$. So the inverse matrix is

$$\mathbf{A}^{-1} = \frac{1}{-6} \begin{bmatrix} 4 & -5 \\ -2 & 1 \end{bmatrix}.$$

You can now confirm that $\mathbf{A}\mathbf{A}^{-1} = \mathbf{A}^{-1}\mathbf{A} = \mathbf{I}$.

Exercise

3.12 Compute the inverse for the following matrices when it exists.

(1) $\mathbf{A} = \begin{bmatrix} -4 & -3 \\ 2 & 6 \end{bmatrix}$.

(2) $\mathbf{B} = \begin{bmatrix} 2 & 1 \\ 10 & 5 \end{bmatrix}$.

(3) $\mathbf{C} = \begin{bmatrix} 13 & 1 \\ -4 & 2 \end{bmatrix}$.

(4) $\mathbf{I} = \begin{bmatrix} 1 & 0 \\ 0 & 1 \end{bmatrix}$.

3.3.11 Matrix Transposition

In some circumstances, you need to flip a matrix around its main diagonal, that is, to exchange rows for columns.

Definition 3.13 (Matrix Transposition) The transpose of a matrix $\mathbf{A}$, written as $\mathbf{A}^T$, is formed by swapping the rows and columns.

If $\mathbf{A}$ is $m \times n$, then $\mathbf{A}^T$ is $n \times m$.

3.3 Matrices

Example 3.23 Suppose a matrix of size 3 by 2 is $A = \begin{bmatrix} 1 & 4 \\ 2 & 5 \\ 3 & 6 \end{bmatrix}$, then its transpose is a matrix of size 2 by 3, that is, $A^T = \begin{bmatrix} 1 & 2 & 3 \\ 4 & 5 & 6 \end{bmatrix}$.

Exercise

3.13 Find the transpose of each of the following matrices:

(1) $A = \begin{bmatrix} 1 & 2 & 10 \\ 4 & 5 & -1 \\ 7 & 0 & -3 \end{bmatrix}$.

(2) $B = \begin{bmatrix} -1 & 4 & -13 \\ 0 & 5 & 8 \end{bmatrix}$.

(3) $C = \begin{bmatrix} 10 \\ -2 \\ 23 \\ -1 \end{bmatrix}$.

(4) $D = \begin{bmatrix} 1, & 0, & -0.7, & 10 \end{bmatrix}$.

Remark 3.4 If $A^T = A$, then A must be square, and it is called a **symmetric matrix**. I is a symmetric matrix. ♦

3.3.11.1 Properties

Let A, B, and C are matrices. It can be shown that:

(1) $(A^T)^T = A$.
(2) $(A + B)^T = A^T + B^T$.
(3) $(AB)^T = B^T A^T$.
(4) $(ABC)^T = C^T B^T A^T$.
(5) If A is a square matrix, then $\det(A) = \det(A^T)$.

Example 3.24 We will illustrate the second and third of properties shown in Sect. 3.3.11.1 using 2 by 2 matrices.

Suppose $\mathbf{A} = \begin{bmatrix} a & b \\ c & d \end{bmatrix}$, and $\mathbf{B} = \begin{bmatrix} e & f \\ g & h \end{bmatrix}$.

Then, $\mathbf{A} + \mathbf{B} = \begin{bmatrix} a+e & b+f \\ c+g & d+h \end{bmatrix}$, and its transpose is $(\mathbf{A} + \mathbf{B})^T = \begin{bmatrix} a+e & c+g \\ b+f & d+h \end{bmatrix}$.

Also, $\mathbf{A}^T + \mathbf{B}^T = \begin{bmatrix} a & c \\ b & d \end{bmatrix} + \begin{bmatrix} e & g \\ f & h \end{bmatrix} = \begin{bmatrix} a+e & c+g \\ b+f & d+h \end{bmatrix}$.

Thus, we observe that $(\mathbf{A} + \mathbf{B})^T = \mathbf{A}^T + \mathbf{B}^T$.

Next, $\mathbf{AB} = \begin{bmatrix} ae+bg & af+bh \\ ce+dg & cf+dh \end{bmatrix}$, and $(\mathbf{AB})^T = \begin{bmatrix} ae+bg & ce+dg \\ af+bh & cf+dh \end{bmatrix}$.

Also, $\mathbf{B}^T \mathbf{A}^T = \begin{bmatrix} e & g \\ f & h \end{bmatrix} \begin{bmatrix} a & c \\ b & d \end{bmatrix} = \begin{bmatrix} ae+bg & ce+dg \\ af+bh & cf+dh \end{bmatrix}$.

Thus, we observe that $(\mathbf{AB})^T = \mathbf{B}^T \mathbf{A}^T$.

In the case of vectors, the transpose of a vector just turns a column (vertical) vector into a row (horizontal) vector and a row vector into a column one.

Example 3.25 If $\mathbf{a} = \begin{bmatrix} 1 \\ 2 \\ 3 \end{bmatrix}$, then $\mathbf{a}^T = (1, 2, 3)$.

In this case, the dot product $\mathbf{a} \cdot \mathbf{a} = \mathbf{a}^T \mathbf{a} = 14$.

Remark 3.5 In practice, the standard vector is assumed to be a column vector. So technically, a row vector is always the transpose of a vector, that is $\mathbf{a}^T$, where $\mathbf{a}$ is a column vector. ♦

Example 3.26 Suppose $\mathbf{A}$ is a matrix of size $m \times n$, $\mathbf{B}$ and $\mathbf{C}$ are matrices of $n \times n$. Prove $\mathbf{A}(\mathbf{B} + \mathbf{C})\mathbf{A}^T = \mathbf{ABA}^T + \mathbf{ACA}^T$.

Solution Applying Eqs. (3.8) and (3.9), we have

$$\mathbf{A}(\mathbf{B} + \mathbf{C})\mathbf{A}^T = (\mathbf{AB} + \mathbf{AC})\mathbf{A}^T$$
$$= \mathbf{ABA}^T + \mathbf{ACA}^T.$$

3.3 Matrices 75

Example 3.27 Suppose the inverse of **A** exists. Prove

$$(\mathbf{A}^{-1})^T = (\mathbf{A}^T)^{-1}. \qquad (3.12)$$

We need to show that $(\mathbf{A}^{-1})^T$ acts like the inverse of $\mathbf{A}^T$. We will make use of the third property of transposition, namely, $(\mathbf{AB})^T = \mathbf{B}^T \mathbf{A}^T$. First:

$$\mathbf{A}^T (\mathbf{A}^{-1})^T = (\mathbf{A}^{-1}\mathbf{A})^T = \mathbf{I}^T = \mathbf{I}.$$

Then:

$$(\mathbf{A}^{-1})^T \mathbf{A}^T = (\mathbf{A}\mathbf{A}^{-1})^T = \mathbf{I}^T = \mathbf{I}.$$

So multiplying $\mathbf{A}^T$ on either side by $(\mathbf{A}^{-1})^T$ gives the identity matrix **I**. This proves that $(\mathbf{A}^{-1})^T = (\mathbf{A}^T)^{-1}$.

3.3.12 Case Study 1 (Continued)

In Sect. 1.3.1 of Chap. 1, we have $\mathbf{X} = \begin{bmatrix} 1 & -5 \\ 1 & 5 \\ 1 & 0 \end{bmatrix}$ and $\mathbf{y} = \begin{bmatrix} 10 \\ 30 \\ 20 \end{bmatrix}$, and we need to compute the coefficients of the least-squares regression method using the following equation:

$$\mathbf{a} = (\mathbf{X}^T \mathbf{X})^{-1} \mathbf{X}^T \mathbf{y}.$$

To compute **a**, we need to obtain the transpose of **X** and the inverse of $\mathbf{X}^T \mathbf{X}$ first and then substitute them to the equation as shown as follows:

- Swap rows and columns of **X** to obtain the transpose of **X**:

$$\mathbf{X}^T = \begin{bmatrix} 1 & 1 & 1 \\ -5 & 5 & 0 \end{bmatrix}.$$

- Compute the matrix multiplication of $\mathbf{X}^T \mathbf{X}$:

$$\mathbf{X}^T \mathbf{X} = \begin{bmatrix} 1 & 1 & 1 \\ -5 & 5 & 0 \end{bmatrix} \begin{bmatrix} 1 & -5 \\ 1 & 5 \\ 1 & 0 \end{bmatrix} = \begin{bmatrix} 3 & 0 \\ 0 & 50 \end{bmatrix}.$$

- Compute the inverse of $\mathbf{X}^T\mathbf{X}$:
  Since the determinant of $\mathbf{X}^T\mathbf{X}$ is 150, the inverse of $\mathbf{X}^T\mathbf{X}$ exists and equals:

  $$(\mathbf{X}^T\mathbf{X})^{-1} = \frac{1}{150}\begin{bmatrix} 50 & 0 \\ 0 & 3 \end{bmatrix}.$$

- Substitute $(\mathbf{X}^T\mathbf{X})^{-1}$, $\mathbf{X}^T$, and $\mathbf{y}$ into the equation of $\mathbf{a}$, and obtain the following:

  $$\mathbf{a} = \frac{1}{150}\begin{bmatrix} 50 & 0 \\ 0 & 3 \end{bmatrix}\begin{bmatrix} 1 & 1 & 1 \\ -5 & 5 & 0 \end{bmatrix}\begin{bmatrix} 10 \\ 30 \\ 20 \end{bmatrix}$$

  $$= \frac{1}{150}\begin{bmatrix} 50 & 50 & 50 \\ -15 & 15 & 0 \end{bmatrix}\begin{bmatrix} 10 \\ 30 \\ 20 \end{bmatrix}$$

  $$= \frac{1}{150}\begin{bmatrix} 3000 \\ 300 \end{bmatrix} = \begin{bmatrix} 20 \\ 2 \end{bmatrix}.$$

  Hence, we get the vector $\mathbf{a}$ with intercept 20 and gradient 2.

3.3.13 Orthogonal Matrix

Orthogonal has been mentioned when we introduce the concept of the dot product in Sect. 3.2.4, where it states that two vectors are orthogonal to each other when their dot product equals zero.

Definition 3.14 (Orthonormal Vectors) Suppose we have m−vectors $x_1, x_2, \cdots, x_m$. These vectors are orthonormal if

- each vector has unit norm, that is $\|x_i\| = 1$, $i = 1, 2, \cdots, m$.
- they are mutually orthogonal, that is $x_i \cdot x_j = 0$, if $i \neq j, i, j = 1, 2, \cdots, m$.

Example 3.28 Given the following three vectors,

$$x_1 = \begin{bmatrix} 0 \\ 0 \\ 1 \end{bmatrix},\ x_2 = \frac{1}{\sqrt{2}}\begin{bmatrix} -1 \\ 1 \\ 0 \end{bmatrix},\ x_3 = \frac{1}{\sqrt{2}}\begin{bmatrix} 1 \\ 1 \\ 0 \end{bmatrix},$$

the norm of each of these vectors is equal to 1, and they are mutually orthogonal to each other. That is, their dot products are all equal to zero. A

(continued)

Example 3.28 (continued)
matrix, denoted as **A**, including these three columns, is an orthogonal matrix as shown as follows:

$$A = \begin{bmatrix} 0 & \frac{-1}{\sqrt{2}} & \frac{1}{\sqrt{2}} \\ 0 & \frac{1}{\sqrt{2}} & \frac{1}{\sqrt{2}} \\ 1 & 0 & 0 \end{bmatrix}.$$

Note that since $x_i \cdot x_i = 1$ for $i = 1, 2, 3$ and $x_i \cdot x_j = 0$ for $i \neq j, i, j = 1, 2, 3$ then $A^T A = A A^T = I$, as you can check for yourself.

Exercise

3.14 Let $m_1^T = \left[\frac{2}{\sqrt{5}}, 0, -\frac{1}{\sqrt{5}}\right]$, $m_2^T = \left[-\frac{1}{\sqrt{5}}, 0, -\frac{2}{\sqrt{5}}\right]$, and $m_3^T = [0, 1, 0]$. Are these three vectors orthogonal to each other?

Definition 3.15 (Orthogonal Matrix) A square matrix **A** with orthonormal columns is called orthogonal. This is equivalent to the following definition:

$$A^T A = A A^T = I,$$

where **I** is the identity matrix.

Remark 3.6 Since $AA^{-1} = I$ if A^{-1} exists (see Sect. 3.3.10.2), and $AA^T = I$ (see Definition 3.15), we obtain $A^T = A^{-1}$, when **A** is a square matrix. ♦

Remark 3.7 Orthogonal vectors (matrix) are useful to build up a new coordinate system. ♦

Exercise

3.15 For the following matrices,

- $Q = \begin{bmatrix} \frac{1}{\sqrt{10}} & \frac{3}{\sqrt{10}} \\ -\frac{3}{\sqrt{10}} & \frac{1}{\sqrt{10}} \end{bmatrix}$ and
- $Q = \begin{bmatrix} \frac{4}{\sqrt{5}} & \frac{3}{\sqrt{5}} \\ -\frac{3}{\sqrt{5}} & \frac{4}{\sqrt{5}} \end{bmatrix},$

(continued)

compute:
(1) $\mathbf{Q}^T$.
(2) $\mathbf{Q}^{-1}$.
(3) Is $\mathbf{Q}$ an orthogonal matrix?

3.4 Linear Combination

3.4.1 Vector Spaces

To motivate this section, let us just start by considering vectors in $\mathcal{R}^2$. Without going into all the details, it is fairly obvious that:

- For any two vectors $\mathbf{x}, \mathbf{w} \in \mathcal{R}^2$ then their sum $\mathbf{x}+\mathbf{w}$ is also $\in \mathcal{R}^2$ [called Closure].
- There is a "zero vector" $\mathbf{0} = (0,0)^T \in \mathcal{R}^2$ that, when added to any vector, does not change it [called Identity].
- For any vector, there is a vector called its *inverse*, that is: for $\mathbf{v} \in \mathcal{R}^2$ then $-\mathbf{v}$ is $\in \mathcal{R}^2$ so that $\mathbf{v} + (-\mathbf{v}) = \mathbf{0}$.

The operation of vector addition is associative and communative; that is, it does not matter what order you add them to; you always get the same result.

Also, in terms of scalar multiplication:

- Multiplying any vector $\mathbf{v} \in \mathcal{R}^2$ by a scalar gives a result which is also $\in \mathcal{R}^2$ [Closure again].
- Multiplying any vector by the scalar 1 leaves it unchanged [Identity again].
- Multiplying a vector by one scalar then another gives the same result as multiplying the scalars together first, that is, $k(j\mathbf{x}) = (kj)\mathbf{x}$ [Associative again].
- Distributivity: $k(\mathbf{x} + \mathbf{w}) = k\mathbf{x} + k\mathbf{w}$ and $(k + j)\mathbf{x} = k\mathbf{x} + j\mathbf{x}$.

A bit more thought says that the same applies to $\mathcal{R}^3$ and in fact to any $\mathcal{R}^n$.
To generalise these ideas, we can define a Vector Space:

Definition 3.16 (Vector Spaces) Let V be a non-empty set with two operations:

- vector addition: this assigns to any $\mathbf{x}, \mathbf{w} \in V$ a sum $\mathbf{x} + \mathbf{w} \in V$. So, vector addition is closed. Vector addition is also associative and communicative and has an identity and inverse.
- scalar multiplication: this assigns to any $\mathbf{x} \in V$, a product $k\mathbf{x} \in V$, where k is a scalar. So, scalar multiplication is closed. Scalar multiplication is also Associative and Distributive and has an Identity.

Lots of structures are vector spaces. For example, $\mathcal{R}^2, \mathcal{R}^3, \mathcal{R}^n$, matrix space, polynomials and function space, and many more are all vector spaces. The advantage of

3.4 Linear Combination

defining an abstract vector space is that once you have proved something for vector spaces in general, then it applies to all the particular vector spaces.

Definition 3.17 (Subspaces) Suppose W is a subset of a vector space V. Then W is a subspace of V if the following two conditions hold:

- the zero vector belongs to W.
- for $\mathbf{x}, \mathbf{w} \in W$
  1. the sum $\mathbf{x} + \mathbf{w} \in W$.
  2. the multiple $k\mathbf{x} \in W$, where k is a scalar.

Example 3.29 Consider the vector space $\mathcal{R}^3$. Let W consist of all vectors whose elements are equal in $\mathcal{R}^3$, such as $(4, 4, 4)$. That is W is the line through the origin O and the point $(1, 1, 1)$. Clearly $O = (0, 0, 0)$ belongs to W, since all entries in O are equal. Further, suppose we have two arbitrary vectors in W, $\mathbf{x} = [a, a, a]$ and $\mathbf{y} = [b, b, b]$. Then, for any real value scalar, k, we have $\mathbf{x} + \mathbf{y} = [a+b, a+b, a+b] \in W$ and $k\mathbf{x} = (ka, ka, ka) \in W$. Thus, W is a subspace of $\mathcal{R}^3$.

Example 3.30 Suppose $S = \left\{ \begin{bmatrix} x_1 \\ x_2 \end{bmatrix} \in \mathcal{R}^2 | x_1 \geq 0 \right\}$. Is S a subspace of R^2? Let us check the conditions.

- S contains the zero vector $\begin{bmatrix} 0 \\ 0 \end{bmatrix}$.
- Suppose $\mathbf{u} = \begin{bmatrix} a \\ b \end{bmatrix}$ and $\mathbf{v} = \begin{bmatrix} c \\ d \end{bmatrix}$ are two vectors in S. We have

$$\begin{bmatrix} a \\ b \end{bmatrix} + \begin{bmatrix} c \\ d \end{bmatrix} = \begin{bmatrix} a+c \\ b+d \end{bmatrix}.$$

Since $a \geq 0$ and $c \geq 0$, we have $a + c \geq 0$. Therefore, the sum of $\mathbf{u} + \mathbf{v}$ is a vector of S.
- However, when multiplying any vector, $\mathbf{u}$, in S, by -1, the direction of the resultant vector, $\mathbf{v} = -\mathbf{u}$, is opposite to $\mathbf{u}$. The vector $\mathbf{v}$ is not in the area covered by S anymore. In Fig. 3.13, the vector $(2, 1)$ is an element of S, since $x_1 = 2 > 0$. After multiplying by -1, the resultant vector, $(-2, -1)$, is not an element of S since the value of $x_1 = -2 < 0$. So S is not a subspace of R^2.

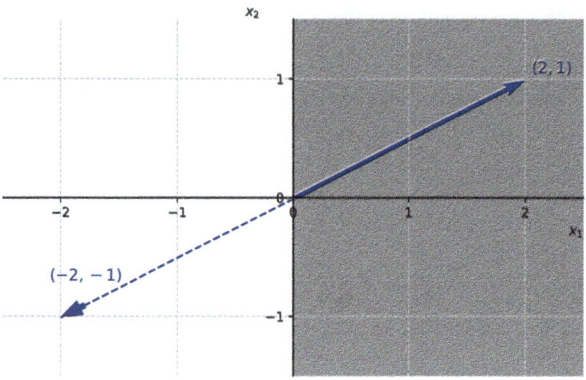

Fig. 3.13 An example of a vector space (with the shaded region representing $x_1 \geq 0$) that is not a subspace of $\mathcal{R}^2$, as explained in Example 3.30

Exercise 3.16

(1) Suppose $S_1 = \left\{ \begin{bmatrix} x_1 \\ x_2 \\ 0 \end{bmatrix} \in \mathcal{R}^3 \right\}$. Is S_1 a subspace of R^3?

(2) Suppose $S_2 = \left\{ \begin{bmatrix} x_1 \\ 0 \\ 0 \end{bmatrix} \in \mathcal{R}^3 \right\}$. Is S_2 a subspace of R^3?

(3) Suppose $S_3 = \left\{ \begin{bmatrix} x_1 \\ 0 \end{bmatrix} \in \mathcal{R}^2 \right\}$. Is S_3 a subspace of R^2?

3.4.2 Linear Combinations and Span

Definition 3.18 (Linear Combination) $\mathbf{v}$ is a linear combination of a set of vectors $\mathbf{u}_1, \mathbf{u}_2, \cdots, \mathbf{u}_d$ if there is a solution to the vector equation

$$\mathbf{v} = x_1 \mathbf{u}_1 + x_2 \mathbf{u}_2 + \cdots + x_d \mathbf{u}_d, \tag{3.13}$$

where $x_1, x_2, \cdots, x_d$ are unknown scalars.

3.4 Linear Combination

Fig. 3.14 An illustration of a linear combination of two vectors, as explained in Example 3.31

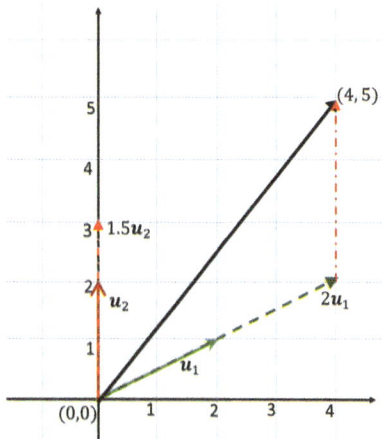

Example 3.31 See Fig. 3.14. We can express $\mathbf{v} = \begin{bmatrix} 4 \\ 5 \end{bmatrix}$ in $\mathcal{R}^2$ as a linear combination of the vectors $\mathbf{u}_1 = \begin{bmatrix} 2 \\ 1 \end{bmatrix}$ and $\mathbf{u}_2 \begin{bmatrix} 0 \\ 2 \end{bmatrix}$ with $x_1 = 2$ and $x_2 = 1.5$, that is,

$$\mathbf{v} = 2\mathbf{u}_1 + 1.5\mathbf{u}_2.$$

$\mathbf{v}$ is the longer diagonal of the parallelogram constructed by $2\mathbf{u}_1$ and $1.5\mathbf{u}_2$.

3.4.2.1 Solving Simultaneous Equations to Find Linear Coefficients

Suppose we want to express $\mathbf{v}^T = (4, -2, 2)$ in $\mathcal{R}^3$ as a linear combination of the vectors $\mathbf{u}_1^T = (1, 2, 1)$, $\mathbf{u}_2^T = (2, 1, 2)$, and $\mathbf{u}_3^T = (-1, 1, 1)$. We seek scalars x_1, x_2, and x_3, such that $\mathbf{v} = x_1\mathbf{u}_1 + x_2\mathbf{u}_2 + x_3\mathbf{u}_3$, that is,

$$\begin{bmatrix} 4 \\ -2 \\ 2 \end{bmatrix} = x_1 \begin{bmatrix} 1 \\ 2 \\ 1 \end{bmatrix} + x_2 \begin{bmatrix} 2 \\ 1 \\ 2 \end{bmatrix} + x_3 \begin{bmatrix} -1 \\ 1 \\ 1 \end{bmatrix}.$$

Since there are three unknown variables, we need to solve three equations:

$$\begin{cases} x_1 + 2x_2 - x_3 = 4 \\ 2x_1 + x_2 + x_3 = -2 \\ x_1 + 2x_2 + x_3 = 2 \end{cases}.$$

We can solve simultaneous equations by elimination and substitution. The solution is $x_1 = \frac{-5}{3}$, $x_2 = \frac{7}{3}$, and $x_3 = -1$, so we can express **v** as follows:

$$\mathbf{v} = \frac{-5}{3}\mathbf{u}_1 + \frac{7}{3}\mathbf{u}_2 - \mathbf{u}_3.$$

3.4.2.2 Span

The next question is: if given a set of vectors, how big is the space of the set of all linear combinations of the vectors? This is called the span of the set of vectors.

Example 3.32 Describe the space formed by the span(**u**), where $\mathbf{u} = \begin{bmatrix} 1 \\ 1 \end{bmatrix}$.

Solution Any linear combination of $\begin{bmatrix} 1 \\ 1 \end{bmatrix}$ is $x_1\mathbf{u}$ which is: $\begin{bmatrix} x_1 \\ x_1 \end{bmatrix}$, where $x_1 \in \mathcal{R}$. Therefore, span(**u**) scales up or scales down along one line, where the two elements in the vector are equal to each other. So the span is a line.

Example 3.33 Describe the space formed by the span(**u**, **v**), where $\mathbf{u} = \begin{bmatrix} 1 \\ 1 \end{bmatrix}$, and $\mathbf{v} = \begin{bmatrix} -2 \\ -2 \end{bmatrix}$.

Solution Any linear combination of $\begin{bmatrix} 1 \\ 1 \end{bmatrix}$ and $\begin{bmatrix} -2 \\ -2 \end{bmatrix}$ is $x_1\mathbf{u} + x_2\mathbf{v}$ which is:

$x_1 \begin{bmatrix} 1 \\ 1 \end{bmatrix} + x_2 \begin{bmatrix} -2 \\ -2 \end{bmatrix} = \begin{bmatrix} x_1 - 2x_2 \\ x_1 - 2x_2 \end{bmatrix}$, where $x_1, x_2 \in \mathcal{R}$. Therefore, span(**u**, **v**) scales up or scales down along one line, where both elements in the vector are equal to $x_1 - 2x_2$. Again, the span is a line.

3.5 Linear Dependence and Independence

Example 3.34 Describe the space formed by the span $\left(\begin{bmatrix} 1 \\ 1 \end{bmatrix}, \begin{bmatrix} 0 \\ 2 \end{bmatrix}\right)$.

Solution Any linear combination of $\begin{bmatrix} 1 \\ 1 \end{bmatrix}$ and $\begin{bmatrix} 0 \\ 2 \end{bmatrix}$ is: $x_1 \begin{bmatrix} 1 \\ 1 \end{bmatrix} + x_2 \begin{bmatrix} 0 \\ 2 \end{bmatrix} = \begin{bmatrix} x_1 \\ x_1 + 2x_2 \end{bmatrix}$, where $x_1, x_2 \in \mathcal{R}$. Therefore, span $\left(\begin{bmatrix} 1 \\ 1 \end{bmatrix}, \begin{bmatrix} 0 \\ 2 \end{bmatrix}\right) = \mathcal{R}^2$, that is, any vector including two elements with arbitrary real numbers. So the span is the whole of $\mathcal{R}^2$.

Definition 3.19 (Spanning Sets) Let V be a vector space. Vectors $\mathbf{u}_1, \mathbf{u}_2, \cdots, \mathbf{u}_d$ are said to form a spanning set of V if every $\mathbf{v} \in V$ is a linear combination of the vectors $\mathbf{u}_1, \mathbf{u}_2, \cdots, \mathbf{u}_d$. That is, the spanning set of $\mathbf{u}_1, \mathbf{u}_2, \cdots, \mathbf{u}_d$ is the whole of V.

In Example 3.34, $\mathbf{u} = \begin{bmatrix} 1 \\ 1 \end{bmatrix}$, and $\mathbf{v} = \begin{bmatrix} 0 \\ 2 \end{bmatrix} \in \mathcal{R}^2$ form a spanning set for $\mathcal{R}^2$. Of course, there are many vectors in $\mathcal{R}^2$ that would form a spanning set for $\mathcal{R}^2$. The simplest example might be the two vectors: $\mathbf{u} = \begin{bmatrix} 1 \\ 0 \end{bmatrix}$ and $\mathbf{v} = \begin{bmatrix} 0 \\ 1 \end{bmatrix}$.

3.5 Linear Dependence and Independence

3.5.1 Linear Dependence and Independence

Intuitively, vectors are linearly dependent if one of them "depends" on the others; that is, it is a linear combination of the others. Vectors are linearly independent if none of them depends on the others; that is, there cannot exist any linear combination of some of the vectors that add up to another one.

Definition 3.20 (Linear Dependence and Independence) Consider the vector equation

$$x_1 \mathbf{u}_1 + x_2 \mathbf{u}_2 + \cdots + x_d \mathbf{u}_d = \mathbf{0}, \tag{3.14}$$

where $\mathbf{u}_1, \mathbf{u}_2$ and $\mathbf{u}_d$ are vectors with n elements, and $x_1, x_2, \cdots, x_d$ are scalars. The vectors $\mathbf{u}_1, \mathbf{u}_2, \cdots, \mathbf{u}_d$ are called linearly independent if $x_1 = 0$, $x_2 = 0$, $\cdots$, $x_d = 0$ is the only solution to Eq. (3.14) The vectors $\mathbf{u}_1, \mathbf{u}_2, \cdots, \mathbf{u}_d$ are linearly dependent if not all $x_1, x_2, \cdots, x_d$ are zeros.

Remark 3.8 When $u_1, u_2, \cdots, u_d$ are linearly independent,

- $x_1 = 0$, $x_2 = 0$, $\cdots, x_d = 0$ is the only solution to Eq. (3.14). On the other hand, when $u_1, u_2, \cdots, u_d$ are linearly dependent, $x_1 = 0$, $x_2 = 0$, $\cdots, x_d = 0$ is not the only solution.
- Any of these vectors cannot be expressed as a linear combination of other elements. That is to say, for example, one cannot rewrite Eq. (3.14) as $u_1 = -\frac{x_2}{x_1}u_2 - \cdots - \frac{x_d}{x_1}u_d$ since $x_1 = 0$. On the other hand, if all the x_i are not zero, then Eq. (3.14) can be rearranged to express one of the vectors in terms of a linear combination of the others, so that they are linearly dependent. For example, suppose $x_2 \neq 0$, then we can write that $u_2 = -\frac{x_1}{x_2}u_1 - \frac{x_3}{x_2}u_3 - \cdots - \frac{x_d}{x_2}u_d$, that is, u_2 is a linear combination of the others.
- Two vectors u and w are linearly dependent if and only if one is a multiple of the other. For example, suppose we have $u = (1, -3)$ and $w = (3, -9)$. u and w are linearly dependent since $w = 3u$. ♦

Example 3.35 Let us consider a 3-dimensional Cartesian coordinate system with X, Y, and Z axes. We have three vectors: the first one is $\begin{bmatrix} 1 \\ 0 \\ 0 \end{bmatrix}$, the second one $\begin{bmatrix} 0 \\ 1 \\ 0 \end{bmatrix}$, and the third one $\begin{bmatrix} 0 \\ 0 \\ 1 \end{bmatrix}$. They are the unit vector along X, Y, and Z axis lines. Now the question is: Can we find a solution for a, b, and c, so that the following linear system is valid?

$$a \begin{bmatrix} 1 \\ 0 \\ 0 \end{bmatrix} + b \begin{bmatrix} 0 \\ 1 \\ 0 \end{bmatrix} + c \begin{bmatrix} 0 \\ 0 \\ 1 \end{bmatrix} = \begin{bmatrix} 0 \\ 0 \\ 0 \end{bmatrix}.$$

The only solution to the above simultaneous equations is $a = b = c = 0$. Therefore, $\begin{bmatrix} 1 \\ 0 \\ 0 \end{bmatrix}$, $\begin{bmatrix} 0 \\ 1 \\ 0 \end{bmatrix}$ and $\begin{bmatrix} 0 \\ 0 \\ 1 \end{bmatrix}$ are linearly independent.

3.5 Linear Dependence and Independence

Example 3.36 Suppose we have three vectors: $\begin{bmatrix} 1 \\ 1 \\ 0 \end{bmatrix}$, $\begin{bmatrix} 1 \\ 3 \\ 2 \end{bmatrix}$ and $\begin{bmatrix} 4 \\ 9 \\ 5 \end{bmatrix}$, and consider the following:

$$a \begin{bmatrix} 1 \\ 1 \\ 0 \end{bmatrix} + b \begin{bmatrix} 1 \\ 3 \\ 2 \end{bmatrix} + c \begin{bmatrix} 4 \\ 9 \\ 5 \end{bmatrix} = \begin{bmatrix} 0 \\ 0 \\ 0 \end{bmatrix},$$

which can be written as follows:

$$\begin{cases} a + b + 4c = 0 \\ a + 3b + 9c = 0 \\ 2b + 5c = 0 \end{cases}.$$

If we subtract the first equation from the second equation, we obtain $2b + 5c = 0$, which is identical to the third one. The solution to these simultaneous equations must satisfy $2b + 5c = 0$. That is, there are many solutions. For example, $a = 3$, $b = 5$, $c = -2$ or $a = -6$, $b = -10$, $c = 4$. In fact, when $a = b = c = 0$, the linear system is also valid. However, this is not the only solution. Therefore, $\begin{bmatrix} 1 \\ 1 \\ 0 \end{bmatrix}$, $\begin{bmatrix} 1 \\ 3 \\ 2 \end{bmatrix}$ and $\begin{bmatrix} 4 \\ 9 \\ 5 \end{bmatrix}$ are linearly dependent.

Exercise

3.17 Determine whether or not the following vectors are linearly dependent:

(1) $\begin{bmatrix} 2 \\ 0 \end{bmatrix}$ and $\begin{bmatrix} 4 \\ 1 \end{bmatrix}$.

(2) $\begin{bmatrix} 1 \\ 0 \\ 2 \end{bmatrix}$, $\begin{bmatrix} 2 \\ 1 \\ 0 \end{bmatrix}$, and $\begin{bmatrix} -1 \\ 1 \\ -2 \end{bmatrix}$.

(3) $\begin{bmatrix} 2 \\ 3 \end{bmatrix}$ and $\begin{bmatrix} 5 \\ 7.5 \end{bmatrix}$.

Fig. 3.15 An example of two vectors linearly independent of each other but not perpendicular, as shown in Example 3.37

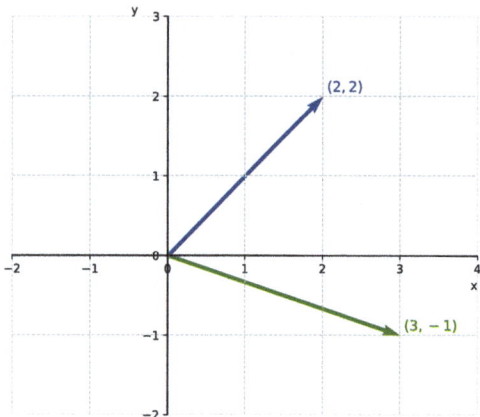

Remark 3.9 Orthogonal vectors are a special case of linear independence. But being linearly independent does not mean the vectors are orthogonal to each other. ♦

Example 3.37 Vectors $\begin{bmatrix} 2 \\ 2 \end{bmatrix}$ and $\begin{bmatrix} 3 \\ -1 \end{bmatrix}$ are not perpendicular, since the dot product of them is 4 and not equal to zero. That they are not perpendicular can also be seen in Fig. 3.15. However, the only solution to the following:

$$x_1 \begin{bmatrix} 2 \\ 2 \end{bmatrix} + x_2 \begin{bmatrix} 3 \\ -1 \end{bmatrix} = \begin{bmatrix} 0 \\ 0 \end{bmatrix},$$

which can be written as simultaneous equations as follows:

$$\begin{cases} 2x_1 + 3x_2 = 0 \\ 2x_1 - x_2 = 0 \end{cases}$$

is $x_1 = 0$ and $x_2 = 0$. Therefore, vectors $\begin{bmatrix} 2 \\ 2 \end{bmatrix}$ and $\begin{bmatrix} 3 \\ -1 \end{bmatrix}$ are linearly independent of each other.

3.5.2 Basis of a Vector Space

We now revisit spanning sets for vector spaces. We saw at the end of Sect. 3.4.2.2 of this chapter an example of two vectors that represented a spanning set for $\mathcal{R}^2$. We now define the **basis** for a vector space.

Definition 3.21 (Basis) A set S of vectors $\mathbf{u}_1, \mathbf{u}_2, \cdots, \mathbf{u}_d$ is a basis of a vector space V if it has the following two properties:

- S is linearly independent;
- S spans V.

This definition says that a basis is a spanning set, and any element in this set cannot be expressed as a linear combination of other elements. In other words, they are linearly independent. Intuitively, the basis is the smallest set of vectors that will generate the whole vector space.

In $\mathcal{R}^2$, we need two non-parallel vectors. In $\mathcal{R}^3$, we need three vectors not in the same plane. In fact, there is a **standard basis** for $\mathcal{R}^2$, namely, the easiest: $\begin{bmatrix} 1 \\ 0 \end{bmatrix}$ and $\begin{bmatrix} 0 \\ 1 \end{bmatrix}$. Similarly in $\mathcal{R}^3$, the **standard basis** is $\begin{bmatrix} 1 \\ 0 \\ 0 \end{bmatrix}, \begin{bmatrix} 0 \\ 1 \\ 0 \end{bmatrix}$, and $\begin{bmatrix} 0 \\ 0 \\ 1 \end{bmatrix}$.

Remark 3.10 Case Study 2
In Sect. 1.3.2 of Chap. 1, we see that the PCA projection plot seems to result from the axes in the original Cartesian coordinate system being rotated. In fact, principal component analysis computes the orthonormal basis, each data projection being a vector that lies in a vector space spanned by this basis. Each element of this orthonormal basis can capture the most crucial information, the variance in the given dataset. We will discuss this further in Chap. 4. ♦

3.6 Connection to Matrices

3.6.1 Determinants and Singular Matrices

A set of n vectors of length n is linearly independent if the matrix with these vectors as columns has a non-zero determinant. For instance, the determinant of $\begin{bmatrix} 1 & 0 & 0 \\ 0 & 1 & 0 \\ 0 & 0 & 1 \end{bmatrix}$ is 1. As said in Sect. 3.3.10.2, such matrices have an inverse. (In this case, the inverse is the same matrix.)

On the other hand, the set of n vectors of length n is linearly dependent if the determinant is zero.

We already know that vectors in the following matrix are linearly dependent:

$$\begin{bmatrix} 1 & 1 & 4 \\ 1 & 3 & 9 \\ 0 & 2 & 5 \end{bmatrix},$$

since it was shown in Example 3.36 of Sect. 3.5.1. And the determinant of it is:

$$\det\left(\begin{bmatrix} 1 & 1 & 4 \\ 1 & 3 & 9 \\ 0 & 2 & 5 \end{bmatrix}\right) = 0.$$

Definition 3.22 (Singular) A square matrix with linearly dependent columns is known as singular.

Since such matrices have a zero determinant, they do not have inverses. So, singular matrices do not have inverses.

Remark 3.11 So why is studying singularity and determinants important in Data Science? This is because many algorithms used in Data Science assume that all features of the data are linearly independent. ♦

3.6.2 Rank

Definition 3.23 (Rank) The maximal number of linearly independent columns of a matrix is called its rank.

Example 3.38 The rank of the matrix, including three vectors shown in Example 3.36 of Sect. 3.5.1, is in fact:

$$\text{rank}\left(\begin{bmatrix} 1 & 1 & 4 \\ 1 & 3 & 9 \\ 0 & 2 & 5 \end{bmatrix}\right) = 2.$$

This can be shown by taking any two columns $\mathbf{u}_i$ and $\mathbf{u}_j$ and solving the equation $x_i \mathbf{u}_i + x_j \mathbf{u}_j = 0$. The only solution is always $x_i = x_j = 0$. So, any two columns are linearly independent, and so the rank is 2.

To find the rank of a general matrix, one can use Gaussian Elimination, which is beyond the content covered in this book. Readers who want to learn more about it can refer to [8]. Note that the rank denoted as r must be less than or equal to the smallest of the two dimensions of the matrix, that is, $r \leq min(m, n)$ for matrix $\mathbf{M}_{m \times n}$.

Exercise

3.18 Let $\mathbf{A} = \begin{bmatrix} 2 & 1 & 4 \\ 3 & 2 & 1 \\ 5 & 3 & 5 \end{bmatrix}$.

(1) Find det(**A**).
(2) Are the columns of **A** linearly independent?
(3) Does **A** have an inverse?
(4) What is the rank of **A**?

Chapter 4
Matrix Decomposition

Matrix decomposition is also called matrix factorisation. Unfortunately, many matrix operations cannot be solved efficiently. However, in the same way, that integers can be decomposed into prime factors to make calculations simpler, and we can use matrix decomposition to reduce a matrix into parts that make it easier to calculate more complex matrix operations. Such parts are, for instance, diagonal matrices and triangular matrices (which only have values in the main diagonal and either the top right half or bottom left half of the matrix). There are many types of matrix decomposition, and in this chapter, you will learn two different matrix decomposition methods: eigendecomposition and singular value decomposition. In addition, you will learn an application of eigendecomposition—principal component analysis.

4.1 Eigendecomposition

We first define eigenvectors and their corresponding eigenvalues of a square matrix A.

Definition 4.1 (Eigendecomposition) Let $\mathbf{A} \in \mathcal{R}^{n \times n}$ be a square matrix. Then $\lambda \in \mathcal{R}$ is an eigenvalue of $\mathbf{A}$ and $\mathbf{u}$ (the non-zero column vector) is the corresponding eigenvector of $\mathbf{A}$ if

$$\mathbf{A}\mathbf{u} = \lambda \mathbf{u}. \tag{4.1}$$

Remark 4.1 Looking at Eq. (4.1), the left-hand side is a matrix multiplication and represents a linear transformation of $\mathbf{u}$; the right-hand side is just a scalar multiplication. A scalar multiplication is just an elongation or shrinking of a vector along its own line. So after the linear transformation, the vector **u** is still along the original line, either in the same or opposite direction (when λ takes a negative value); the length of the vector is changed if the absolute value of the scalar λ does not equal 1. ♦

The equation $A\mathbf{u} = \lambda\mathbf{u}$ has non-zero solutions for the vector **u** if and only if the matrix $\mathbf{A} - \lambda\mathbf{I}$ has a zero determinant, that is, $\det(\mathbf{A} - \lambda\mathbf{I}) = 0$.

To see why it needs $\det(\mathbf{A} - \lambda\mathbf{I}) = 0$, let us rewrite Eq. (4.1) as follows:

$$A\mathbf{u} = \lambda\mathbf{u} \Leftrightarrow A\mathbf{u} - \lambda\mathbf{u} = \mathbf{0} \Leftrightarrow A\mathbf{u} - \lambda\mathbf{I}\mathbf{u} = \mathbf{0} \Leftrightarrow (\mathbf{A} - \lambda\mathbf{I})\mathbf{u} = \mathbf{0}. \qquad (4.2)$$

Looking at $(\mathbf{A} - \lambda\mathbf{I})\mathbf{u} = \mathbf{0}$, then $(\mathbf{A} - \lambda\mathbf{I})$ is just a matrix formed by subtracting two matrices. Now suppose the matrix $(\mathbf{A} - \lambda\mathbf{I})$ is invertible. This means $\det(\mathbf{A} - \lambda\mathbf{I})$ is non-zero and $(\mathbf{A} - \lambda\mathbf{I})^{-1}$ exists.

Then if we multiply $(\mathbf{A} - \lambda\mathbf{I})^{-1}$ on both sides of $(\mathbf{A} - \lambda\mathbf{I})\mathbf{u} = \mathbf{0}$, we obtain $(\mathbf{A} - \lambda\mathbf{I})^{-1}(\mathbf{A} - \lambda\mathbf{I})\mathbf{u} = (\mathbf{A} - \lambda\mathbf{I})^{-1}\mathbf{0} \Rightarrow \mathbf{I}\mathbf{u} = \mathbf{0} \Rightarrow \mathbf{u} = \mathbf{0}$.

This contradicts Definition 4.1, which says **u** is non-zero. Therefore, we cannot assume $(\mathbf{A} - \lambda\mathbf{I})$ is invertible, which means $\det(\mathbf{A} - \lambda\mathbf{I}) = 0$.

Remark 4.2 $\det(\mathbf{A} - \lambda I)$ is called the characteristic polynomial of **A**. It is a polynomial of degree n in λ and has n roots. $\det(\mathbf{A} - \lambda I) = 0$ is called the characteristic equation of **A**. Sometimes finding the roots of the characteristic equation is just referred to as finding the roots of the characteristic polynomial. ♦

4.1.1 Computing Eigenvalues and Eigenvectors

Suppose **A** is an n-square matrix. The following shows the procedure to compute eigenvalues and eigenvectors:

- Step 1: Find the characteristic polynomial of **A**.
- Step 2: Find the roots of the characteristic equation of **A** to obtain the eigenvalues of **A**.
- Step 3: Repeat the following two steps for each eigenvalue λ:
  1. Form the matrix $\mathbf{M} = \mathbf{A} - \lambda\mathbf{I}$.
  2. Find the solution of $\mathbf{M}\mathbf{u} = \mathbf{0}$.

These non-zero vectors **u** are linearly independent eigenvectors of **A**. Each of them has its corresponding eigenvalue λ.

4.1 Eigendecomposition

Example 4.1 Find eigenvalues and eigenvectors of the following 2×2 matrix

$$\mathbf{A} = \begin{bmatrix} 4 & 2 \\ 3 & -1 \end{bmatrix}.$$

Solution

- Step 1: Find the characteristic polynomial of $\mathbf{A}$.

$$\mathbf{A} - \lambda \mathbf{I} = \begin{bmatrix} 4-\lambda & 2 \\ 3 & -1-\lambda \end{bmatrix}.$$

$$|\mathbf{A} - \lambda \mathbf{I}| = (4-\lambda)(-1-\lambda) - 6 = \lambda^2 - 3\lambda - 10.$$

- Step 2: Find the roots of the characteristic equation of $\mathbf{A}$ to obtain the eigenvalues of $\mathbf{A}$.

$$\text{Set } \lambda^2 - 3\lambda - 10 = 0.$$

$$(\lambda - 5)(\lambda + 2) = 0.$$

The roots $\lambda_1 = 5$ and $\lambda_2 = -2$ are the eigenvalues of $\mathbf{A}$.
- Step 3 (1): Form the matrix $\mathbf{M} = \mathbf{A} - \lambda \mathbf{I}$ for $\lambda_1 = 5$.

$$\mathbf{M} = \begin{bmatrix} 4-5 & 2 \\ 3 & -1-5 \end{bmatrix} = \begin{bmatrix} -1 & 2 \\ 3 & -6 \end{bmatrix}.$$

- Step 3 (2): Find the solution of $M\mathbf{u} = \mathbf{0}$.

$$M\mathbf{u} = \begin{bmatrix} -1 & 2 \\ 3 & -6 \end{bmatrix} \begin{bmatrix} u_1 \\ u_2 \end{bmatrix} = \begin{bmatrix} 0 \\ 0 \end{bmatrix}.$$

$$\begin{cases} -u_1 + 2u_2 = 0 \\ 3u_1 - 6u_2 = 0 \end{cases} \rightarrow u_1 - 2u_2 = 0.$$

The system has only one free variable. Any non-zero solution of this one variable is an eigenvector of $\lambda_1 = 5$. For example, $\mathbf{u} = \begin{bmatrix} 6 \\ 3 \end{bmatrix}$. Another way to do it is to find a unit eigenvector. From $u_1 - 2u_2 = 0$, we have $u_1 = $

(continued)

Example 4.1 (continued)

$2u_2$. According to the definition of a unit vector, we have $\sqrt{u_1^2 + u_2^2} = \sqrt{(2u_2)^2 + u_2^2} = \sqrt{5u_2^2} = 1$. Possible solutions are $\mathbf{u} = \begin{bmatrix} \frac{2}{\sqrt{5}} \\ \frac{1}{\sqrt{5}} \end{bmatrix}$, or $\mathbf{u} = \begin{bmatrix} -\frac{2}{\sqrt{5}} \\ -\frac{1}{\sqrt{5}} \end{bmatrix}$. Note these two vectors are just in opposite directions.

- Repeat Steps 3 (1) and 3 (2) for $\lambda_2 = -2$.

$$\mathbf{M} = \begin{bmatrix} 4-(-2) & 2 \\ 3 & -1-(-2) \end{bmatrix} = \begin{bmatrix} 6 & 2 \\ 3 & 1 \end{bmatrix}.$$

$$\mathbf{Mu} = \begin{bmatrix} 6 & 2 \\ 3 & 1 \end{bmatrix} \begin{bmatrix} u_1 \\ u_2 \end{bmatrix} = \begin{bmatrix} 0 \\ 0 \end{bmatrix}.$$

$$\begin{cases} 6u_1 + 2u_2 = 0 \\ 3u_1 + u_2 = 0 \end{cases} \to 3u_1 + u_2 = 0.$$

Again the system has only one free variable, and any non-zero solution of this variable is an eigenvector of $\lambda_2 = -2$. For example, $\mathbf{u} = \begin{bmatrix} 1 \\ -3 \end{bmatrix}$, whose unit vector is $\mathbf{u} = \begin{bmatrix} \frac{1}{\sqrt{10}} \\ -\frac{3}{\sqrt{10}} \end{bmatrix}$; or $\mathbf{u} = \begin{bmatrix} -1 \\ 3 \end{bmatrix}$, whose unit vector is $\mathbf{u} = \begin{bmatrix} -\frac{1}{\sqrt{10}} \\ \frac{3}{\sqrt{10}} \end{bmatrix}$.

Example 4.2 Find eigenvalues and eigenvectors of the following 3×3 matrix

$$\mathbf{A} = \begin{bmatrix} 3 & 2 & 2 \\ 2 & 3 & 2 \\ 2 & 2 & 3 \end{bmatrix}.$$

(continued)

4.1 Eigendecomposition

Example 4.2 (continued)
Solution

- Step 1: Find the characteristic polynomial of **A**

$$A - \lambda I = \begin{bmatrix} 3-\lambda & 2 & 2 \\ 2 & 3-\lambda & 2 \\ 2 & 2 & 3-\lambda \end{bmatrix}.$$

Applying the method described in Sect. 3.3.9 of Chap. 3, we can produce Fig. 4.1. From Fig. 4.1, we can obtain the following:

$$|A - \lambda I| = (3-\lambda)^3 + 2 \times 2^3 - 3 \times (2 \times 2 \times (3-\lambda)) = -\lambda^3 + 9\lambda^2 - 15\lambda + 7.$$

- Step 2: Find the roots of the characteristic equation of **A** to obtain the eigenvalues of **A**.

$$\text{Set } -\lambda^3 + 9\lambda^2 - 15\lambda + 7 = 0.$$

$$(\lambda - 1)(\lambda - 1)(\lambda - 7) = 0.$$

The roots $\lambda_1 = \lambda_2 = 1$ and $\lambda_3 = 7$ are the eigenvalues of **A**.
- Step 3 (1): Form the matrix $\mathbf{M} = \mathbf{A} - \lambda \mathbf{I}$ for $\lambda_1 = \lambda_2 = 1$.

$$M = \begin{bmatrix} 3-1 & 2 & 2 \\ 2 & 3-1 & 2 \\ 2 & 2 & 3-1 \end{bmatrix} = \begin{bmatrix} 2 & 2 & 2 \\ 2 & 2 & 2 \\ 2 & 2 & 2 \end{bmatrix}.$$

- Step 3 (2): Find the solution of $\mathbf{Mu} = \mathbf{0}$.

$$Mu = \begin{bmatrix} 2 & 2 & 2 \\ 2 & 2 & 2 \\ 2 & 2 & 2 \end{bmatrix} \begin{bmatrix} u_1 \\ u_2 \\ u_3 \end{bmatrix} = \begin{bmatrix} 0 \\ 0 \\ 0 \end{bmatrix} \rightarrow u_1 + u_2 + u_3 = 0.$$

Any non-zero solution is an eigenvector of $\lambda = 1$. For example, $\mathbf{u} = \begin{bmatrix} 1 \\ -1 \\ 0 \end{bmatrix}$, whose unit vector is $\mathbf{u} = \begin{bmatrix} \frac{1}{\sqrt{2}} \\ \frac{-1}{\sqrt{2}} \\ 0 \end{bmatrix}$.

(continued)

Example 4.2 (continued)
- Repeat Steps 3 (1) and 3 (2) for $\lambda_3 = 7$.

$$\mathbf{M} = \begin{bmatrix} 3-7 & 2 & 2 \\ 2 & 3-7 & 2 \\ 2 & 2 & 3-7 \end{bmatrix} = \begin{bmatrix} -4 & 2 & 2 \\ 2 & -4 & 2 \\ 2 & 2 & -4 \end{bmatrix}.$$

$$\mathbf{Mu} = \begin{bmatrix} -4 & 2 & 2 \\ 2 & -4 & 2 \\ 2 & 2 & -4 \end{bmatrix} \begin{bmatrix} u_1 \\ u_2 \\ u_3 \end{bmatrix} = \begin{bmatrix} 0 \\ 0 \\ 0 \end{bmatrix}.$$

$$\begin{cases} -4u_1 + 2u_2 + 2u_3 = 0 \\ 2u_1 - 4u_2 + 2u_3 = 0 \\ 2u_1 + 2u_2 - 4u_3 = 0 \end{cases} \rightarrow u_1 = u_2 = u_3.$$

Any non-zero solution is an eigenvector of $\lambda_3 = 7$. For example, $\mathbf{u} = \begin{bmatrix} 1 \\ 1 \\ 1 \end{bmatrix}$, whose unit vector is $\mathbf{u} = \begin{bmatrix} \frac{1}{\sqrt{3}} \\ \frac{1}{\sqrt{3}} \\ \frac{1}{\sqrt{3}} \end{bmatrix}.$

Remark 4.3 In many applications of eigendecomposition, eigenvectors are unit vectors, which means that their length or magnitude is equal to 1. ♦

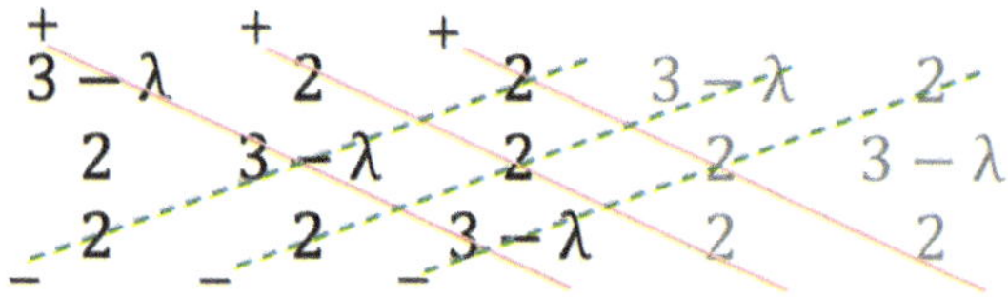

Fig. 4.1 An illustration of calculating the determinant of $\mathbf{A} - \lambda \mathbf{I}$ in Example 4.2

Exercises

4.1 Compute the eigenvalues and eigenvectors for the following matrices.

(1) $A = \begin{bmatrix} 3 & 1 \\ 3 & 5 \end{bmatrix}$.

(2) $B = \begin{bmatrix} 2 & 1 \\ 4 & 5 \end{bmatrix}$.

(3) $C = \begin{bmatrix} -4 & -3 \\ 2 & 3 \end{bmatrix}$.

(4) $D = \begin{bmatrix} 1 & -3 \\ 2 & 6 \end{bmatrix}$.

(5) $E = \begin{bmatrix} -1 & 2 & 2 \\ 2 & 2 & 2 \\ -3 & 6 & -6 \end{bmatrix}$.

4.2 Can you find eigenvalues and eigenvectors for the matrix $F = \begin{bmatrix} 0 & 2 \\ -2 & 0 \end{bmatrix}$?

4.1.2 Diagonalisation

Finally, we can decompose a matrix A into simpler parts. If $n \times n$ matrix A has n eigenvectors $u_1, u_2, \cdots, u_n$ with associated eigenvalues $\lambda_1, \lambda_2, \cdots, \lambda_n$, then A can be written in a diagonalised form

$$A = UDU^{-1}, \qquad (4.3)$$

where $U = [u_1, u_2, \cdots, u_n]$, and D are a diagonal matrix with $\lambda_1, \lambda_2, \cdots, \lambda_n$ as the main diagonal elements.

If A can be expressed this way, it is said to be diagonalisable. Since it is not always possible to find eigenvectors and eigenvalues for a matrix, then not all matrices are diagonalisable.

Remark 4.4 Since each u_i is a column vector, then $U = [u_1, u_2, \cdots, u_n]$ is just a normal matrix with each of its columns equal to one of the u_i in turn. It is just another way of describing a matrix. ◆

To prove that A can be diagonalised in the way described in Eq. (4.3), first we multiply U from right on both sides of Eq. (4.3) and then multiply U^{-1} from left on both sides, we have

$$D = U^{-1}AU. \qquad (4.4)$$

Now we have to show that $\mathbf{U}^{-1}\mathbf{A}\mathbf{U}$ is the diagonal matrix with $\lambda_1, \lambda_2, \cdots, \lambda_n$ on the main diagonal.

Proof First note that $\mathbf{U}^{-1}\mathbf{u_i}$ is the ith column of $\mathbf{U}^{-1}\mathbf{U}$. That is $\mathbf{U}^{-1}\mathbf{u_i}$ is the ith column of the identity matrix $\mathbf{I}$ with a size of $n \times n$ since $\mathbf{U}^{-1}\mathbf{U} = \mathbf{I}$. Also remember that $\mathbf{U} = [\mathbf{u_1}, \mathbf{u_2}, \cdots, \mathbf{u_n}]$ and that from the definition of eigenvectors that $\mathbf{Au_i} = \lambda\mathbf{u_i}$. Then:

$$\begin{aligned}\mathbf{U}^{-1}\mathbf{A}\mathbf{U} &= \mathbf{U}^{-1}\mathbf{A}[\mathbf{u_1}, \mathbf{u_2}, \cdots, \mathbf{u_n}] \\ &= \mathbf{U}^{-1}[\mathbf{Au_1}, \mathbf{Au_2}, \cdots, \mathbf{Au_n}] \\ &= \mathbf{U}^{-1}[\lambda_1\mathbf{u_1}, \lambda_2\mathbf{u_2}, \cdots, \lambda_n\mathbf{u_n}] \\ &= [\lambda_1\mathbf{U}^{-1}\mathbf{u_1}, \lambda_2\mathbf{U}^{-1}\mathbf{u_2}, \cdots, \lambda_n\mathbf{U}^{-1}\mathbf{u_n}] \\ &= \begin{bmatrix} \lambda_1 & 0 & \ldots & 0 \\ 0 & \lambda_2 & \ldots & 0 \\ \vdots & \vdots & & \vdots \\ 0 & 0 & \ldots & \lambda_n \end{bmatrix}.\end{aligned} \quad (4.5)$$

$\square$

Example 4.3 Let us revisit Example 4.1. After performing the eigendecomposition, one possible solution for $\mathbf{U} = [\mathbf{u_1}, \mathbf{u_2}]$ is $\mathbf{U} = \begin{bmatrix} \frac{2}{\sqrt{5}} & \frac{1}{\sqrt{10}} \\ \frac{1}{\sqrt{5}} & \frac{-3}{\sqrt{10}} \end{bmatrix}$. The determinant of $\mathbf{U}$ equals to $\frac{-7}{\sqrt{50}}$ and $\mathbf{U}^{-1} = \frac{1}{\frac{-7}{\sqrt{50}}} \begin{bmatrix} \frac{-3}{\sqrt{10}} & \frac{-1}{\sqrt{10}} \\ \frac{-1}{\sqrt{5}} & \frac{2}{\sqrt{5}} \end{bmatrix}$. Substituting $\mathbf{U}^{-1}$, $\mathbf{A}$, and $\mathbf{U}$ into Eq. (4.4), we have

$$\mathbf{D} = \frac{1}{\frac{-7}{\sqrt{50}}} \begin{bmatrix} \frac{-3}{\sqrt{10}} & \frac{-1}{\sqrt{10}} \\ \frac{-1}{\sqrt{5}} & \frac{2}{\sqrt{5}} \end{bmatrix} \begin{bmatrix} 4 & 2 \\ 3 & -1 \end{bmatrix} \begin{bmatrix} \frac{2}{\sqrt{5}} & \frac{1}{\sqrt{10}} \\ \frac{1}{\sqrt{5}} & \frac{-3}{\sqrt{10}} \end{bmatrix} = \begin{bmatrix} 5 & 0 \\ 0 & -2 \end{bmatrix},$$

where the two elements along the main diagonal are the eigenvalues of $\mathbf{A}$.

Remark 4.5 If the $n \times n$ matrix $\mathbf{A}$ is symmetric, then it is diagonalisable. Moreover, its eigenvectors corresponding to different eigenvalues are orthogonal to each other, which is a special case of linear independence. Hence, $\mathbf{U}$ in Eq. (4.3) is an orthogonal matrix. ♦

Exercise

4.3 Diagonalisation. Find a matrix $\mathbf{U}$, so that $\mathbf{D} = \mathbf{U}^{-1}\mathbf{A}\mathbf{U}$ is diagonal.

(1) $\mathbf{A} = \begin{bmatrix} 3 & 1 \\ 3 & 5 \end{bmatrix}$ (The first matrix in Exercise 4.1).

(2) $\mathbf{A} = \begin{bmatrix} 3 & 2 \\ 2 & 6 \end{bmatrix}$. Since $\mathbf{A}$ is symmetric, check whether $\mathbf{U}$ is orthogonal.

4.2 Principal Component Analysis

Principal Component Analysis (PCA) is widely used in many Data Science applications. It extracts important information from the data. This important information relates to the total variation contained in the data. One can use PCA to compress the size of the dataset by keeping only the information relating to the most variance in the data. One can also use PCA to visualise the structure of the data.

4.2.1 Mathematics Behind PCA

First, we need to define some concepts from Statistics. In each case, we will give the definitions in their general form but give an example in terms of a small number that might be easier to visualise.

Suppose $\mathbf{X}$ is a data matrix including n data observations with d dimensions (or variables, features, or attributes). Each element of $\mathbf{X}$ is denoted as $x_{i,j}$, where $i = 1, \ldots, n$ and $j = 1, \ldots, d$. For example, we could have 6 data points with 3 features (dimensions), for instance, height, width, and depth.

- Define the sample mean for each dimension

$$\bar{x}_j = \frac{1}{n} \sum_{i=1}^{n} x_{i,j}. \tag{4.6}$$

In our example

$$\bar{x}_j = \frac{1}{6} \sum_{i=1}^{6} x_{i,j}.$$

So $\bar{x}_1$ would be the mean height.

- Define the sample standard deviation for each dimension

$$s(x_j) = \sqrt{\frac{\sum_{i=1}^{n}(x_{i,j} - \bar{x}_j)^2}{n-1}}. \tag{4.7}$$

The sample standard deviation can be considered as an average distance from the centre of the data within a specific dimension.

In our example,

$$s(x_1) = \sqrt{\frac{\sum_{i=1}^{6}(x_{i,1} - \bar{x}_1)^2}{6-1}}.$$

is the average distance in the data to the mean height.

The squared standard deviation is called variance:

$$var(x_j) = (s(x_j))^2. \tag{4.8}$$

- The degree to which a pair of variables is linearly related is referred to as the correlation between the two variables. Here we define the sample covariance, which measures the correlation between two dimensions (the hth dimension and kth dimension) in the data.

$$cov(x_h, x_k) = \frac{\sum_{i=1}^{n}(x_{i,h} - \bar{x}_h)(x_{i,k} - \bar{x}_k)}{n-1}. \tag{4.9}$$

The sign of a covariance value can tell us whether two features are positively correlated or negatively correlated.

In our example, the correlation between the 2nd (width) and 3rd (depth) over all six data points is:

$$cov(x_2, x_3) = \frac{\sum_{i=1}^{6}(x_{i,2} - \bar{x}_2)(x_{i,3} - \bar{x}_3)}{6-1}.$$

- A particular correlation coefficient is the Pearson correlation coefficient: r

$$r = \frac{cov(x_h, x_k)}{\sqrt{var(x_h)var(x_k)}}. \tag{4.10}$$

Pearson correlation coefficient has a value between -1 and 1. It usually evaluates the linear relationship between two continuous variables. If we want to check the strength of the correlation between two features, then we need to consider the absolute value of the Pearson correlation coefficient.

4.2 Principal Component Analysis

- The covariance matrix is a $d \times d$ square matrix, Σ given by:

$$\Sigma = \begin{bmatrix} cov(x_1, x_1) & cov(x_1, x_2) & \cdots & cov(x_1, x_d) \\ cov(x_2, x_1) & cov(x_2, x_2) & \cdots & cov(x_2, x_d) \\ \vdots & \vdots & \cdots & \vdots \\ cov(x_d, x_1) & cov(x_d, x_2) & \cdots & cov(x_d, x_d) \end{bmatrix}.$$

For example, for our data, the covariance matrix is a 3×3 matrix:

$$\Sigma = \begin{bmatrix} cov(x_1, x_1) & cov(x_1, x_2) & cov(x_1, x_3) \\ cov(x_2, x_1) & cov(x_2, x_2) & cov(x_2, x_3) \\ cov(x_3, x_1) & cov(x_d, x_2) & cov(x_3, x_3) \end{bmatrix}.$$

The covariance matrix is symmetrical, where $cov(x_h, x_k) = cov(x_k, x_h)$. Each element of the main diagonal is the variance of a specific dimension.

For our data, for example, $cov(x_1, x_2) = cov(x_2, x_1)$ since the correlation between height and width is the same as the correlation between width and height.

Also, on the main diagonal, the first number is

$$cov(x_1, x_1) = \frac{\sum_{i=1}^{6}(x_{i,1} - \bar{x}_1)(x_{i,1} - \bar{x}_1)}{6-1} = \frac{\sum_{i=1}^{6}(x_{i,1} - \bar{x}_1)^2}{6-1}.$$

This is the variance $var(x_1)$ of the heights.

- The covariance matrix is symmetric, so we can do eigendecomposition on the covariance matrix, since there exist $\mathbf{u}_i$ eigenvectors of Σ such that

$$\Sigma \mathbf{u}_i = \lambda_i \mathbf{u}_i.$$

Recall that if the matrix is symmetric, then eigenvectors corresponding to different eigenvalues must be orthogonal to each other.

4.2.2 The Definition of PCA

The d principal components of data $\mathbf{X}$ ($n \times d$) are the d eigenvectors $\mathbf{u}_1, \mathbf{u}_2, \ldots, \mathbf{u}_d$ corresponding to the d ordered eigenvalues $\lambda_1 \geq \lambda_2 \geq \ldots \geq \lambda_d$ of the covariance of $\mathbf{X}$, Σ.

Remember that there are d dimensions (or features) of each data item. Suppose there are just three dimensions, like height, width, and depth. These can be plotted in a three-dimensional space using height as the x-axis, width as the y-axis, and depth as the z-axis. What we are doing with PCA is replacing these three mutually perpendicular axes with three different mutually perpendicular ones. The first of

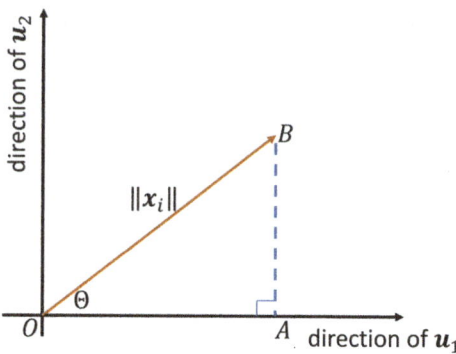

Fig. 4.2 An illustration of projecting the data onto u_1

these new axes, $\mathbf{u}_1$, will be the direction of the most variance in the data, the second axis, $\mathbf{u}_2$, will have the next most variance and so on.

So the first principal component of the data $\mathbf{X}$ is the vector $\mathbf{u}_1$, such that the projection of the data onto $\mathbf{u}_1$, that is, $\mathbf{X}\mathbf{u}_1$, has the largest variance, subject to the normalising constraint $\mathbf{u}_1^T \mathbf{u}_1 = 1$. This normalising constraint means that the principal component is a unit eigenvector, and its variance can be measured by the corresponding eigenvalue.

The second principal component is always orthogonal to the first component, and the third principal component is orthogonal to both the first two, etc. Up to d principal components can be found by this method. Projections of the data on each principal component are obtained as linear combinations of the original variables, that is, $\mathbf{X}\mathbf{u}_i$.

Remark 4.6 When calculating $\mathbf{X}\mathbf{u}_1$, each data vector $\mathbf{x}_i$ in the matrix $\mathbf{X}$ forms a dot product with $\mathbf{u}_1$. Also $\mathbf{u}_1^T \mathbf{u}_1 = 1$, that is the vector $\mathbf{u}_1$ is a unit vector of length 1. So using the second definition of dot product, the geometric definition, we get $\mathbf{x}_i \cdot \mathbf{u}_1 = \|\mathbf{x}_i\| \cos\theta$, where θ is the angle between them. Looking at Fig. 4.2, which as a two-dimensional representation only has $\mathbf{u}_1$ and $\mathbf{u}_2$, we see that $\cos\theta = \overrightarrow{OA}/\overrightarrow{OB}$ and so $\overrightarrow{OA} = \overrightarrow{OB}\cos\theta = \|\mathbf{x}_i\|\cos\theta$. This can be seen as laying the length of $\mathbf{x}_i$ onto the vector $\mathbf{u}_1$, and this is what is meant by projecting the data onto $\mathbf{u}_1$. ◆

4.2.3 PCA in Practice

The following procedure shows how to perform PCA in a real-world setting:

1. Pre-process the given dataset with n data points and d attributes: For example, normalise the data, so that they have zero means and unit standard deviations. We use $\mathbf{X}$ to denote the normalised data matrix with a size of n by d.
2. Calculate the covariance matrix. The size of the covariance matrix is d by d.

4.2 Principal Component Analysis

3. Compute the eigenvectors $\mathbf{u}_i$ and eigenvalues λ_i of the covariance matrix. The size of the eigenvector matrix is d by d, where each column vector is an eigenvector. The number of eigenvalues is d.
4. Select k principal components, where $k \leq d$. If we are trying to illustrate the data then we usually just pick $k = 2$, since it is easy to plot and visualise. Usually these are the first two principle components, since we want to visualise the greatest variance. If we are trying to compress the data, we may use any value $k < d$ and again pick the first k components so as to retain the most variation.
5. Derive the new dataset. That is, project the normalised data onto the selected k principal components. This step involves a very basic operation: matrix multiplication.

$$\text{projected\_data} = \text{normalised\_data} \times \text{selected\_principal\_components}.$$

For example, the projections along the first principal component are given by:

$$\text{projected\_data} = \mathbf{X}_{n \times d} \mathbf{u}_{d \times 1} = u_{11}\mathbf{x}_{,1} + u_{21}\mathbf{x}_{,2} + \cdots + u_{d1}\mathbf{x}_{,d},$$

where $u_{11}, u_{21}, \ldots$ and u_{d1} are elements of $\mathbf{u}_1$ and $\mathbf{x}_{,1}, \mathbf{x}_{,2}, \ldots$, and $\mathbf{x}_{,d}$ denote the individual attributes of the data, that is the column vectors of the data in the matrix $\mathbf{X}$. It can be further written as follows:

$$\text{projected\_data} = u_{11} \begin{bmatrix} x_{11} \\ \vdots \\ x_{n1} \end{bmatrix} + u_{21} \begin{bmatrix} x_{12} \\ \vdots \\ x_{n2} \end{bmatrix} + \cdots + u_{d1} \begin{bmatrix} x_{1d} \\ \vdots \\ x_{nd} \end{bmatrix}.$$

As can be seen, the projection of each data point on the first principal component is a weighted sum of all attributes or features, where the weights are elements in the corresponding eigenvector. In general, projections of data points on a specific principal component are a linear combination of all attributes or features of the dataset.

Since PCA is a method to extract the total variation information of the data, it is important to report how much selected principal components have captured this information. We have:

- Total variation in the original data is $tr(\mathbf{\Sigma})$.
- Total variation of principal components is $\sum \lambda_i$.
- When performing a PCA, the variation information of $\mathbf{\Sigma}$ is kept in λ's, so we have

$$\sum \lambda_i = tr(\mathbf{\Sigma}).$$

- $\frac{\lambda_i}{\sum \lambda_j}$ is the amount of information contained in the ith principal component.

- $\frac{(\lambda_1+\lambda_2+...+\lambda_k)}{\sum \lambda_j}$ is proportion information in first k principal components;
- We can use PCA to do feature extraction. That is to select the first k principal components. When doing feature extraction, we want $\frac{(\lambda_1+\lambda_2+...+\lambda_k)}{\sum \lambda_j}$ large but also want k small.

4.2.4 Case Study 2: Continued (1)

Consider a small data set $\mathbf{X} = \begin{bmatrix} 1 & 5 \\ 2 & 2 \\ 3 & 3 \\ 4 & 4 \\ 5 & 1 \end{bmatrix}$.

This data has five data points, each with dimension 2, so n is 5 and d is 2.

- Pre-process the given dataset. In this example, we remove the mean value from each dimension. Since $\bar{x}_1 = \frac{1+2+3+4+5}{5} = 3$ and $\bar{x}_2 = \frac{5+2+3+4+1}{5} = 3$, the datasets having zero means are shown as follows: $\mathbf{newX} = \begin{bmatrix} -2 & 2 \\ -1 & -1 \\ 0 & 0 \\ 1 & 1 \\ 2 & -2 \end{bmatrix}$.

Here the variance of both columns is the same (and is 2.5). In this case, since the variance is the same, there is no need to normalise by dividing by the standard deviation. This would not be the case with realistic examples. More will be said about this in Chap. 7, Sect. 7.4.3.

- Calculate the covariance matrix using Eqs. (4.7), (4.8), and (4.9). We have:

$$var(newx_1) = (s(newx_1))^2$$
$$= \left(\sqrt{\frac{(-2)^2 + (-1)^2 + 0 + 1^2 + 2^2}{5-1}}\right)^2$$
$$= 2.5,$$

$$var(newx_2) = (s(newx_2))^2$$
$$= \left(\sqrt{\frac{(2)^2 + (-1)^2 + 0 + 1^2 + (-2)^2}{5-1}}\right)^2$$
$$= 2.5,$$

4.2 Principal Component Analysis

and

$$cov(newx_1, newx_2) = cov(newx_2, newx_1)$$

$$= \frac{\sum_{i=1}^{5}(x_{i,1} - 0)(x_{i,2} - 0)}{5 - 1}$$

$$= \frac{(-2) \times (2) + (-1) \times (-1) + 0 \times 0 + 1 \times 1 + 2 \times (-2)}{4}$$

$$= -1.5.$$

The covariance matrix is $\Sigma = \begin{bmatrix} 2.5 & -1.5 \\ -1.5 & 2.5 \end{bmatrix}$. Since the data is two-dimensional, the size of the covariance matrix is 2 by 2.

- Compute the eigenvectors and eigenvalues of the covariance matrix. We obtain the following:
  - $\lambda_1 = 4; \mathbf{u}_1^T = [\frac{1}{\sqrt{2}} \frac{-1}{\sqrt{2}}]$.
  - $\lambda_2 = 1; \mathbf{u}_2^T = [\frac{1}{\sqrt{2}} \frac{1}{\sqrt{2}}]$.

  The sum of eigenvalues $4 + 1 = 5$ equals the sum of elements along the covariance matrix $2.5 + 2.5 = 5$. Readers are encouraged to check the results by following the steps shown in Sect. 4.1.1 of this chapter.

- Select principal components. We shall use both the first principal component and the second principal component to visualise the data in this example. The first principal component is $\mathbf{u}_1^T = [\frac{1}{\sqrt{2}} \frac{-1}{\sqrt{2}}]$ having the largest eigenvalue of 4. The first principal component captures $\frac{\lambda_1}{\lambda_1 + \lambda_2} = \frac{4}{4+1} = 80\%$ of the total variation in the dataset.

- Derive the new dataset using **XU**, where **U** is the matrix formed with columns equal to the two eigenvectors $\mathbf{u}_1$ and $\mathbf{u}_2$.

  Doing this for each data point, in turn, we can see where each point is moved to in the PCA space. So taking the first data point $[-2, 2]$ on the first principal component, we have

  $$\text{projected\_data\_pc1} = [-2, 2] \times \begin{bmatrix} \frac{1}{\sqrt{2}} \\ -\frac{1}{\sqrt{2}} \end{bmatrix} = -2 \times \frac{1}{\sqrt{2}} + 2 \times -\frac{1}{\sqrt{2}} = -\frac{4}{\sqrt{2}}.$$

  To project the data point $[-2, 2]$ on the second principal component, we have

  $$\text{projected\_data\_pc2} = [-2, 2] \times \begin{bmatrix} \frac{1}{\sqrt{2}} \\ \frac{1}{\sqrt{2}} \end{bmatrix} = -2 \times \frac{1}{\sqrt{2}} + 2 \times \frac{1}{\sqrt{2}} = 0.$$

Therefore, the projection of data point $[-2, 2]$ in the PCA space is $[-\frac{4}{\sqrt{2}}, 0]$. Projections of the other four data points can be calculated in the same way. The final PCA plot is shown in Fig. 1.5 of Chap. 1. This PCA plot has captured all variation information in the original dataset.

This example is unrealistic, since it only has two dimensions. It was picked, so that all the stages could be calculated by hand and the working can be explained. In the next section, we illustrate a realistic data set.

4.2.5 A Principal Component Analysis on the Sparrow Dataset

In this example, we demonstrate PCA on a female sparrows dataset. The data includes 49 sparrows with five body measurements which are *total length, alar extent, length of beak and head, length of humerus* and *length of keel of sternum*. After a severe storm, about half of the 49 birds died. The researcher wanted to know whether they could find any support for Charles Darwin's theory of natural selection [9].

First, we normalise the data, so that each feature has a zero mean and a unit standard deviation. We then obtain the covariance matrix of the normalised data, equivalent to the correlation-coefficient matrix, which is shown as follows:

$$\Sigma = \begin{bmatrix} 1.0000 & 0.7350 & 0.6618 & 0.6453 & 0.6051 \\ 0.7350 & 1.0000 & 0.6737 & 0.7685 & 0.5290 \\ 0.6618 & 0.6737 & 1.0000 & 0.7632 & 0.5263 \\ 0.6453 & 0.7685 & 0.7632 & 1.0000 & 0.6066 \\ 0.6051 & 0.5290 & 0.5263 & 0.6066 & 1.0000 \end{bmatrix}.$$

As can be seen, $tr(\Sigma) = 5.0000$. After doing the eigendecomposition, we have the following eigenvalues:

- $\lambda_1 = 3.6160, \lambda_2 = 0.5315, \lambda_3 = 0.3864, \lambda_4 = 0.3016, \lambda_5 = 0.1645$,
- $\sum \lambda_i = 5.0000$,

and eigenvectors are shown in Table 4.1.

Table 4.1 Eigenvectors and eigenvalues of the covariance matrix of the sparrow dataset

	u_1	u_2	u_3	u_4	u_5
	0.4518	−0.0507	−0.6905	0.4204	0.3739
	0.4617	0.2996	−0.3405	−0.5479	−0.5301
	0.4505	0.3246	0.4545	0.6063	−0.3428
	0.4707	0.1847	0.4109	−0.3883	0.6517
	0.3977	−0.8765	0.1785	−0.0689	−0.1924
λ_i	3.6160	0.5315	0.3864	0.3016	0.1645

4.2 Principal Component Analysis

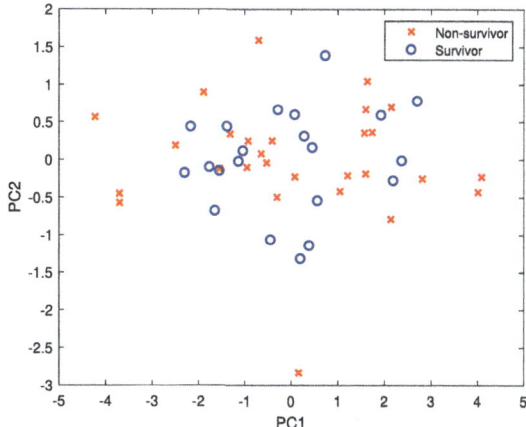

Fig. 4.3 A PCA visualisation plot of the sparrow dataset, where the first two principal components capture about 82.95% of the total variance in the data

Figure 4.3 shows the first principal component plotted against the second principal component. The difference along the first principal component (denoted as $PC1$) axis between the rightmost point and the leftmost point is about 8, and the difference along the second principal component (denoted as $PC2$) axis between the highest point and the lowest point is greater than 4 but much less than the 8 along the $PC1$ axis. In fact, from Table 4.1, it can be seen that the variance along $PC1$ is 3.616, and the variance along $PC2$ is 0.5315. The first two principal components have captured $(3.6160 + 0.5315)/5.0 = 82.95\%$ of the total variation among the data.

PCA is often used to help visualise the most important aspects of multiple dimensional data, since multiple dimensional data is not visualisable in itself. It can often show relationships that are not obvious in the original data. For the sparrow dataset, we can see that the two classes cannot be linearly separated in the PCA space. However, it does illustrate that there are some outliers in the non-survival class.

Projections (denoted as $proj\_pc1$) along $PC1$ are calculated using the following equation:

$$proj\_pc1 = \mathbf{X}\mathbf{u}_1$$

$$= [\mathbf{x}_1, \mathbf{x}_2, \mathbf{x}_3, \mathbf{x}_4, \mathbf{x}_5] \begin{bmatrix} 0.4518 \\ 0.4617 \\ 0.4505 \\ 0.4707 \\ 0.3977 \end{bmatrix} \quad (4.11)$$

$$= 0.4518\mathbf{x}_1 + 0.4617\mathbf{x}_2 + 0.4504\mathbf{x}_3 + 0.4707\mathbf{x}_4 + 0.3977\mathbf{x}_5.$$

where $\mathbf{x}_1, \mathbf{x}_2, \cdots, \mathbf{x}_5$ are the five columns of $\mathbf{X}$.

Projections (denoted as $proj\_pc2$) along $PC2$ are calculated using the following equation:

$$proj\_pc2 = \mathbf{X}\mathbf{u}_2$$

$$= [\mathbf{x}_1, \mathbf{x}_2, \mathbf{x}_3, \mathbf{x}_4, \mathbf{x}_5] \begin{bmatrix} -0.0507 \\ 0.2996 \\ 0.3246 \\ 0.1847 \\ -0.8765 \end{bmatrix} \quad (4.12)$$

$$= -0.0507\mathbf{x}_1 + 0.2.996\mathbf{x}_2 + 0.3246\mathbf{x}_3 + 0.1847\mathbf{x}_4 - 0.8765\mathbf{x}_5.$$

where $\mathbf{x}_1, \mathbf{x}_2, \cdots, \mathbf{x}_5$ are the five columns of $\mathbf{X}$.

Researchers can trace back to the original data based on the PCA projection plot to look into more details. For instance, the lowest point in Fig. 4.3 clearly looks like an outlier. So we can look at $proj\_pc1$ being positive with a value just above zero and $proj\_pc2$ being negative and less than -2.5. In fact, this specific sparrow has a *total length* of 162 millimeters (mm), an *alar extent* of 239 mm, a *length of beak and head* of 30.3 mm, a *length of humers* of 18.0 mm, and a *length of keel of sternum* of 23.1 mm. After removing the mean and converting each feature to the unit standard deviation, we have $x_1 \approx 1.11$, $x_2 \approx -0.46$, $x_3 \approx -1.47$, $x_4 \approx -0.84$, and $x_5 \approx 2.32$. It tells us that this sparrow's *total length* and *length of keel of sternum* are greater than their corresponding mean feature values, while the other three features are less than the corresponding mean feature values. Researchers may further work out why this sparrow is an outlier by comparing its body structural information with other sparrows' body structural information.

Exercise

4.4 Do a principal component analysis on the following small dataset $\mathbf{Y}$ involving five data points: $\mathbf{Y} = \begin{bmatrix} 3 & 3 \\ 0 & 0 \\ -3 & -3 \\ -1 & 1 \\ 1 & -1 \end{bmatrix}$.

(1) Compute the mean value of each variable.
(2) Compute the standard deviation of each variable.
(3) Compute the covariance between two variables.
(4) Write down the covariance matrix.
(5) Find the eigenvalues and eigenvectors of the covariance matrix.
(6) Find the percentage of variance captured by each principal component.
(7) Compute the projection for the first data point [3, 3] in the PCA space.

4.3 Singular Value Decomposition

Singular value decomposition (SVD) is one of the most known and widely used matrix decomposition methods. SVD can be used as a method for dealing with large, high-dimensional data and finding important dimensions in the data.

Definition 4.2 (Singular Value Decomposition) Let M be the real $n \times d$ matrix that we want to decompose. The SVD theorem states:

$$\mathbf{M}_{n \times d} = \mathbf{U}_{n \times n} \mathbf{S}_{n \times d} \mathbf{V}^T_{d \times d}, \qquad (4.13)$$

where

- $\mathbf{U}$ is a column-orthonormal matrix; the columns of the $\mathbf{U}$ matrix are called the left-singular vectors of $\mathbf{M}$;
- $\mathbf{S}$ is
  - a $n \times d$ diagonal matrix;
  - the diagonal values in the $\mathbf{S}$ matrix are known as the singular values of the original matrix $\mathbf{M}$;
  - the singular values are stored in descending order along the main diagonal in $\mathbf{S}$;
  - the number of non-zero values in $\mathbf{S}$ is equal to the rank of matrix $\mathbf{M}$.
- $\mathbf{V}$ is a column-orthonormal matrix; the columns of $\mathbf{V}$ are called the right-singular vectors of $\mathbf{M}$.

The SVD described in Eq. (4.13) is called a full SVD. Some of the non-zero singular values may be significant, while others may be very small and not significant.

The singular value decomposition can also be done as follows:

$$\mathbf{M}_{n \times d} = \mathbf{U}_{n \times k} \mathbf{S}_{k \times k} \mathbf{V}^T_{k \times d}, \qquad (4.14)$$

where $k \leq \min(n, d)$. This is often called compact SVD, or economy SVD.

If $k = r$, where r is the rank of $\mathbf{M}$, then $\mathbf{M}_{n \times d} = \mathbf{U}_{n \times k} \mathbf{S}_{k \times k} \mathbf{V}^T_{k \times d}$; otherwise, if $k < r$, then $\mathbf{M}_{n \times d} \approx \mathbf{U}_{n \times k} \mathbf{S}_{k \times k} \mathbf{V}^T_{k \times d}$ which is useful in data compression as we will see later (Sect. 4.3.6 of this chapter).

Note that the SVD of a matrix is not a unique solution.

4.3.1 Intuitive Interpretations

Let $\mathbf{M} = \begin{bmatrix} 0 & -2 \\ 2 & 0 \end{bmatrix}$, and $\triangle pqr$ denotes a triangle constructed with $\vec{op} = \begin{bmatrix} 1 \\ 3 \end{bmatrix}$, $\vec{oq} = \begin{bmatrix} 2 \\ 3 \end{bmatrix}$, and $\vec{or} = \begin{bmatrix} 1 \\ 1 \end{bmatrix}$ (see the red triangle in Fig. 4.4a).

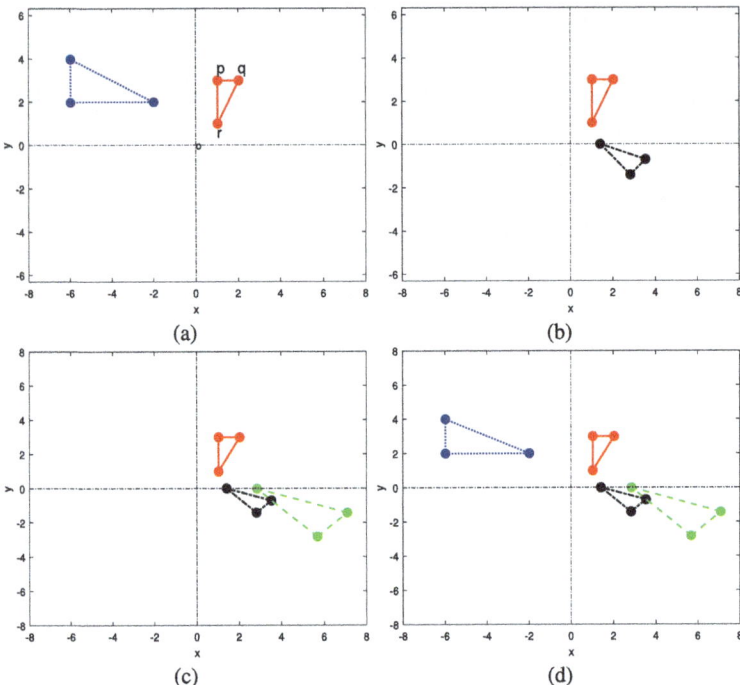

Fig. 4.4 An illustration of how SVD works. (**a**) The original triangle, $\triangle pqr$, is shown in red, and its transformed version, after the linear transformation defined by matrix **M**, is shown in blue. (**b**) The triangle is shown in black after the linear transformation defined by matrix $\mathbf{V}^T$. (**c**) The triangle is shown in green after the linear transformation defined by matrix $\mathbf{SV}^T$. (**d**) The triangle is shown in blue after the linear transformation defined by matrix $\mathbf{USV}^T$. This is the same as the transformation defined by **M** shown in (**a**)

After doing SVD on **M** (Sect. 4.3.3 of this chapter shows a possible method to perform SVD), we have $\mathbf{U} = \begin{bmatrix} -\frac{1}{\sqrt{2}} & \frac{1}{\sqrt{2}} \\ \frac{1}{\sqrt{2}} & \frac{1}{\sqrt{2}} \end{bmatrix}$, $\mathbf{S} = \begin{bmatrix} 2 & 0 \\ 0 & 2 \end{bmatrix}$, and $\mathbf{V}^T = \begin{bmatrix} \frac{1}{\sqrt{2}} & \frac{1}{\sqrt{2}} \\ \frac{1}{\sqrt{2}} & -\frac{1}{\sqrt{2}} \end{bmatrix}$.

As mentioned in Sect. 3.3.4 of Chap. 3, a vector can be linearly transformed through matrix multiplication. Figure 4.4a shows the triangle $\triangle pqr$ in red and the triangle (in blue) after the linear transformation given by matrix **M**, which is the result of **M** multiplying each edge of $\triangle pqr$. As can be seen, the triangle has been rotated and made bigger. Panel (b) displays, in black, the resultant triangle after rotating the triangle $\triangle pqr$ via a linear transformation given by matrix $\mathbf{V}^T$, that is, $\mathbf{V}^T$ multiplies each edge of $\triangle pqr$. As can be seen, the size of the triangle remains the same. The black triangle is then stretched after another linear transformation given by **S**; that is, **S** multiplies each edge of the black triangle, as shown in panel (c) in green. This time there is no rotation involved. In panel (d), a further rotation has happened, and the final triangle is shown in blue, which is the result after **U** is multiplied by the green triangle. The size of the blue triangle is the same as the green one. The blue triangles in panels (a) and (d) are the same. This shows that the linear transformation represented by multiplying by **M** can be decomposed into three simpler, linear transformations represented by multiplying by matrix $\mathbf{V}^T$, matrix **S**, and matrix **U** in that order.

4.3.2 Properties of the SVD

If $\mathbf{M} = \mathbf{USV}^T$, where the size of **S** is $n \times d$, then

1. $\mathbf{U}^T\mathbf{U} = \mathbf{I}$. This can be obtained by Definition 4.2, since $\mathbf{U}^T$ has rows which are orthonormal and **U** has columns which are the same orthonormal vectors and matrix multiplication multiplies rows from the first matrix by columns of the second.
2. $\mathbf{V}^T\mathbf{V} = \mathbf{I}$. This can be obtained by Definition 4.2 for the same reason.
3. $\mathbf{M}^T = \mathbf{VS}^T\mathbf{U}^T$. This can be obtained by applying the fourth property of Sect. 3.3.11.1 of Chap. 3.
4. $\mathbf{US}(:, i) = \mathbf{MV}(:, i)$, where $(:, i)$ is the ith column of each matrix. This can be obtained by multiplying the right-hand side of both sides of Eq. (4.13) by **V** and considering each column separately.

4.3.3 Find a Singular Value Decomposition of a Matrix

Let **M** be a matrix. One can find a singular value decomposition on **M** by following the procedure shown as follows:

1. Compute $\mathbf{A} = \mathbf{M}^T\mathbf{M}$.
2. Find the eigenvalues and eigenvectors of **A**:

    a. sort the eigenvalues in descending order;
    b. the square roots of the eigenvalues are the singular values;
    c. the corresponding unit eigenvectors are the right-singular vectors **V** of **M**.

3. Find the left-singular vectors **U** one column at a time by using the property $\mathbf{US}(:, i) = \mathbf{MV}(:, i)$.

Alternatively, one can do the singular value decomposition on **M** starting with the calculation of $\mathbf{A} = \mathbf{MM}^T$. In this way, the unit eigenvectors obtained from the eigendecomposition are the left-singular vectors of **M**. To find the right-singular vectors **V**, one can use $\mathbf{U}^T\mathbf{M} = \mathbf{SV}^T$, which can be obtained by multiplying the left-hand side of both sides of Eq. (4.13) of this chapter by $\mathbf{U}^T$. We can then find $\mathbf{V}^T$ one row at a time.

4.3.4 Case Study 2: Continued (2)

Find a singular value decomposition of

$$\mathbf{newX} = \begin{bmatrix} -2 & 2 \\ -1 & -1 \\ 0 & 0 \\ 1 & 1 \\ 2 & -2 \end{bmatrix}.$$

Note that **newX** is a 5×2 matrix, so $n = 5$ and $d = 2$. We are actually going to find the compact or economy version of SVD, using Eq. (4.14) of this chapter, where $k = d = 2$, where 2 is the rank of **newX**.

Solution (Note we have the particular matrix **newX** instead of **M** as used in the theory):

1. Compute $\mathbf{newX}^T \mathbf{newX}$.

$$\mathbf{A} = \mathbf{newX}^T \mathbf{newX} = \begin{bmatrix} 10 & -6 \\ -6 & 10 \end{bmatrix}.$$

2. Do the eigendecomposition on **A** by following the procedure introduced in Sect. 4.1.1 of this chapter, and we obtain $\lambda_1 = 16$, $\lambda_2 = 4$, and two possible corresponding eigenvectors are $\mathbf{v}_1 = \begin{bmatrix} -\frac{1}{\sqrt{2}} \\ \frac{1}{\sqrt{2}} \end{bmatrix}$ and $\mathbf{v}_2 = \begin{bmatrix} \frac{1}{\sqrt{2}} \\ \frac{1}{\sqrt{2}} \end{bmatrix}$, respectively.

   a. Sort the eigenvalues in descending order.
   b. The square roots of the eigenvalues are the singular values. In this case, they are $s_1 = \sqrt{16} = 4$ and $s_2 = \sqrt{4} = 2$, and they are elements along the main diagonal of the singular-value matrix:

$$\mathbf{S} = \begin{bmatrix} 4 & 0 \\ 0 & 2 \end{bmatrix}.$$

   c. The corresponding unit eigenvectors are the right-singular vectors of **M**. Then **V** is the matrix with these two eigenvectors as columns, that is

$$\mathbf{V} = \begin{bmatrix} -\frac{1}{\sqrt{2}} & \frac{1}{\sqrt{2}} \\ \frac{1}{\sqrt{2}} & \frac{1}{\sqrt{2}} \end{bmatrix}.$$

4.3 Singular Value Decomposition

3. Find the left-singular vectors **U** by using the property $\mathbf{US}(:, i) = \mathbf{newXV}(:, i)$; that is, we find **U** one column at a time:

$$\mathbf{u}_1 = \frac{1}{s_1}\mathbf{newXv}_1 = \frac{1}{4}\begin{bmatrix} -2 & 2 \\ -1 & -1 \\ 0 & 0 \\ 1 & 1 \\ 2 & -2 \end{bmatrix}\begin{bmatrix} -\frac{1}{\sqrt{2}} \\ \frac{1}{\sqrt{2}} \end{bmatrix} = \begin{bmatrix} \frac{1}{\sqrt{2}} \\ 0 \\ 0 \\ 0 \\ -\frac{1}{\sqrt{2}} \end{bmatrix},$$

$$\mathbf{u}_2 = \frac{1}{s_2}\mathbf{newXv}_2 = \frac{1}{2}\begin{bmatrix} -2 & 2 \\ -1 & -1 \\ 0 & 0 \\ 1 & 1 \\ 2 & -2 \end{bmatrix}\begin{bmatrix} \frac{1}{\sqrt{2}} \\ \frac{1}{\sqrt{2}} \end{bmatrix} = \begin{bmatrix} 0 \\ -\frac{1}{\sqrt{2}} \\ 0 \\ \frac{1}{\sqrt{2}} \\ 0 \end{bmatrix}.$$

Thus

$$\mathbf{U} = [\mathbf{u}_1, \mathbf{u}_2] = \begin{bmatrix} \frac{1}{\sqrt{2}} & 0 \\ 0 & -\frac{1}{\sqrt{2}} \\ 0 & 0 \\ 0 & \frac{1}{\sqrt{2}} \\ -\frac{1}{\sqrt{2}} & 0 \end{bmatrix}.$$

To conclude, we have found a singular value decomposition of **newX**:

$$\mathbf{newX} = \begin{bmatrix} -2 & 2 \\ -1 & -1 \\ 0 & 0 \\ 1 & 1 \\ 2 & -2 \end{bmatrix} = \begin{bmatrix} \frac{1}{\sqrt{2}} & 0 \\ 0 & -\frac{1}{\sqrt{2}} \\ 0 & 0 \\ 0 & \frac{1}{\sqrt{2}} \\ -\frac{1}{\sqrt{2}} & 0 \end{bmatrix}\begin{bmatrix} 4 & 0 \\ 0 & 2 \end{bmatrix}\begin{bmatrix} -\frac{1}{\sqrt{2}} & \frac{1}{\sqrt{2}} \\ \frac{1}{\sqrt{2}} & \frac{1}{\sqrt{2}} \end{bmatrix}^T.$$

Remark 4.7 If, instead of finding the compact or economy SVD, we had found the full SVD, then **S** would be a 5×2 matrix:

$$\mathbf{S} = \begin{bmatrix} 4 & 0 \\ 0 & 2 \\ 0 & 0 \\ 0 & 0 \\ 0 & 0 \end{bmatrix},$$

and **U** would be a 5×5 matrix. But since the last three rows of **S** are all zeros, the last three columns of **U** are irrelevant. Hence, finding the compact or economy SVD with $k = d = r = 2$, where $r = 2$ is the rank of **newX**, finds all the relevant information of the full SVD, since we can only get 2 non-zero singular values anyway. ♦

Exercise

4.5 Find a singular value decomposition for $k = d = r = 2$ of

$$Y = \begin{bmatrix} 3 & 3 \\ 0 & 0 \\ -3 & -3 \\ -1 & 1 \\ 1 & -1 \end{bmatrix}.$$

4.3.5 An Example of the Interpretation of SVD on a Small Dataset

Results of SVD can provide insights to explain concepts included in a dataset. Suppose four children have ranked five books written by Louis Sachar and Philip Pullman, respectively. The ranking is shown in Table 4.2, where the maximum score is 5:

We have the ranking matrix shown as follows:

$$M = \begin{bmatrix} 5 & 5 & 4 & 0 & 0 \\ 4 & 4 & 3 & 0 & 0 \\ 0 & 0 & 0 & 5 & 5 \\ 0 & 0 & 0 & 3 & 3 \end{bmatrix},$$

Table 4.2 The rankings of five books reviewed by four children

	Louis Sacha			Philip Pullman	
	Holes	Small steps	Fuzzy mud	Serpentine	Northern lights
Mary	5	5	4	0	0
Jack	4	4	3	0	0
Tim	0	0	0	5	5
Ann	0	0	0	3	3

4.3 Singular Value Decomposition

where each row shows rankings for all five books by one child and each column shows rankings for one specific book reviewed by all four children. We perform an SVD using Eq. (4.14) of this chapter and set $k = 2$ (note that the rank of $\mathbf{M}$ is 3). Keeping two decimal places for each value, we have

$$\mathbf{M} \approx \mathbf{U}\mathbf{S}\mathbf{V}^T = \begin{bmatrix} 0.79 & 0 \\ 0.62 & 0 \\ 0 & 0.86 \\ 0 & 0.51 \end{bmatrix} \begin{bmatrix} 10.3 & 0 \\ 0 & 8.2 \end{bmatrix} \begin{bmatrix} 0.62 & 0.62 & 0.48 & 0 & 0 \\ 0 & 0 & 0 & 0.71 & 0.71 \end{bmatrix}. \quad (4.15)$$

In this example, the concepts are two authors. The $\mathbf{U}$ matrix connects children to the author they like, where the first column corresponds to Louis Sachar and the second column corresponds to Philip Pullman. For example, Jack likes Louis Sachar's books and has not ranked Philip Pullman's books. Scores corresponding to Jack in the $\mathbf{U}$ matrix are in the second row with values of 0.62 and 0, respectively. Scores corresponding to Mary are 0.79 and 0, respectively, shown in the first row of $\mathbf{U}$. Mary has not ranked Philip Pullman's books either. The score from Mary to Louis Sachar is higher than the one from Jack to Louis Sachar, because Mary has given a higher rank to each of those three books written by Louis Sachar than Jack has.

The matrix $\mathbf{S}$ tells us the strength of concepts in the dataset. 10.3 is the strength of Louis Sachar, while 8.2 is the strength of Philip Pullman. The information about Louis Sachar is stronger, because there is more information in the dataset about books written by Louis Sachar.

Finally, the $\mathbf{V}$ matrix connects books to authors. The first three books in the first row are written by Louis Sachar, and the last two books in the second row are written by Philip Pullman. Interestingly, one cannot compare values in the $\mathbf{V}$ across two authors. For example, it does not make sense to compare 0.62 to 0.71. However, one may compare values within each specific author. For instance, 0.62 is bigger than 0.48, and it suggests that both *Holes* and *Small Steps* have a better ranking than *Fuzzy Mud* overall for Louis Sachar in this small review dataset.

Now let us swap the first two rows of $\mathbf{M}$ and keep values in $\mathbf{N}$ as shown as follows:

$$\mathbf{N} = \begin{bmatrix} 4 & 4 & 3 & 0 & 0 \\ 5 & 5 & 4 & 0 & 0 \\ 0 & 0 & 0 & 5 & 5 \\ 0 & 0 & 0 & 3 & 3 \end{bmatrix}.$$

After performing an SVD on **N**, we have

$$\mathbf{N} \approx \mathbf{USV}^T = \begin{bmatrix} -0.62 & 0 \\ -0.79 & 0 \\ 0 & 0.86 \\ 0 & 0.51 \end{bmatrix} \begin{bmatrix} 10.3 & 0 \\ 0 & 8.2 \end{bmatrix} \begin{bmatrix} -0.62 & -0.62 & -0.48 & 0 & 0 \\ 0 & 0 & 0 & 0.71 & 0.71 \end{bmatrix}.$$

Comparing with the SVD output in the mathematical expression (4.15), one can see that **S** is the same, and the absolute values of the two $\mathbf{V}^T$ matrices are equal. However, the first two rows of the **U** matrix have been swapped, and the signs of values in these two rows have been changed.

Furthermore, let us swap the first column and the last column of **M** and keep values in **Z** as shown as follows:

$$\mathbf{Z} = \begin{bmatrix} 0 & 5 & 4 & 0 & 5 \\ 0 & 4 & 3 & 0 & 4 \\ 5 & 0 & 0 & 5 & 0 \\ 3 & 0 & 0 & 3 & 0 \end{bmatrix}.$$

After performing an SVD on **Z**, we have

$$\mathbf{Z} \approx \mathbf{USV}^T = \begin{bmatrix} 0.79 & 0 \\ 0.62 & 0 \\ 0 & -0.86 \\ 0 & -0.51 \end{bmatrix} \begin{bmatrix} 10.3 & 0 \\ 0 & 8.2 \end{bmatrix} \begin{bmatrix} 0 & 0.62 & 0.48 & 0 & 0.62 \\ -0.71 & 0 & 0 & -0.71 & 0 \end{bmatrix}.$$

Comparing with the SVD output in the mathematical expression (4.15), one can see that **S** is the same, and the absolute values of the two **U** matrices are equal. But the two $\mathbf{V}^T$ matrices are different. The absolute values seem the same; however, the first column and the last column have been swapped, and the signs of some values have been changed.

Remark 4.8 **U** is a matrix that holds important information about rows of a given data matrix; $\mathbf{V}^T$ is a matrix that holds important information about columns of the given data matrix. ♦

4.3.5.1 One More Property of SVD

If $\mathbf{M}_{n \times d} = \mathbf{U}_{n \times n} \mathbf{S}_{n \times d} \mathbf{V}^T_{d \times d}$, then $s_i = ||\mathbf{Mv}_i||$, where s_i is the ith value along the main diagonal of **S**, and $\mathbf{v}_i$ is the ith column of **V**.

For example, if we calculate $\mathbf{Mv}_1$, where $\mathbf{v}_1$ as shown in the mathematical expression (4.15), we have

$$\mathbf{Mv}_1 = \begin{bmatrix} 5 & 5 & 4 & 0 & 0 \\ 4 & 4 & 3 & 0 & 0 \\ 0 & 0 & 0 & 5 & 5 \\ 0 & 0 & 0 & 3 & 3 \end{bmatrix} \begin{bmatrix} 0.62 \\ 0.62 \\ 0.48 \\ 0 \\ 0 \end{bmatrix} = \begin{bmatrix} 8.12 \\ 6.4 \\ 0 \\ 0 \end{bmatrix}.$$

The norm of $\mathbf{Mv}_1$ is $\sqrt{8.12^2 + 6.4^2} \approx 10.3$, which is approximately equal to the first element of $\mathbf{S}$ in the mathematical expression (4.15). The approximation is caused by the fact that we have kept only two decimal places in (4.15). As mentioned previously, s_1 is the strength of the first author in the data matrix. $\mathbf{Mv}_1$ shows each child's overall rating to the first author. It is an average of all five books weighted by the first row in $\mathbf{V}^T$. Therefore, $||\mathbf{Mv}_1||$ may be considered as a score over all four children, weighted by the first row of $\mathbf{V}^T$, which connects books to Louis Sachar.

4.3.6 An Example of Image Compression Using SVD

Let us do image compression using SVD on a cat image[1] as shown in Fig. 4.5. The size of the image $n \times d$ is 668×640. That is, the image has $n = 668$ row and $d = 640$ column pixels. An image can be treated as a matrix of pixels with corresponding colour values and can be decomposed using SVD with a smaller number of singular values that retain only the essential information that comprises the image which results in a smaller image file size.

Figure 4.6 shows the sorted singular values after performing a full SVD on the cat image. As can be seen, singular values decrease dramatically from the first one to the 50th and converge to about 0 after the 100th singular value.

We can compress the image by using a smaller number (k) of singular values on the right-hand side of Eq. (4.14) in this chapter to reconstruct the image. Figure 4.7 shows three compressed images with the number of singular values equal to 50, 20, and 5, respectively.

The quality of a compressed image can be measured using the following equation (assuming the size of the original image is $n \times d$.):

$$\frac{s_1^2 + \cdots + s_k^2}{s_1^2 + \cdots + s_d^2} \times 100\%, \tag{4.16}$$

that is, the sum of the squares of the retained singular values divides by the sum of the squares of all of the singular values. In this example, the image quality is

---

[1] This image was sourced from Pexels: https://www.pexels.com/search/cats/.

Fig. 4.5 The black-and-white version of the original colour image of a cat

Fig. 4.6 A plot of singular values sorted in descending order

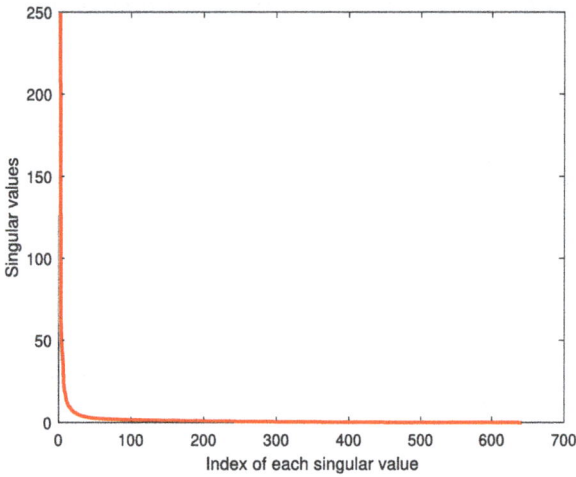

99.37%, 98.73%, and 95.57% for $k = 50, 20,$ and 5, respectively, calculated from Python programming. The compressed image with $k = 5$ is not good, as can be seen in Fig. 4.7, although a percentage value of 95.57 seems a lot.

The compression ratio of an image can be calculated using the following equation:

$$\frac{n \times d}{k \times (n + 1 + d)}. \tag{4.17}$$

4.3 Singular Value Decomposition

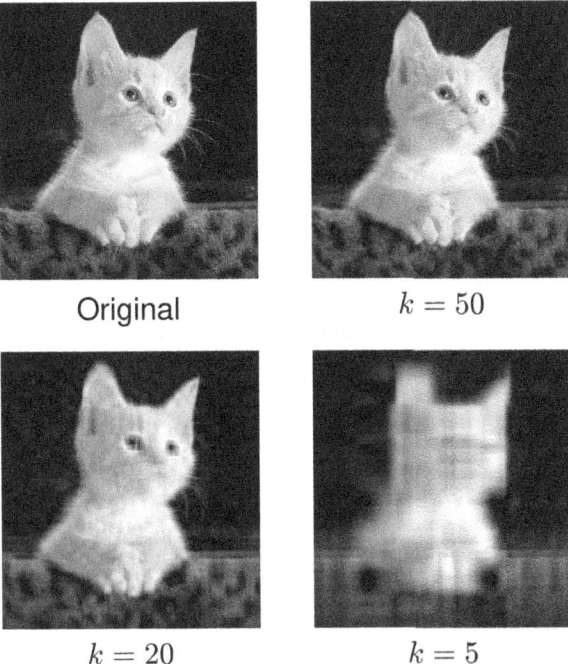

Fig. 4.7 The original image of a cat and its reconstructions using different numbers of singular values

The top line is the size of the original matrix = $n \times d$. The bottom line is the sum of three small matrices after SVD, namely, $n \times k + k \times k + k \times d$. However, since the singular values are kept in the main diagonal of $\mathbf{S}$, one can just save the top k values along the main diagonal rather than the whole $k \times k$ matrix. Hence the bottom line is $n \times k + k \times 1 + k \times d = k \times (n + 1 + d)$.

Remark 4.9 When performing an image compression task, one needs to consider both the image quality and the compression ratio. That is to have a compression ratio as large as possible while keeping the compressed image as good as the original one. ♦

Example 4.4 Compute the compression ratio of the cat image for $k = 50$ (the second image in the first row of Fig. 4.7).

Solution Compression ratio = $\frac{668 \times 640}{50 \times (668 + 1 + 640)} \approx 6.53$.

Exercise

4.6 Compute the compression ratio in Example 4.4 with $k = 20$.

4.4 The Relationship Between PCA and SVD

If $M = USV^T$, then columns of V are principal directions (or axes). Singular values are related to the eigenvalues of the covariance matrix via the following equation, where n is the number of data points:

$$\lambda_i = s_i^2/(n-1), \tag{4.18}$$

Now let us compare results obtained in Sects. 4.2.4 and 4.3.4 of this chapter. As can be seen, the two columns of V in the SVD calculation are equal to the two eigenvectors in the PCA calculation. If we substitute $s_1 = 4$ and $s_2 = 2$ to Eq. (4.18), we have $\frac{4^2}{5-1} = 4$ and $\frac{2^2}{5-1} = 1$, respectively, which are eigenvalues of PCA in Case Study 2 (Sect. 4.2.4).

Exercise

4.7 Exercises 4.4 and 4.5 work on the same data matrix using PCA and SVD, respectively. Apply Eq. (4.18) to results obtained from Exercise 4.5, and compare these eigenvalues with what you have obtained in Exercise 4.4.

Chapter 5
Calculus

This chapter introduces calculus. Calculus deals with the way in which quantities grow or change in relationship with each other. This chapter includes finding the derivative of a function, finding an integral, and some applications of derivatives, such as finding the local minimum and maximum of a function. Many readers will have covered this material before; this chapter will, therefore, represent a reminder for such readers. Doing the many exercises will help with that revision. For others, the many examples and exercises will aid in the learning process.

5.1 Limits of Functions

The principles behind both differentiation and integration in calculus are based on the concept of *limits*. So, before introducing the derivative of a function, we need to have an understanding of limits.

The limiting value of something is the value you get as you approach it ever and ever closer. To find the limiting value of a function of x at a point x_0, if such exists, you need to look at the value of the function as you approach ever closer to the point. If the limiting value of the function A exists, then we need to show that the function gets closer to A as x approaches x_0. This is formalised in the following definition.

Definition 5.1 (Limits) Let $f(x)$ be a function defined at all values of x with the possible exception of $x = x_0$. If for any positive number ϵ (however small), there exists a positive number δ so that whenever $0 < |x - x_0| < \delta$, the function $f(x)$ satisfies $|f(x) - A| < \epsilon$. We say A is the limit of $f(x)$ as x approaches x_0 ($x \to x_0$) and denote it as

$$\lim_{x \to x_0} f(x) = A.$$

Fig. 5.1 An example of a function, $f(x) = \frac{x^2-4}{x-2}$, that is undefined at $x = 2$ (indicated by the hollow circle) but well-behaved nearby

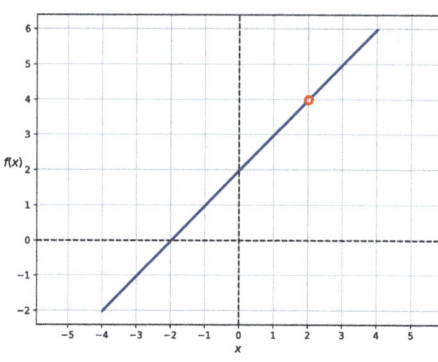

Table 5.1 As $x \to 1$, we have $f(x) = 2x \to 2$.

x	0	0.9	0.99	0.999	0.9999
f(x)	0	1.8	1.98	1.998	1.9998

Why do we need the concept of limits? Some functions are not defined at a point but are well-behaved nearby. For example, see Fig. 5.1, where function $f(x) = \frac{x^2-4}{x-2}$ is undefined at $x = 2$, since this value makes the bottom line zero, and so $f(x)$ is undefined. However, as can be seen, as $x \to 2$, we have $f(x) = 4$. This means when x is near 2 but not equal to it, the values of $f(x)$ are near 4. That is,

$$\lim_{x \to 2} f(x) = 4.$$

How near can it be? The answer is that it can be as near as we want it to be. For example, if $f(x) = 2x$, then as $x \to 1$, we have $f(x) \to 2$, as shown in Table 5.1.

5.1.1 Left- and Right-Hand Limits

Let us consider $f(x) = \frac{|x-3|}{x-3}$. The function is undefined at $x = 3$ (see Fig. 5.2). Suppose we imagine that x is moving. Then, it can approach 3 either from the right or from the left. We indicate these by writing $x \to 3^+$ and $x \to 3^-$, respectively. In this example, as $x \to 3^-$, we have $f(x) = -1$. On the other hand, as $x \to 3^+$, we have $f(x) = 1$. We can write these as

$$\lim_{x \to 3^-} f(x) = -1, \text{ and } \lim_{x \to 3^+} f(x) = 1.$$

We say that

$$\lim_{x \to x_0^-} f(x) = A_1, \text{ if } f(x) \to A_1 \text{ as } x \to x_0^-,$$

5.1 Limits of Functions

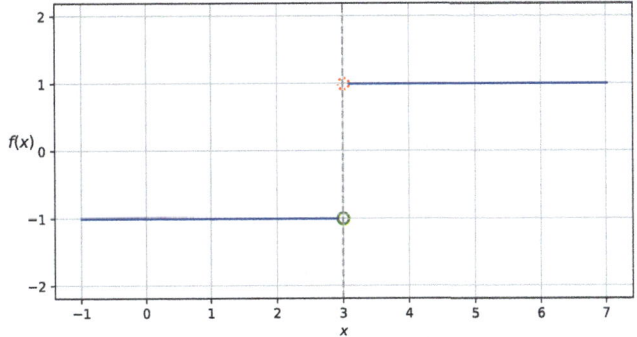

Fig. 5.2 An illustration of the limits of the function $f(x) = \frac{|x-3|}{x-3}$ as $x \to 3$, approaching from the right (dotted circle) and from the left (solid circle)

and

$$\lim_{x \to x_0^+} f(x) = A_2, \text{ if } f(x) \to A_2 \text{ as } x \to x_0^+.$$

If $A_1 = A_2 = A$, then

$$\lim_{x \to x_0^-} f(x) = \lim_{x \to x_0^+} f(x) = A,$$

that is, it does not matter which side x approaches x_0 from, then the limit of the function exists and we say that

$$\lim_{x \to x_0} f(x) = A.$$

This can be seen in the example above, $f(x) = \frac{x^2 - 4}{x - 2}$, where the value $f(x) = 4$ is obtained if you approach 2 from either side.

5.1.2 Theorems on Limits

Suppose $g(x)$ and $h(x)$ are two functions. If

$$\lim_{x \to x_0} g(x) = A \text{ and } \lim_{x \to x_0} h(x) = B,$$

then

-
$$\lim_{x \to x_0} (g(x) \pm h(x)) = A \pm B = \lim_{x \to x_0} g(x) \pm \lim_{x \to x_0} h(x).$$

-
$$\lim_{x \to x_0} (g(x)h(x)) = AB = \lim_{x \to x_0} g(x) \lim_{x \to x_0} h(x).$$

-
$$\lim_{x \to x_0} \frac{g(x)}{h(x)} = \frac{A}{B} = \frac{\lim_{x \to x_0} g(x)}{\lim_{x \to x_0} h(x)}, \text{ if } B \neq 0.$$

Sometimes it happens that as $x \to x_0$, the limit of either $g(x)$ or $h(x)$, or both does not exist. Finding the limits of such functions is beyond the scope of this book. Readers may want to view details from [6].

Example 5.1 Find the limit for the following function as $x \to 1$:

$$3x - 1.$$

Solution

$$\lim_{x \to 1}(3x - 1) = \lim_{x \to 1} 3x - \lim_{x \to 1} 1 = 3 \lim_{x \to 1} x - \lim_{x \to 1} 1 = 3 \times 1 - 1 = 2.$$

Example 5.2 Find the limit for the following function as $x \to 2$:

$$(x^3)(x^2).$$

Solution

$$\lim_{x \to 2}((x^3)(x^2)) = \lim_{x \to 2} x^3 \lim_{x \to 2} x^2 = (\lim_{x \to 2} x)^3 (\lim_{x \to 2} x)^2 = 2^3 \times 2^2 = 8 \times 4 = 32.$$

(continued)

Example 5.2 (continued)
Note that

$$(x^3)(x^2) = x^5,$$

and

$$\lim_{x \to 2} (x^5) = (\lim_{x \to 2} x)^5 = 2^5 = 32,$$

which gives us confidence that the rule is correct.

Example 5.3 Find the limit for the following function as $x \to 1$:

$$\frac{x^3 - 1}{x^2 - 4x + 2}.$$

Solution

$$\lim_{x \to 1} \frac{x^3 - 1}{x^2 - 4x + 2} = \frac{\lim_{x \to 1}(x^3 - 1)}{\lim_{x \to 1}(x^2 - 4x + 2)}$$

$$= \frac{\lim_{x \to 1} x^3 - \lim_{x \to 1} 1}{\lim_{x \to 1} x^2 - 4\lim_{x \to 1} x + \lim_{x \to 1} 2}$$

$$= \frac{(\lim_{x \to 1} x)^3 - 1}{(\lim_{x \to 1} x)^2 - 4 \times \lim_{x \to 1} x + 2}$$

$$= \frac{1^3 - 1}{1^2 - 4 \times 1 + 2}$$

$$= 0.$$

It can be seen that one can substitute the value of x_0 into a rational function to find the limit of the function. However, if a rational function's denominator equals zero or approaches ∞ after substituting, it needs to be treated differently. Let us consider the following two examples.

Example 5.4 Find the limit for the following function as $x \to 3$:

$$\frac{x-3}{x^2-9}.$$

Solution When $x \to 3$, both limits of the numerator and the denominator are zeros, so we cannot take the limit of the numerator and denominator separately. Instead, we can cancel the common factor $x-3$ from the numerator and the denominator. Therefore,

$$\lim_{x \to 3} \frac{x-3}{x^2-9} = \lim_{x \to 3} \frac{x-3}{(x-3)(x+3)} = \lim_{x \to 3} \frac{1}{x+3} = \frac{\lim_{x \to 3} 1}{\lim_{x \to 3}(x+3)} = \frac{1}{6}.$$

Example 5.5 Find the limit for the following function as $x \to \infty$:

$$\frac{2x^3+3x^2+1}{6x^3+4x^2-2}.$$

Solution When $x \to \infty$, both the numerator and the denominator approach ∞. Therefore, we cannot apply theorems of limits directly. Instead, let us divide the numerator and the denominator by x^3 first, then we can find the limit:

$$\lim_{x \to \infty} \frac{2x^3+3x^2+1}{6x^3+4x^2-2} = \lim_{x \to \infty} \frac{2+\frac{3}{x}+\frac{1}{x^3}}{6+\frac{4}{x}-\frac{2}{x^3}} = \frac{1}{3}.$$

Exercise

5.1 Find the following limits.

(1)
$$\lim_{x \to 2}(x^2-3),$$

(2)
$$\lim_{x \to \infty}\left(1+\frac{1}{x}\right)\left(3-\frac{1}{x^3}\right),$$

(continued)

(3)
$$\lim_{x \to 2} \frac{x-2}{\sqrt{x+2}},$$

(4)
$$\lim_{x \to \frac{1}{2}} \frac{8x^3 - 1}{6x^2 - 5x + 1},$$

(5)
$$\lim_{x \to 0} \frac{4x^3 - 2x^2 + x}{3x^2 + 2x}.$$

5.1.3 Continuity

Let $f(x)$ be defined for all values of x, near $x = x_0$ as well as $x = x_0$. The function $f(x)$ is called continuous at $x = x_0$, if $\lim_{x \to x_0} f(x) = f(x_0)$. Consider the following:

$$f(x) = \begin{cases} x^3, & x \neq 1 \\ 0, & x = 1 \end{cases} \quad \text{and} \quad \lim_{x \to 1} f(x) = 1.$$

$f(x)$ is not continuous at $x = 1$ since $f(1) = 0$ and $\lim_{x \to 1} f(x) = 1 \neq f(1)$.

5.2 Derivatives

We can apply the concept of limit in many applications.

For example, the position of an object with uniform motion (constant velocity) along the x-axis at time t is $x = f(t)$. Its velocity is the ratio of the distance it travels to the time it takes, which is the same at all points. However, if the motion is not uniform, its velocity is different at different time intervals. For any time interval,

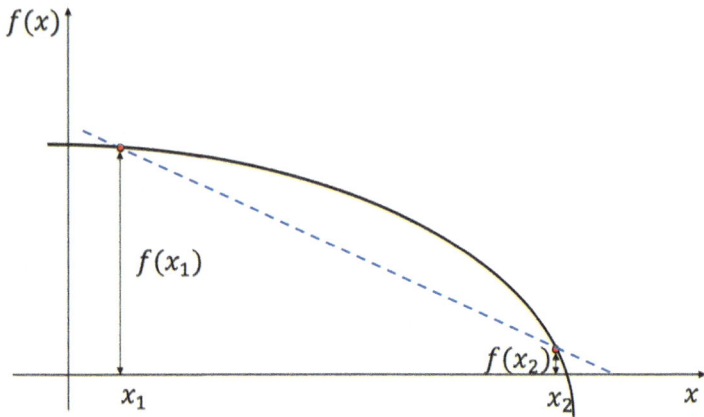

Fig. 5.3 An illustration of the slope of a secant line, represented by the dashed line

denoting the starting time point as t_0, when the object is at point x_0, the velocity can be found by calculating the following:

$$\frac{x - x_0}{t - t_0} = \frac{f(t) - f(t_0)}{t - t_0}, \tag{5.1}$$

which is the average velocity within the time interval. If $t \to t_0$ and the limit of Eq. (5.1) exists, then

$$\lim_{t \to t_0} \frac{f(t) - f(t_0)}{t - t_0}$$

is the velocity of the object at the instant in time $t = t_0$.

Let us consider another example. The slope of the line joining two points on a curve, called a secant line, as shown in Fig. 5.3, can be calculated as follows:

$$\text{slope} = \frac{\text{change in } f(x)}{\text{change in } x} = \frac{f(x_2) - f(x_1)}{x_2 - x_1}. \tag{5.2}$$

If $x_1 \to x_2$ and the limit of Eq. (5.2) exists, then

$$\lim_{x_1 \to x_2} \frac{f(x_2) - f(x_1)}{x_2 - x_1}$$

is the slope of the line just touching the curve, called the tangent, at the point $(x_2, f(x_2))$.

5.2 Derivatives

In the above two examples, we tried to calculate the limit of the rate of change of a function: the change of the dependent variable related to the change of the independent variable, which can be written in general as follows:

$$\lim_{\Delta x \to 0} \frac{f(x_0 + \Delta x) - f(x_0)}{\Delta x}, \tag{5.3}$$

where Δx and $f(x_0 + \Delta x) - f(x_0)$ are the increase of the independent variable and the dependent variable of the function $y = f(x)$, respectively. This limit is called the *derivative* of the function $f(x)$ at x_0.

Definition 5.2 (Derivative and Differential) The ratio defined in Eq. (5.3) is called the derivative of the function $y = f(x)$ at its domain value x_0. If the derivative can be formed at each point of a subdomain of f, then f is said to be differentiable at a general point x on that subdomain, and a new function f' has been constructed, denoted as $f'(x)$, $\frac{dy}{dx}$, or $\frac{df(x)}{dx}$. That is,

$$f'(x) = \frac{dy}{dx} = \frac{df(x)}{dx} = \lim_{\Delta x \to 0} \frac{f(x + \Delta x) - f(x)}{\Delta x}.$$

It is also convenient to define the *differential*, particularly in applications where a linear approximation to a function is required. The *differential*, dy, or principal part of the change in a function with respect to changes in the independent variable, is defined as:

$$dy = f'(x)dx,$$

which is the differential of y or $f(x)$.

Remark 5.1 Note that in general $dy \neq \Delta y$, where $\Delta y = f(x + \Delta x) - f(x)$. $\frac{dy}{dx}$ is not actually a fraction at all. It is the limit of the fraction $\frac{\Delta y}{\Delta x}$ as $\Delta x \to 0$. ♦

Example 5.6 Let $f(x) = 6x + 5$, and use the derivative definition to find $f'(x)$ at any point x.

Solution

$$f(x + \Delta x) = 6(x + \Delta x) + 5 = 6x + 6\Delta x + 5$$

$$f(x + \Delta x) - f(x) = 6\Delta x$$

$$\lim_{\Delta x \to 0} \frac{6\Delta x}{\Delta x} = 6.$$

(continued)

Example 5.6 (continued)
Since we know from Sect. 2.3.2.1 of Chap. 2 that $y = f(x) = 6x + 5$ is a straight line and that its gradient is 6 everywhere, then the result that $f'(x) = 6$, as found here, is correct.

Example 5.7 Let $f(x) = |x|$. Compute $f'(x)$ at $x = 0$ using the definition of derivative.

Solution

$$\lim_{\Delta x \to 0} \frac{f(0 + \Delta x) - f(0)}{\Delta x} = \lim_{\Delta x \to 0} \frac{|\Delta x| - 0}{\Delta x} = \lim_{\Delta x \to 0} \frac{|\Delta x|}{\Delta x}.$$

$$\text{If } \Delta x < 0, \lim_{\Delta x \to 0^-} \frac{f(0 + \Delta x) - f(0)}{\Delta x} = -1;$$

$$\text{if } \Delta x > 0, \lim_{\Delta x \to 0^+} \frac{f(0 + \Delta x) - f(0)}{\Delta x} = 1.$$

$$\text{Since } \lim_{\Delta x \to 0^-} \frac{f(0 + \Delta x) - f(0)}{\Delta x} \neq \lim_{\Delta x \to 0^+} \frac{f(0 + \Delta x) - f(0)}{\Delta x},$$

the derivative of $f(x) = |x|$ at $x = 0$ does not exist. There is no one unique tangent for $f(x) = |x|$ at $x = 0$. For example, some are shown as dotted lines in Fig. 5.4.

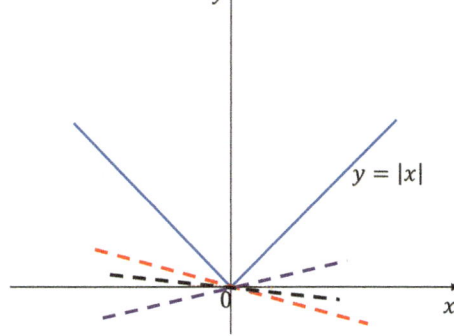

Fig. 5.4 An illustration of $f(x) = |x|$. Dotted lines are three examples of possible tangents passing through $x = 0$

Exercise

5.2 From the definition of the derivative, find whether $f'(x)$ exists at $x = 0$. If it exists, calculate $f'(x = 0)$. Also, calculate $f'(x)$ at a general point x.

(1) $f(x) = x^2$.
(2) $f(x) = x^2 + x$.

5.2.1 Derivatives of Some Elementary Functions

This is not a Mathematics textbook, so we just list the derivatives of some of the most common functions. In the following, we assume each function is a differentiable function of x or θ, where θ is measured in radians. c and a denote constants.

$$\frac{d}{dx} c = 0 \tag{5.4}$$

$$\frac{d}{dx} ax^c = cax^{c-1} \tag{5.5}$$

$$\frac{d}{d\theta} \sin\theta = \cos\theta \tag{5.6}$$

$$\frac{d}{d\theta} \cos\theta = -\sin\theta \tag{5.7}$$

$$\frac{d}{dx} e^{ax} = ae^{ax} \tag{5.8}$$

$$\frac{d}{dx} \ln x = \frac{1}{x} \tag{5.9}$$

Readers may read [6] to find derivatives for more trigonometrical and hyperbolic functions.

5.2.2 Rules for Differentiation

Again, we just list these rules. If $f(x)$, $g(x)$, and $h(x)$ are differentiable functions, then the following differentiation rules are valid.

- Addition Rule

$$\frac{d}{dx}(g(x)+h(x)) = \frac{d}{dx}g(x) + \frac{d}{dx}h(x).$$

-

$$\frac{d}{dx}(g(x)-h(x)) = \frac{d}{dx}g(x) - \frac{d}{dx}h(x).$$

-

$$\frac{d}{dx}Cg(x) = C\frac{d}{dx}g(x),$$

where C is any constant.
- Product Rule

$$\frac{d}{dx}(g(x)h(x)) = g(x)\frac{d}{dx}h(x) + h(x)\frac{d}{dx}g(x).$$

- Quotient Rule

$$\frac{d}{dx}\frac{g(x)}{h(x)} = \frac{h(x)\frac{d}{dx}g(x) - g(x)\frac{d}{dx}h(x)}{(h(x))^2}.$$

- If $y = g(x)$, and $x = g^{-1}(y)$, then $\frac{dy}{dx}$ and $\frac{dx}{dy}$ are related by

$$\frac{dy}{dx} = \frac{1}{\frac{dx}{dy}}.$$

That is, the derivative of an inverse function is the reciprocal of the derivative of the function.
- Chain Rule
In calculus, the chain rule is a formula for computing the derivative of the composition of two or more functions. For two functions: If $y = f(u)$ and $u = g(x)$, then

$$\frac{dy}{dx} = \frac{dy}{du}\frac{du}{dx}.$$

This generalises to three or more functions, for instance: If $y = f(z), z = g(w)$ and $w = h(x)$, then

$$\frac{dy}{dx} = \frac{dy}{dz}\frac{dz}{dw}\frac{dw}{dx}.$$

5.2 Derivatives

Example 5.8 Let $f(x) = x^2$. Find $f'(x)$.

Solution Applying Eq. (5.5), we have:

$$f'(x) = 2x^{2-1} = 2x.$$

Example 5.9 Let $h(x) = 3x^2 + 5x$. Find $\frac{d}{dx}h(x)$.

Solution Applying the addition rule and Eq. (5.5), we have:

$$\frac{d}{dx}h(x) = \frac{d}{dx}(3x^2) + \frac{d}{dx}5x = 6x + 5.$$

Example 5.10 Let $h(x) = x(3x + 5)$. Find $h'(x)$.

Solution Consider $f(x) = x$ and $g(x) = 3x + 5$ and applying the product rule, Eq. (5.5) and the addition rule, we have:

$$h'(x) = f(x)\frac{d}{dx}g(x) + g(x)\frac{d}{dx}f(x)$$
$$= x \times 3 + (3x + 5) \times 1$$
$$= 6x + 5.$$

Since $h(x) = x(3x + 5) = 3x^2 + 5x$, this is the same as Example 5.9 and has the same solution. This gives us confidence that the product rule is correct.

Example 5.11 Let $h(x) = e^x \cos x$. Find $\frac{d}{dx}h(x)$.

Solution Consider $f(x) = e^x$ and $g(x) = \cos x$, then applying the product rule and Eqs. (5.7), with independent variable x rather than θ, and (5.8), we have:

$$\frac{d}{dx}h(x) = f(x)\frac{d}{dx}g(x) + g(x)\frac{d}{dx}f(x)$$

(continued)

Example 5.11 (continued)

$$= e^x \times (-\sin x) + \cos x \times e^x$$
$$= e^x(\cos x - \sin x).$$

Example 5.12 Let $h(x) = \tan x = \frac{\sin x}{\cos x}$. Find $\frac{d}{dx}h(x)$.

Solution Consider $f(x) = \sin x$ and $g(x) = \cos x$, then applying the quotient rule and Eqs. (5.6) and (5.7), we have:

$$\frac{d}{dx}h(x) = \frac{g(x)\frac{d}{dx}f(x) - f(x)\frac{d}{dx}g(x)}{(g(x))^2}$$

$$= \frac{\cos x \times \cos x - \sin x \times (-\sin x)}{\cos^2 x}$$

$$= \frac{\cos^2 x + \sin^2 x}{\cos^2 x}$$

$$= \frac{1}{\cos^2 x}$$

$$= \sec^2 x.$$

Example 5.13 Let $\sigma(x)$ be a sigmoid function, defined as: $\sigma(x) = \frac{1}{1+e^{-x}}$. Find $\frac{d}{dx}\sigma(x)$.

Solution Applying the quotient rule, we have:

$$\frac{d}{dx}\sigma(x) = \frac{(1+e^{-x}) \times 0 - 1 \times (-e^{-x})}{(1+e^{-x})^2}$$

$$= \frac{e^{-x}}{(1+e^{-x})^2}$$

$$= \frac{1}{1+e^{-x}} \times \frac{(1+e^{-x}) - 1}{1+e^{-x}}$$

$$= \sigma(x)(1 - \sigma(x)).$$

(continued)

Example 5.13 (continued)
Note that in line one of the above, we have applied (1) the derivative of a constant is zero; and (2) Eq. (5.8). Figure 5.5 shows a sigmoid function and its derivative in the domain of $[-10, 10]$. The function itself is bounded between 0 and 1, and its derivative is symmetrical about the vertical axis of $x = 0$. Values of the derivative are convergent to 0 as x approaches either ∞ or $-\infty$.

Example 5.14 Let $y = \ln x$ and $x = e^y$. Find $\frac{dy}{dx}$ and $\frac{dx}{dy}$.

Solution Using Eqs. (5.9) and (5.8), we have: $\frac{dy}{dx} = \frac{1}{x}$ and $\frac{dx}{dy} = e^y = x$. Since the functions $y = \ln x$ and $x = e^y$ are inverse functions (see Example 2.23 in Sect. 2.3.4 of Chap. 2), we have shown that

$$\frac{dy}{dx} = \frac{1}{\frac{dx}{dy}}.$$

Example 5.15 Let $y = \sin^{-1} x$, and $y \in [-\frac{\pi}{2}, \frac{\pi}{2}]$. Find $\frac{dy}{dx}$.

Solution We can apply the rule of calculating the derivative for an inverse function in this case. Since $x = \sin y$ and $\frac{dx}{dy} = \frac{d}{dy} \sin y = \cos y$, we have:

$$\frac{dy}{dx} = \frac{1}{\cos y} = \frac{1}{\sqrt{1 - (\sin y)^2}} = \frac{1}{\sqrt{1 - x^2}}.$$

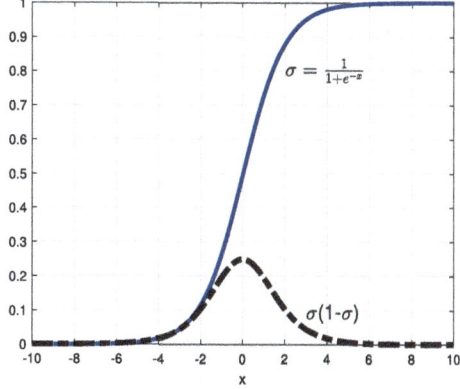

Fig. 5.5 An illustration of a sigmoid function (solid line) and its derivative (dash-dotted line)

Example 5.16 Let $f(x) = \ln(\cos x)$. Find $\frac{dy}{dx}$.

Solution Let $y = \ln z$ and $z = \cos x$. Applying the chain rule, we have $\frac{dy}{dz} = \frac{1}{z}$, $\frac{dz}{dx} = -\sin x$, and

$$\frac{dy}{dx} = \frac{dy}{dz}\frac{dz}{dx} = \frac{1}{z} \times (-\sin x) = -\frac{\sin x}{\cos x} = -\tan x.$$

Example 5.17 Find the derivative of $f(x) = \max(0, x)$.

Solution If $x \leq 0$, the function value is 0. The derivative of any constant value is 0. If $x > 0$, the function value is the value of x. The derivative of x is 1. Figure 5.6 shows the function (the left panel) and its derivative (the right panel) in the domain of $[-5, 5]$. This function is widely used as an activation function in neural networks, and its name is the Rectified Linear Unit, or ReLU for short. Technically, the derivative is undefined when the input is 0. In practice, if we assume that the derivative is zero here, there are no problems. Many real-world applications using neural networks have empirically proved that models with ReLU as the activation function are easier to train and can have a better performance.

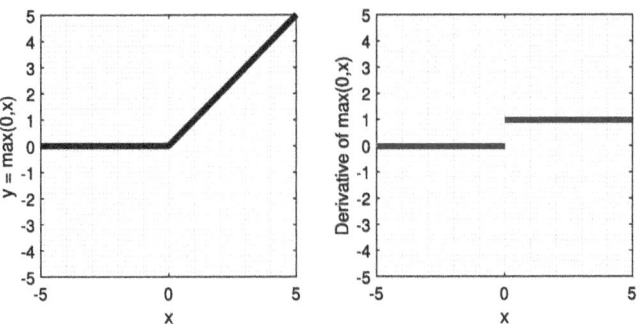

Fig. 5.6 An illustration of a ReLU function (on the left) and its derivative (on the right)

Exercises

5.3 Find $\frac{d}{dx} f(x)$ of the following functions.

(1) $f(x) = x + 6$.
(2) $f(x) = x^6$.
(3) $f(x) = 10e^x$.
(4) $f(x) = 5\ln(x)$.
(5) $f(x) = \ln(x) \sin x$.
(6) $f(x) = \frac{e^x}{\cos x}$.
(7) $f(x) = e^{10x+1}$.
(8) $f(x) = 4(e^x)^2 + 5$.
(9) $f(x) = e^{3x} \sin 5x$.
(10) $f(x) = \frac{\ln(8x)}{e^{(x^2)}}$.

5.2.3 The Second Derivative

In general, the derivative y' or $f'(x)$ of a function $y = f(x)$ in an interval is still a function of x. For example, see the derivative in Figs. 5.5 and 5.6, respectively. If $f'(x)$ is also differentiable in the interval, we call the derivative of $y' = f'(x)$ the second-order derivative of $y = f(x)$, denoted as y'', $f''(x)$, or $\frac{d^2 y}{dx^2}$. That is,

$$y'' = (y')' \text{ or } \frac{d^2 y}{dx^2} = \frac{d}{dx} \frac{dy}{dx}.$$

Similarly, the nth derivative of $f(x)$ if it exists, is denoted as $y^{(n)}$, $f^{(n)}(x)$, or $\frac{d^n y}{dx^n}$. This book considers the first and the second derivative of a function only.

Example 5.18 Find the second derivative of the following function:

$$y = 5x + 8.$$

Solution

$$y' = 5, \quad y'' = (y')' = 0.$$

Example 5.19 Find the second derivative of the following function:

$$y = \sin(\omega x).$$

Solution

$$y' = \omega\cos(\omega x), \ y'' = (\omega\cos(\omega x))' = -\omega^2\sin(\omega x) = -\omega^2 y.$$

Exercise

5.4 Find the second derivative of the following functions.
(1) $y = x^3 \ln x$.
(2) $y = ae^{-\alpha x}$,   express the answer in terms of y.
(3) $y = ae^{-\alpha x} + be^{\alpha x}$,   express the answer in terms of y.

5.3 Finding Local Maxima and Minima Using Derivatives

There is a close relationship between the function monotony and the sign of its derivative. Suppose function $y = f(x)$ is continuous in the interval of $[a, b]$ and is differentiable in (a, b). Recall that the derivative of a function at a specific point can be considered as the slope of the tangent line of the function passing through that point. If $f'(x) > 0$ for all $x \in (a, b)$, then $y = f(x)$ monotonically increases in the interval of $[a, b]$ (Fig. 5.7a). On the other hand, if $f'(x) < 0$ for all $x \in (a, b)$, then $y = f(x)$ monotonically decreases in the interval of $[a, b]$ (Fig. 5.7b).

We can use the relationship between the function monotony and the sign of its derivative to find extreme points, the local maxima and minima, of a function. Looking at Fig. 5.8, we see function $f(x)$ has four local minima, $f(x_1)$, $f(x_4)$, $f(x_6)$, and $f(x_8)$, respectively, and three local maxima, $f(x_2)$, $f(x_5)$, and $f(x_7)$, respectively, in the interval of $[a, b]$. Among them, the local maximum value $f(x_2)$ is smaller than the local minimum value $f(x_6)$. In fact, $f(x_1)$ is the overall minimum, and $f(x_7)$ is the overall maximum of the function in the interval of $[a, b]$. We can also see that all tangents at either local minima or local maxima are horizontal. However, when a tangent is horizontal, the corresponding function value is not necessarily a local minimum or a local maximum, such as the point $(x_3, f(x_3))$. $(x_3, f(x_3))$ is called a point of inflection. Any point $(x, f(x))$ at which $f'(x) = 0$ is called a critical point.

5.3 Finding Local Maxima and Minima Using Derivatives

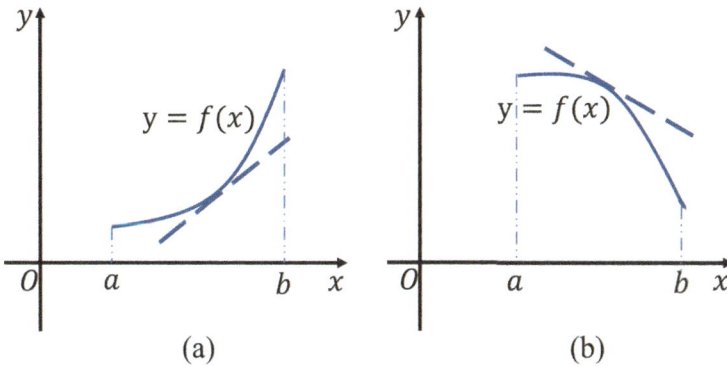

Fig. 5.7 An illustration of the relationship between the sign of the derivative and the monotonicity of a function, where the solid line represents the function and the dashed line represents the tangent line

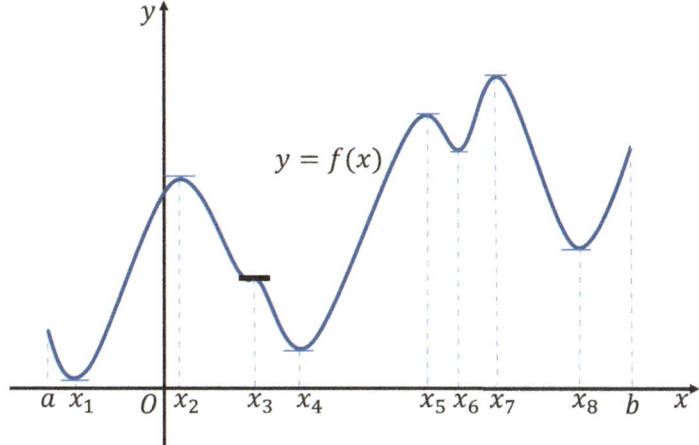

Fig. 5.8 An illustration of local maxima and minima of a function

Now let us have a look at Fig. 5.9. The top panel shows a diagram of a function, the middle shows the first derivative of the function, and the bottom shows the second derivative. We can see that the local minimum point of the function has a zero first derivative and a positive second derivative; the local maximum point of the function has a zero first derivative and a negative second derivative.

The general idea to find local maxima and minima of a function $y = f(x)$ using derivatives is:

- Step 1: to find critical values x using the condition $f'(x) = 0$.
- Step 2: to determine the exact nature of the function at a critical point $(x, f(x))$, $f''(x)$ needs to be calculated.

Fig. 5.9 An illustration of the relationship between local maxima or minima and their corresponding first and second derivatives

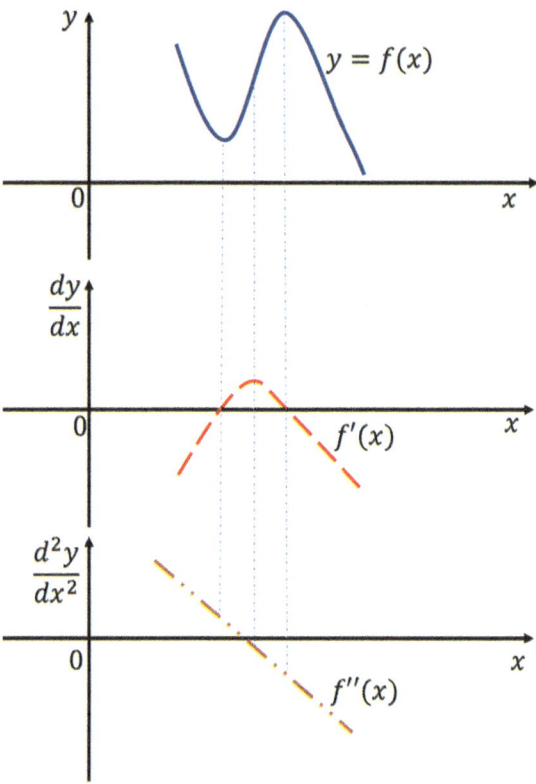

- If $f''(x) > 0$, the critical point is a local minimum point.
- If $f''(x) < 0$, the critical point is a local maximum point.
- If $f''(x) = 0$, it needs further investigation.

Example 5.20 Find the local minima and maxima of function $f(x) = (x^2 - 1)^3 - 1$.

Solution

- Find critical values using the condition $f'(x) = 0$.
  - First, find $f'(x)$ by applying the chain rule (see Sect. 5.2.2 of this chapter). Set $u = (x^2 - 1)$, then we have:

$$\frac{df(x)}{dx} = \frac{df(u)}{du}\frac{du}{dx} = 3u^2 2x = 6x(x^2 - 1)^2.$$

(continued)

5.3 Finding Local Maxima and Minima Using Derivatives

Example 5.20 (continued)

– Set $6x(x^2 - 1)^2 = 0$, we have: $x_1 = -1$, $x_2 = 0$, and $x_3 = 1$.

Substitute these values to the function, we obtain three critical points: $(-1, -1)$, $(0, -2)$, and $(1, -1)$.

- Calculate $f''(x)$, that is, calculate the derivative of $(6x(x^2 - 1)^2)$. Consider $g(x) = 6x$ and $h(x) = (x^2 - 1)^2$, and apply the product rule to $\frac{d}{dx}(g(x)h(x))$ and the chain rule to $h(x)$, then we have:

$$f''(x) = 6(x^2 - 1)^2 + 6x \cdot 2(x^2 - 1) \cdot 2x$$
$$= 30x^4 - 36x^2 + 6$$
$$= 6(5x^2 - 1)(x^2 - 1).$$

- Substitute $x_1 = -1$, $x_2 = 0$, and $x_3 = 1$ into $f''(x) = 6(5x^2 - 1)(x^2 - 1)$, separately.

  – Since $f''(0) = 6 > 0$, $f(x)$ has the minimum value at $x = 0$, which is $f(0) = -2$.
  – Since $f''(-1) = 0$ and $f''(1) = 0$, each critical point needs further investigation. When taking a close value from the left side of -1, for example, -1.01, we have $f'(-1.01) < 0$; taking a close value from the right side of -1, for example, -0.9, we have $f'(-0.9) < 0$. Since there is no sign change to $f'(x)$, then we conclude there is no maximum or minimum at $x = -1$. Similarly, there is no maximum or minimum at $x = 1$. Both points are, in fact, points of inflection with zero gradients.

Remark 5.2 A point of inflection is a point where the gradient line at a point is above the curve on one side and below the curve on the other side of the point. Or where the curve changes from being concave downward to being concave upward or vice versa. At these points $f''(x) = 0$. If $f'(x) = 0$ as well, then we have a point of inflection with a zero gradient like point $(x_3, f(x_3))$ on Fig. 5.8. Other points of inflection are like the point where $f''(x) = 0$ but $f'(x) \neq 0$ on Fig. 5.9. This is where the middle vertical dash-dotted line goes down from the middle of the upward-sloping part of the graph in the top part of the figure to the bottom part of the figure, where it shows that $f''(x) = 0$ (in fact the gradient itself, $f'(x)$ has a maximum at that point as seen in the middle part of the Fig. 5.9). ♦

Exercise

5.5 Find any local maximums and minimums of the following functions.
(1) $f(x) = x^3 - 3x^2 - 24x + 3$.
(2) $f(x) = 4x^3 - 3x^2 + 1$.
(3) $f(x) = 24x - 2x^3$.
(4) $f(x) = 4x^2 - \frac{1}{x}$.
(5) $f(x) = x^4 - 4x^3 - 2x^2 + 12x + 4$.
(6) for $x \in [0, 2\pi]: f(x) = e^x(\cos x + \sin x)$.
(7) $f(x) = xe^{-x}$.
(8) $f(x) = x^2 e^{-x}$.

5.4 Integrals

Earlier, we introduced how to find the derivative of a differentiable function. In this section, we will discuss the inverse operation of finding derivatives. That is, given a function $f(x)$, and we will see how we can find a differentiable function $F(x)$ so that the derivative of $F(x)$ equals $f(x)$. First, we consider the area under a curve by summing up (integrating) small areas. This again will draw on the concept of limits.

Consider the area of the region A shown in Fig. 5.10. That is the area under $f(x)$ bounded in the interval of $[a, b]$. Suppose we divide the interval into n sub-intervals by inserting arbitrarily $n - 1$ points $x_1, x_2, \ldots, x_{n-1}$, and the length of each subinterval is:

$$\Delta x_1 = a - x_1, \Delta x_2 = x_2 - x_1, \ldots, \Delta x_n = b - x_{n-1}.$$

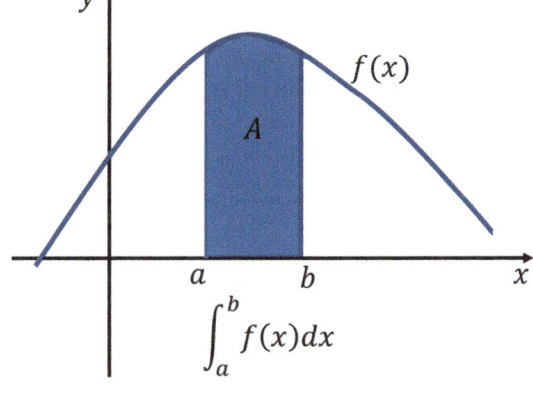

Fig. 5.10 An illustration of the definite integral of $f(x)$ over the interval $[a, b]$, represented as the shaded area under the curve

Randomly choose a point in each subinterval ξ_i ($x_{i-1} \leq \xi_i \leq x_i$), where $i = 1, \ldots, n$, $x_0 = a$ and $x_n = b$, and take the product of the function value at ξ_i and the length of its corresponding subinterval Δx_i. This product represents the area of a rectangle of height $f(\xi_i)$ and width Δx_i and will be an approximation to the area under the curve in that subinterval Δx_i. Then, the sum of these products can be written as:

$$S = \sum_{i=1}^{n} f(\xi_i) \Delta x_i. \tag{5.10}$$

Definition 5.3 (Definite Integrals) Let the number of subintervals n increase so that the lengths $\Delta x_i \to 0$. Denote $\lambda = \max\{\Delta x_1, \Delta x_2, \cdots, \Delta x_n\}$. If the sum of Eq. (5.10) approaches a limit that does not depend on how we divide the interval, then we denote this limit by the following:

$$\int_a^b f(x)dx = \lim_{\lambda \to 0} \sum_{i=1}^{n} f(\xi_i) \Delta x_i. \tag{5.11}$$

This is called the definite integral of $f(x)$ between a and b. Finding the summation is known as integration. $f(x)$ is called the integrand; a and b are the limits of integration or the endpoints of integration; dx tells us x is the variable of integration.

Geometrically, the integral that is the limit of the sum (Eq. 5.10) represents the total area of all rectangles (defined by subintervals) under a bounded function. For example, $\int_0^4 x dx = 8$. This can be viewed in Fig. 5.11. The line represents the function of $f(x) = x$. The area in the interval of $[0, 4]$ under the line of $f(x) = x$ is a triangle, whose area is computed as $\frac{1}{2} \times 4 \times 4 = 8$.

5.4.1 First Fundamental Theorem of Calculus

For f continuous on $[a, b]$, define a function F by the following:

$$F(x) = \int_a^x f(t)dt \quad \text{for } x \text{ in } [a, b], \tag{5.12}$$

then F is differentiable on (a, b) and $F'(x) = f(x)$. That is, differentiating F gives us back the original function $f(x)$. We call $F(x)$ an antiderivative of $f(x)$.

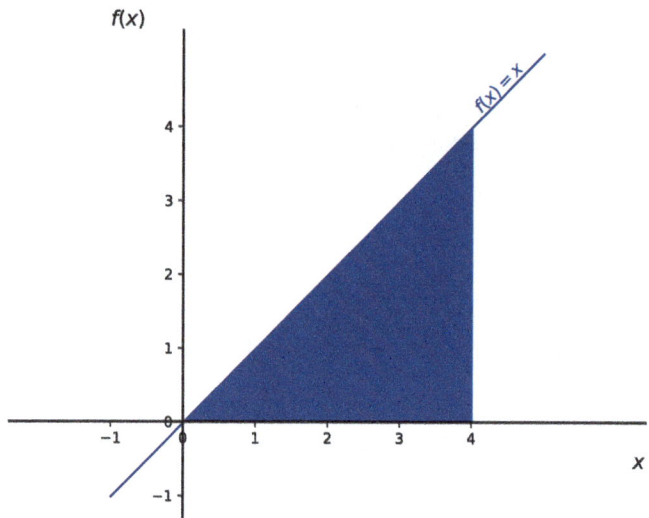

Fig. 5.11 An illustration of the definite integral under a line of $f(x) = x$ in the interval of [0, 4]

Example 5.21 Suppose $f(x) = x^5$. It means $F'(x) = \frac{d}{dx}F(x) = f(x) = x^5$. Can you think of some functions whose derivatives are x^5? Consider $\frac{d}{dx}c = 0$ and $\frac{d}{dx}ax^c = cax^{c-1}$ (Eqs. 5.4 and 5.5). Some possible functions can be $F(x) = \frac{x^6}{6}$, $F(x) = \frac{x^6}{6} + 100$, and $F(x) = \frac{x^6}{6} + 34.5$. In fact, any function of x of the form $F(x) = \frac{x^6}{6} + C$ for some constant C is an antiderivative of $f(x) = x^5$.

Remark 5.3 Not all functions $f(x)$ have antiderivatives $F(x)$ in terms of elementary functions (polynomials, exponentials, logarithms, trigonometric functions, etc.), but the definite integral, understood as the limit of a summation, can still exist. ♦

5.4.2 Indefinite Integrals

The family of all antiderivatives of $f(x)$, denoted as $\int f(x)dx$, are called indefinite integrals.

Remark 5.4 A definite integral, written as $\int_a^b f(x)$, is a number. An indefinite integral, written as $\int f(x)dx$, is a family of functions. It is a family of functions, because the constant C can take any value. ♦

5.4.3 Second Fundamental Theorem of Calculus

If f is continuous on $[a, b]$ and F is any of the family of antiderivative of f with respect to x, then

$$\int_{x=a}^{b} f(x)dx = F(b) - F(a). \tag{5.13}$$

Remark 5.5 Care should be taken when finding an area using the equation

$$\int_{x=a}^{b} f(x)dx = F(b) - F(a). \tag{5.14}$$

When $f(x)$ is below the axis for all of the interval $[a, b]$, then $F(b) - F(a)$ is negative, so if the function $f(x)$ is both above and below the axis in the interval $[a, b]$, then $F(b) - F(a)$ is the difference between the upper area and the lower area. This could be zero if the areas are identical. ♦

5.4.4 Integrals of Some Elementary Functions

We use a, c, and C to denote constants in the following set of formulas. These can all be checked by differentiating the right-hand side to get back to the left-hand side.

$$\int adx = ax + C, \tag{5.15}$$

$$\int ax^c dx = \frac{ax^{c+1}}{c+1} + C, \ (c \neq -1), \tag{5.16}$$

$$\int \frac{a}{x} dx = a \ln|x| + C, \tag{5.17}$$

$$\int ce^{ax} dx = \frac{ce^{ax}}{a} + C, \tag{5.18}$$

$$\int ca^x dx = \frac{ca^x}{\ln a} + C, \tag{5.19}$$

$$\int a\cos x\,dx = a\sin x + C, \tag{5.20}$$

$$\int a\sin x\,dx = -a\cos x + C. \tag{5.21}$$

Readers may find integral formula for more elementary functions in [6].

Example 5.22 Find the following integral:

$$\int 6x^2\,dx.$$

Solution Applying Eq. (5.16), we have:

$$\int 6x^2\,dx = \frac{6x^3}{3} + C = 2x^3 + C.$$

Example 5.23 Find the following integral:

$$\int \frac{dx}{x\sqrt[3]{x}}.$$

Solution Applying Eq. (5.16), we have:

$$\int \frac{dx}{x\sqrt[3]{x}} = \int x^{-\frac{4}{3}}\,dx = \frac{x^{-\frac{4}{3}+1}}{-\frac{4}{3}+1} + C = -3x^{-\frac{1}{3}} + C. = -\frac{3}{\sqrt[3]{x}} + C.$$

Exercise

5.6 Calculate the following integrals.

(1) $\int e^{2x}\,dx$,
(2) $\int -8x^3\,dx$,
(3) $\int \frac{6}{x}\,dx$.
(4) $\int 3e^{-x}\,dx$,
(5) $\int 5\sin x\,dx$,

5.4.5 Two Properties of Integrals

If $f(x)$ and $g(x)$ are integrable in $[a, b]$, then

$$\int_a^b (f(x) + g(x))dx = \int_a^b f(x)dx + \int_a^b g(x)dx, \quad (5.22)$$

$$\int_a^b cf(x)dx = c \int_a^b f(x)dx, \quad c \text{ is a constant.} \quad (5.23)$$

Similarly, for indefinite integrals, the following two properties are valid:

$$\int (f(x) + g(x))dx = \int f(x)dx + \int g(x)dx, \quad (5.24)$$

$$\int cf(x)dx = c \int f(x)dx, \quad c \text{ is a constant.} \quad (5.25)$$

Example 5.24 Find the following integral:

$$\int \frac{(x-1)^3}{x^2} dx.$$

Solution

$$\int \frac{(x-1)^3}{x^2} dx = \int \frac{x^3 - 3x^2 + 3x - 1}{x^2} dx$$

$$= \int x \, dx - 3 \int dx + 3 \int \frac{dx}{x} - \int \frac{dx}{x^2}$$

$$= \frac{x^2}{2} - 3x + 3 \ln |x| + \frac{1}{x} + C.$$

Example 5.25 Find the following integral:

$$\int 10^x e^x dx.$$

(continued)

Example 5.25 (continued)
Solution

$$\int 10^x e^x dx = \int (10e)^x = \frac{(10e)^x}{\ln(10e)} + C = \frac{10^x e^x}{\ln 10 + 1} + C.$$

Exercise

5.7 Calculate the following integrals.

(1) $\int (4\sin x + e^x)dx$,
(2) $\int_{\theta=0}^{\frac{\pi}{3}} (\theta + \cos\theta)d\theta$,
(3) $\int \frac{e^x + e^{-x}}{2} dx$

5.5 Further Integration Techniques

In many real-world applications, we need to integrate more complex functions. The methods shown in the previous two sections are not enough. We will introduce two more new techniques in this subsection. More examples can be found in [6].

5.5.1 Integration by Substitution

This method transforms an integral over one variable x to an integral over a different variable u by making a substitution. That is:

$$\int_{x_1}^{x_2} f(x)dx = \int_{u_1}^{u_2} g(u)du.$$

The idea behind this integration method is that by making a substitution, you produce a new integral that is simpler to evaluate. This can often be achieved by substituting for the "most difficult part" of the integral, or recognising that the integrand is of the form that you would get having differentiated a composite function using the chain rule.

The following shows the procedure when applying the substitution method to an integral $\int_{x_1}^{x_2} f(x)dx$:

- Step 1: Think of a substitution $u = h(x)$ that will make the integral simpler.
- Step 2: Differentiate the substitution $u = h(x)$, and write dx in terms of du.

5.5 Further Integration Techniques

- Step 3: For a definite integral, we must also determine the new limits of integration.
- Step 4: Make the following substitutions into the original integral:
  - Replace x with the equation from Step 1.
  - Replace dx with the equation from Step 2.
  - Replace limits x_1 and x_2 with the new limits from Step 3.
- Step 5: Do the new integral in terms of u.
- Step 6: Write the answer in terms of x.

Example 5.26 Perform the following integral:

$$\int (2x+5)^4 dx.$$

Solution This integral could be done by expanding out the bracket, but it will give a more convenient solution if we replace the "difficult bit" in the bracket. So set $u = 2x + 5$.

Differentiate the equation $u = 2x + 5$, and obtain $\frac{du}{dx} = 2$ or equivalently $dx = \frac{du}{2}$.

Substituting $dx = \frac{du}{2}$ and $u = 2x + 5$ into the original integral gives

$$\int (2x+5)^4 dx = \int u^4 \frac{du}{2} = \int \frac{u^4}{2} du = \frac{u^5}{2(5)} + C = \frac{(2x+5)^5}{10} + C.$$

Example 5.27 Perform the following integral:

$$\int x^3 \sqrt{x^4 + 3}\, dx.$$

Solution The key to this substitution is recognising that this is the sort of result that you could get by using the chain rule of differentiation. The "difficult bit" in the bracket, when differentiated, gives the other part of the integrand (apart from a constant). That is, $\frac{d(x^4+3)}{dx} = 4x^3$. So set $u = x^4 + 3$.

(continued)

Example 5.27 (continued)

Differentiate the equation $u = x^4 + 3$, and obtain $dx = \frac{du}{4x^3}$.
Substituting $dx = \frac{du}{4x^3}$ and $u = x^4 + 3$ into the original integral gives:

$$\int x^3 \sqrt{x^4 + 3}\, dx = \int x^3 \sqrt{u}\, \frac{du}{4x^3}$$

$$= \int \frac{\sqrt{u}}{4}\, du$$

$$= \frac{u^{\frac{1}{2}+1}}{4(\frac{1}{2}+1)} + C$$

$$= \frac{u^{\frac{3}{2}}}{6} + C$$

$$= \frac{(x^4 + 3)^{\frac{3}{2}}}{6} + C.$$

Example 5.28 Evaluate:

$$\int_0^{\frac{\pi}{2}} \cos^5 x \sin x\, dx.$$

Solution Again, $-\sin x$ is what you get when differentiating $\cos x$, so the chain rule is implicated. Hence, we set $u = \cos x$.

Differentiate the equation $u = \cos x$, and obtain $dx = -\frac{du}{\sin x}$.
When $x = 0$, $u = 1$; when $x = \frac{\pi}{2}$, $u = 0$.

Substituting $dx = -\frac{du}{\sin x}$, $u = \cos x$ and new limits into the original integral gives:

$$\int_0^{\frac{\pi}{2}} \cos^5 x \sin x\, dx = -\int_1^0 u^5\, du = \int_0^1 u^5\, du = \frac{u^6}{6}\bigg|_0^1 = \frac{1}{6}.$$

5.5 Further Integration Techniques

Example 5.29 Perform the following integral:

$$\int \frac{x+2}{\sqrt{x+3}}dx.$$

Solution Here there is no obvious chain rule, but the denominator is the "difficult bit". So try $u = \sqrt{x+3}$.
Hence $u^2 = x+3$ and $x = u^2 - 3$ and obtain $dx = 2u\,du$.
Substituting $dx = 2u\,du$ and $u = \sqrt{x+3}$ into the original integral gives:

$$\int \frac{x+2}{\sqrt{x+3}}dx = \int \frac{u^2-1}{u} 2u\,du$$

$$= \int (2u^2 - 2)du$$

$$= \frac{2u^3}{3} - 2u + C$$

$$= \frac{2}{3}(x+3)^{\frac{3}{2}} - 2(x+3)^{\frac{1}{2}} + C.$$

You can check the result is correct by differentiating the answer! Note that the substitution of $u = x+3$ also works as you can check.

Exercise

5.8 Calculate the following integrals.

(1) $\int x\sqrt{(2-x^2)}dx.$
(2) $\int \frac{dx}{(x-1)^2}.$
(3) $\int_0^{\frac{\pi}{8}} \sin(4x)dx.$
(4) $\int \frac{x^2}{(1+x^3)^2}dx.$
(5) $\int_0^1 \frac{x}{(1+x)^3}dx.$
(6) $\int_{\frac{\pi}{6}}^{\frac{\pi}{2}} \frac{\cos x}{\sin^3 x}dx.$
(7) $\int \frac{x}{1+x^2}dx.$
(8) $\int xe^{x^2}dx.$

5.5.2 Integration by Parts

The formula of integration by parts can be derived from the product rule for differentiation (see Sect. 5.2.2 of this chapter). Suppose u and v are functions of x and differentiable:

$$\frac{d}{dx}(uv) = u\frac{dv}{dx} + v\frac{du}{dx}.$$

Integrating the above equation with respect to x, we get:

$$uv = \int u\frac{dv}{dx}dx + \int v\frac{du}{dx}dx.$$

Rearranging this gives us the integration by parts formula:

$$\int u\frac{dv}{dx}dx = uv - \int v\frac{du}{dx}dx. \tag{5.26}$$

The key to this method of integration is to treat the integral as a product. One part of the product is represented by $\frac{dv}{dx}$ on the left-hand side of Eq. (5.26) and is integrated to give v on the right-hand side. This part needs to be something that you can, therefore, integrate. The other part of the integral is represented by u and is differentiated, giving $\frac{du}{dx}$ on the right-hand side. In this way, only part of the original integral is integrated, and hopefully, the new integral is made simpler.

Example 5.30 Perform the following integral:

$$\int x \sin x \, dx.$$

Solution The integrand is a product, both of which can be integrated separately. But if u is taken as x, then the new integrand will contain the differential of x, which is just 1, and so will be simpler. Hence consider $u = x$ and $\frac{dv}{dx} = \sin x$. Then we have $\frac{du}{dx} = 1$ and $v = -\cos x$. Applying the integration by parts formula (Eq. 5.26) gives:

$$\int x \sin x \, dx = x(-\cos x) - \int (-\cos x) 1 \, dx$$

$$= -x \cos x + \int \cos x \, dx$$

$$= -x \cos x + \sin x + C.$$

5.5 Further Integration Techniques

Example 5.31 Perform the following integral:

$$\int x \ln x \, dx.$$

Solution The only part we can integrate is the x, so we set $u = \ln x$ and $\frac{dv}{dx} = x$. Then we have $\frac{du}{dx} = \frac{1}{x}$ and $v = \frac{1}{2}x^2$. Applying the integration by parts formula (Eq. 5.26) gives:

$$\int x \ln x \, dx = \ln x \times \frac{1}{2}x^2 - \int \frac{1}{2}x^2 \frac{1}{x} dx = \frac{1}{2}x^2 \ln x$$

$$- \int \frac{1}{2} x \, dx = \frac{1}{2}x^2 \ln x - \frac{1}{4}x^2 + C.$$

Example 5.32 Evaluate:

$$\int_0^{\frac{\pi}{2}} x^2 \cos x \, dx.$$

Solution Again, both parts of the integrand can be integrated, but using u as x^2 will lead to a simpler integral. So set $u = x^2$ and $\frac{dv}{dx} = \cos x$. Then $\frac{du}{dx} = 2x$ and $v = \sin x$. Applying the integration by parts formula (Eq. 5.26) gives:

$$\int_0^{\frac{\pi}{2}} x^2 \cos x \, dx = (x^2 \sin x) \Big|_0^{\frac{\pi}{2}} - 2 \int_0^{\frac{\pi}{2}} x \sin x \, dx.$$

Now, the new integral can again be evaluated using integration by parts. But, in fact, we have just done it (Example 5.30). So we get:

$$(x^2 \sin x) \Big|_0^{\frac{\pi}{2}} - 2(-x \cos x + \sin x) \Big|_0^{\frac{\pi}{2}} = \frac{\pi^2}{4} - 2.$$

Exercise

5.9 Calculate the following integrals.

(1) $\int xe^x dx$.
(2) $\int x^2 \sin x \, dx$.
(3) $\int x^3 \ln x \, dx$.
(4) $\int_1^2 \ln x \, dx$,   Hint: treat $\ln x$ as $1 \ln x$.
(5) $\int x \sin 4x \, dx$.
(6) $\int 2x \ln 5x \, dx$.
(7) $\int \sin^2 x \, dx$.   Hint: treat $\sin^2 x$ as $\sin x \sin x$ in the integration by parts and then use $\cos^2 x = 1 - \sin^2 x$ and rearrange!

Chapter 6
Advanced Calculus

This chapter takes the study of calculus forward into more advanced topics involving multiple variable functions. In general, most functions that are found in the machine learning field are ones with many variables rather than just one. Quite often, we are trying to maximise some value or minimise some error function, so the ability to differentiate such functions and find their maxima and minima will be essential. This leads us to the methods of partial differentiation that enable us to find gradients in different planes as described in Sect. 6.1. We also briefly look at multiple integrals that will be needed when we look at probability distributions of multiple continuous random variables in Chap. 11.

6.1 Partial Derivatives

6.1.1 The First Partial Derivatives

In functions with two or more variables, the partial derivative is the derivative with respect to one of those variables, keeping all other variables constant. For example, consider a function $f(x, y)$ with two variables (such as $f(x, y) = x^2 + 2xy + y^3$). Partial derivatives of $f(x, y)$ with respect to x and y are denoted by $\frac{\partial f}{\partial x}$ and $\frac{\partial f}{\partial y}$, respectively.

Let $\Delta x = dx$ and $\Delta y = dy$ be increments given to x and y of $f(x, y)$, respectively. Δf is then the subsequent incremental change of the function f. By Eq. (5.3) (in Sect. 5.2 of Chap. 5), if the corresponding limits exist, we have

$$\frac{\partial f}{\partial x} = \lim_{\Delta x \to 0} \frac{f(x + \Delta x, y) - f(x, y)}{\Delta x},$$

and

$$\frac{\partial f}{\partial y} = \lim_{\Delta y \to 0} \frac{f(x, y + \Delta y) - f(x, y)}{\Delta y}.$$

When evaluating these partial derivatives at a particular point (x_0, y_0), they can be denoted as $\frac{\partial f}{\partial x}\big|_{\substack{x=x_0 \\ y=y_0}}$ and $\frac{\partial f}{\partial y}\big|_{\substack{x=x_0 \\ y=y_0}}$, respectively.

Again, we can define the *differential* as we did for ordinary differentiation. The following expression

$$df = \frac{\partial f}{\partial x} dx + \frac{\partial f}{\partial y} dy \tag{6.1}$$

is called the total differential of f, or the principal part of the change in the function f with respect to changes in the independent variables.

Remark 6.1 Note that in general, $df \neq \Delta f$. If $\Delta x = dx$ and $\Delta y = dy$ are small, then df is a close approximation of Δf. ♦

Thought of visually or graphically, a function of two variables is a surface in three dimensions. Assume $z = f(x, y)$. Then, $\frac{\partial f}{\partial x}$ keeps y constant, and so is a gradient (see Sect. 6.1.4 of this chapter) on the curve where the surface meets a plane parallel to the $x - z$ plane. Similarly, $\frac{\partial f}{\partial y}$ keeps x constant, and so is a gradient on the curve where the surface meets a plane parallel to the $y - z$ plane. The same considerations apply to more variables, but the graph is no longer possible to visualise.

Example 6.1 Suppose $f(x, y) = x^2 + 2xy + y^3$. Find the partial derivatives at the point of $(2, 1)$.

Solution Consider y as a constant:

$$\frac{\partial f}{\partial x} = 2x + 2y.$$

Consider x as a constant,

$$\frac{\partial f}{\partial y} = 2x + 3y^2.$$

Substitute the point of $(2, 1)$ into the two partial derivative results, then we have

$$\frac{\partial f}{\partial x}\bigg|_{\substack{x=2 \\ y=1}} = 2 \times 2 + 2 \times 1 = 6,$$

(continued)

Example 6.1 (continued)
and

$$\left.\frac{\partial f}{\partial y}\right|_{\substack{x=2\\y=1}} = 2 \times 2 + 3 \times 1^2 = 7.$$

Example 6.2 A cylinder is being made that is 30 cm radius and 60 cm high. The tolerances in construction are that the radius is $\pm 0.05\%$ and the height is $\pm 0.01\%$. Find the approximate maximum error in volume to the nearest integer and hence find the percentage error that this represents.

Solution Radius error: $\pm 0.05\%$ of 30 cm is ± 0.015 cm.
Height error: $\pm 0.01\%$ of 60 cm is ± 0.006 cm.
Set the cylinder's radius, height, and volume as r, h, and V. We have

$$V = \pi r^2 h.$$

Let Δr, Δh, and ΔV denote increments of r, h, and V. Applying Eq. (6.1), it gives us the following:

$$\Delta V \approx dV = \frac{\partial V}{\partial r}dr + \frac{\partial V}{\partial h}dh = 2\pi rh \Delta r + \pi r^2 \Delta h.$$

To get an estimate of the maximum error, both Δr and Δh should be positive (or negative).

Substitute $r = 30$, $h = 60$, $\Delta r = 0.015$, and $\Delta h = 0.006$, and then we have

$$\Delta V \approx 2\pi \times 30 \times 60 \times 0.015 + \pi \times 30^2 \times 0.006 = 187 (\text{cm}^3).$$

The actual volume should be:

$$V = \pi r^2 h = \pi \times 30^2 \times 60 = 169,646 (\text{cm}^3).$$

So the error represents

$$187 \div 169,646 \times 100\% = 0.11\%.$$

Exercises

6.1 Find the following partial derivatives:

(1) At the point $(3, -1)$ for the function $f(x, y) = x^3 y + 5x^2 y^2 + 2xy^3$
(2) At the point $(1, \frac{\pi}{2})$ for the function $f(x, y) = x^2 \sin y - 3x \cos y$
(3) At the point $(0, 2)$ for the function $f(x, y) = y^3 e^{2x} + y^2 e^{3x} + y e^{4x}$.

6.2 A triangle is being stamped out of a sheet of metal. Its height is 5 cm, and its base is 10 cm. The tolerances in this process are that the height is $\pm 0.2\%$ and the base is $\pm 0.1\%$. Find the approximate maximum error in the area to three decimal places and hence the percentage error that this represents.

6.1.2 The Second Partial Derivatives

The second partial derivatives are the partial derivatives of the first derivative function. For example, let us consider a function $f(x, y)$ with two variables. If its first derivatives $\frac{\partial f}{\partial x}$ and $\frac{\partial f}{\partial y}$ are continuous and the partial derivatives of $\frac{\partial f}{\partial x}$ and $\frac{\partial f}{\partial y}$ all exist, then the second derivatives of $f(x, y)$ can be denoted as follows:

- $\frac{\partial^2 f}{\partial x^2}$ — the partial derivative of $\frac{\partial f}{\partial x}$ with respect to x
- $\frac{\partial^2 f}{\partial y \partial x}$ — the partial derivative of $\frac{\partial f}{\partial x}$ with respect to y
- $\frac{\partial^2 f}{\partial y^2}$ — the partial derivative of $\frac{\partial f}{\partial y}$ with respect to y
- $\frac{\partial^2 f}{\partial x \partial y}$ — the partial derivative of $\frac{\partial f}{\partial y}$ with respect to x

Remark 6.2 For most well-behaved functions (ones where the two second partial derivatives involved are continuous), we have $\frac{\partial^2 f}{\partial y \partial x} = \frac{\partial^2 f}{\partial x \partial y}$. ♦

Example 6.3 Find the second partial derivatives of the function $f(x, y) = x^3 + 6xy + 3y^3$.

Solution Consider y as a constant: $\frac{\partial f}{\partial x} = 3x^2 + 6y$.
Consider x as a constant, $\frac{\partial f}{\partial y} = 6x + 9y^2$.
Consider y as a constant in $\frac{\partial f}{\partial x}$: $\frac{\partial^2 f}{\partial x^2} = 6x$.
Consider y as a constant in $\frac{\partial f}{\partial y}$: $\frac{\partial^2 f}{\partial x \partial y} = 6$.
Consider x as a constant in $\frac{\partial f}{\partial x}$: $\frac{\partial^2 f}{\partial y \partial x} = 6$.
Consider x as a constant in $\frac{\partial f}{\partial y}$: $\frac{\partial^2 f}{\partial y^2} = 18y$.

Exercise

6.3 Find the first and second partial derivatives of the following functions:
(1) $f(x, y) = 3x^4 y + 6x^3 y^2 - 4x^2 y^3 + xy^4$.
(2) $f(x, y) = x^2 \sin y + 6x^3 \cos y$.
(3) $f(x, y) = e^{(x^2+y^2)}$.
(4) $f(x, y) = (x^3 + y^3) ln(y^3 + x^3)$.
(5) $f(x, y) = \frac{x^3+3y}{x}$.

6.1.3 Differentiation of Composite Functions with Two Variables

Let us consider function $z = f(u, v)$ with two variables u and v, where both u and v are functions with one variable t, that is $u = \varphi(t)$ and $v = \psi(t)$. If both u and v are differentiable functions of t, the function z is continuous, and the partial derivatives exist with respect to u and v, and then the differentiation of the composite function can be computed as follows:

$$\frac{dz}{dt} = \frac{\partial z}{\partial u}\frac{du}{dt} + \frac{\partial z}{\partial v}\frac{dv}{dt}. \tag{6.2}$$

t is often time and expresses the dependence of each of u and v on the passing of time. Note that we can find $\frac{dz}{dt}$ as full differentiation since, in reality, z can be expressed as a function of t.

Example 6.4 Suppose $z = \sin u \cos v$, where $u = e^t$ and $v = \ln t$. Find $\frac{dz}{dt}$.

Solution

$$\frac{dz}{dt} = \frac{\partial z}{\partial u}\frac{du}{dt} + \frac{\partial z}{\partial v}\frac{dv}{dt}$$

$$= \cos u \cos v (e^t) + \frac{\sin u(-\sin v)}{t}$$

$$= \cos(e^t) \cos(\ln t)(e^t) - \frac{\sin(e^t) \sin(\ln t)}{t}.$$

The same result would be found if you substituted t for u and v first, but the differentiation is more complicated.

Similarly, consider both u and v as functions of two variables, s and t. If the partial derivatives of functions $u = \varphi(s,t)$ and $v = \psi(s,t)$ exist with respect to s and t, and function $z = f(u,v)$ is continuous, and the partial derivatives exist with respect to u and v, then the differentiation of the composite function can be computed as follows:

$$\frac{\partial z}{\partial s} = \frac{\partial z}{\partial u}\frac{\partial u}{\partial s} + \frac{\partial z}{\partial v}\frac{\partial v}{\partial s}, \tag{6.3}$$

$$\frac{\partial z}{\partial t} = \frac{\partial z}{\partial u}\frac{\partial u}{\partial t} + \frac{\partial z}{\partial v}\frac{\partial v}{\partial t}. \tag{6.4}$$

Example 6.5 Suppose $z = e^u \cos v$, where $u = st$ and $v = s + t$. Find $\frac{\partial z}{\partial s}$ and $\frac{\partial z}{\partial t}$.

Solution

$$\frac{\partial z}{\partial s} = \frac{\partial z}{\partial u}\frac{\partial u}{\partial s} + \frac{\partial z}{\partial v}\frac{\partial v}{\partial s}$$

$$= e^u \cos v \cdot t - e^u \sin v$$

$$= e^{st}(t\cos(s+t) - \sin(s+t)).$$

This is the same result that you would get if you first substituted s and t for u and v into the formula for z and then found the partial derivative $\frac{\partial z}{\partial s}$. Usually, though, having lots of simple functions to differentiate is easier than having a complicated composite function:

$$\frac{\partial z}{\partial t} = \frac{\partial z}{\partial u}\frac{\partial u}{\partial t} + \frac{\partial z}{\partial v}\frac{\partial v}{\partial t}$$

$$= e^u \cos v \cdot s - e^u \sin v$$

$$= e^{st}(s\cos(s+t) - \sin(s+t)).$$

Again, the same result would be found by substituting s and t for u and v as before.

6.1 Partial Derivatives

Exercise

6.4 Differentiation of composite functions:

(1) $z = e^u \ln v$, where $u = \sin t$ and $v = \cos t$. Find $\frac{dz}{dt}$.
(2) $z = (1 + u^2) \sin v$, where $u = s^2 + t^2$ and $v = st^2$. Find $\frac{\partial z}{\partial s}$ and $\frac{\partial z}{\partial t}$.

6.1.4 Gradient

Armed with the definitions of partial derivatives in coordinate directions, we can define a vector representing the total gradient in the full space concerned.

Suppose a function f is differentiable in a region. The gradient of the function, denoted by $\operatorname{grad} f$, is a vector function where each element is a partial derivative with respect to one of the variables. For example, the gradient of $f(x, y, z)$ can be written as the following vector:

$$(\operatorname{grad} f)^T = \left[\frac{\partial f(x, y, z)}{\partial x}, \frac{\partial f(x, y, z)}{\partial y}, \frac{\partial f(x, y, z)}{\partial z} \right].$$

Since the gradient is a vector, it can provide information on the magnitude and direction of the vector. Suppose the gradient of a function $f(x, y)$ is given by the vector

$$(\operatorname{grad} f)^T = \left[\frac{\partial f(x, y)}{\partial x}, \frac{\partial f(x, y)}{\partial y} \right], \tag{6.5}$$

then the magnitude is calculated as follows:

$$|\operatorname{grad} f| = \sqrt{\left(\frac{\partial f(x, y)}{\partial x} \right)^2 + \left(\frac{\partial f(x, y)}{\partial y} \right)^2}.$$

If $\frac{\partial f(x,y)}{\partial x} \neq 0$, then the tangent of the angle θ from the x-axis to the gradient is given by

$$\tan \theta = \frac{\frac{\partial f(x,y)}{\partial y}}{\frac{\partial f(x,y)}{\partial x}}.$$

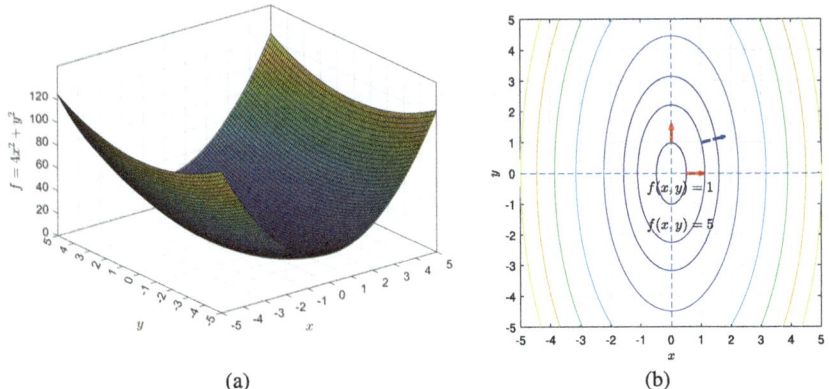

Fig. 6.1 A geometric description of the gradient vector for the function $f(x, y) = 4x^2 + y^2$ used in Example 6.6

Example 6.6 Suppose $f(x, y) = 4x^2 + y^2$. Figure 6.1a shows function values as heights above a grid in the $x - y$ plane. Applying Eq. (6.5) for $f(x, y)$, we have $(\text{grad} f)^T = [8x, 2y]$. Figure 6.1b shows the contour lines of the function for constant levels (heights, function values) over the interval $[-5, 5]$ for x and y, respectively. For example, the innermost oval line contains all the pairs of points (x, y) that have the same function value of 1. The gradient at a general point is given by $[8x, 2y]$. Each gradient vector is perpendicular to the corresponding tangent line. Three particular gradient vectors are plotted at the points $(\frac{1}{2}, 0)$, $(0, 1)$, and $(1, 1)$, respectively. The first two are shown as solid-line arrows perpendicular to the tangents of $f(x, y) = 1$, and the third as a dashed-line arrow perpendicular to the tangent of $f(x, y) = 5$.

Remark 6.3 The gradient vector points in the direction of the maximum rate of increase of the function. Or it points in the opposite direction of the maximum rate of decrease of the function. Readers interested in this may want to learn more about directional derivatives and the gradient in Section 2.E of [11]. ♦

Exercise

6.5 Find the gradient vectors at the points indicated for:

(1) $f(x, y) = x^3 y^3$ at points $(1, 1)$, $(2, 1)$, $(1, 2)$
(2) $f(x, y) = x^2 \sin y + y^2 \cos x$, at points $(0, 1)$, $(1, 0)$, $(\frac{\pi}{2}, \frac{\pi}{2})$, $(\frac{\pi}{4}, \frac{\pi}{4})$

6.1.5 Jacobian Matrix

If $f(x, y)$ and $g(x, y)$ are differentiable in a region, the Jacobian matrix of f and g with respect to x and y can be defined by

$$J = \begin{bmatrix} \frac{\partial f}{\partial x} & \frac{\partial f}{\partial y} \\ \frac{\partial g}{\partial x} & \frac{\partial g}{\partial y} \end{bmatrix}.$$

Similarly, if $f(x, y, z)$, $g(x, y, z)$, and $h(x, y, z)$ are differentiable in a region, the Jacobian matrix of f, g, and h with respect to x, y, and z can be defined by

$$J = \begin{bmatrix} \frac{\partial f}{\partial x} & \frac{\partial f}{\partial y} & \frac{\partial f}{\partial z} \\ \frac{\partial g}{\partial x} & \frac{\partial g}{\partial y} & \frac{\partial g}{\partial z} \\ \frac{\partial h}{\partial x} & \frac{\partial h}{\partial y} & \frac{\partial h}{\partial z} \end{bmatrix}.$$

Extensions are easily made. That is, each row of the Jacobian matrix includes partial derivatives of a specific function with respect to all variables. A Jacobian matrix can be either a rectangular matrix or a square matrix. Essentially, the Jacobian matrix collects all the information about first derivatives together in one place and is useful for translating between coordinate systems (see Sect. 6.3.1 of this chapter) and tells us about the local behaviour of a function in terms of its gradient.

Example 6.7 If $f_1(x, y) = e^x \cos y$ and $f_2(x, y) = e^x \sin y$, determine the Jacobian matrix.

(continued)

Example 6.7 (continued)
Solution Note that the question can also be written in vector form by collecting the component functions together as define function $F : \mathbb{R}^2 \to \mathbb{R}^2$ given by $F(x, y) = (f_1(x, y), f_2(x, y))$.

$$J = \begin{bmatrix} \frac{\partial f_1}{\partial x} & \frac{\partial f_1}{\partial y} \\ \frac{\partial f_2}{\partial x} & \frac{\partial f_2}{\partial y} \end{bmatrix} = \begin{bmatrix} e^x \cos y & -e^x \sin y \\ e^x \sin y & e^x \cos y \end{bmatrix}.$$

Note that the determinant of J is $e^{2x} \cos^2 y + e^{2x} \sin^2 y = e^{2x}$.

Example 6.8 Determine the Jacobian matrix of the function $F : \mathbb{R}^3 \to \mathbb{R}^2$ given by $F(x, y, z) = (4xy + 3xz^3, 3xyz^2)$.

Solution Set $f_1(x, y, z) = 4xy + 3xz^3$ and $f_2(x, y, z) = 3xyz^2$:

$$J = \begin{bmatrix} \frac{\partial f_1}{\partial x} & \frac{\partial f_1}{\partial y} & \frac{\partial f_1}{\partial z} \\ \frac{\partial f_2}{\partial x} & \frac{\partial f_2}{\partial y} & \frac{\partial f_2}{\partial z} \end{bmatrix} = \begin{bmatrix} 4y + 3z^3 & 4x & 9xz^2 \\ 3yz^2 & 3xz^2 & 6xyz \end{bmatrix}.$$

Exercise

6.6 Determine the Jacobian matrix for the following:
(1) $f_1(x, y) = x^2 \sin y$ and $f_2(x, y) = y^3 \sin x$.
(2) $F : \mathbb{R}^2 \to \mathbb{R}^2$ given by $F(r, \theta) = (r \cos \theta, r \sin \theta)$.
(3) $F : \mathbb{R}^3 \to \mathbb{R}^3$ given by $F(r, \theta, \phi) = (r \sin \theta \cos \phi, r \sin \theta \sin \phi, r \cos \theta)$.

6.1.6 Hessian Matrix

A Hessian matrix of a function is a square matrix of the second partial derivatives of the function. It collects together all the information about the second derivatives and tells us about the curvature of a function at a point. For a two-variable function $f(x, y)$, its Hessian matrix is defined by the following:

$$\mathcal{H}f(x, y) = \begin{bmatrix} H_{11}(x, y) & H_{12}(x, y) \\ H_{21}(x, y) & H_{22}(x, y) \end{bmatrix} = \begin{bmatrix} \frac{\partial^2 f}{\partial x^2} & \frac{\partial^2 f}{\partial x \partial y} \\ \frac{\partial^2 f}{\partial y \partial x} & \frac{\partial^2 f}{\partial y^2} \end{bmatrix}.$$

6.1 Partial Derivatives

Extensions are easily made to a function with more than two variables. Each row of a Hessian matrix includes partial derivatives of the first derivative with respect to a specific variable. For example, the Hessian matrix of a three-variable function $f(x, y, z)$ is defined by the following:

$$\mathcal{H}f(x, y, z) = \begin{bmatrix} H_{11}(x, y, z) & H_{12}(x, y, z) & H_{13}(x, y, z) \\ H_{21}(x, y, z) & H_{22}(x, y, z) & H_{23}(x, y, z) \\ H_{31}(x, y, z) & H_{32}(x, y, z) & H_{33}(x, y, z) \end{bmatrix} = \begin{bmatrix} \frac{\partial^2 f}{\partial x^2} & \frac{\partial^2 f}{\partial x \partial y} & \frac{\partial^2 f}{\partial x \partial z} \\ \frac{\partial^2 f}{\partial y \partial x} & \frac{\partial^2 f}{\partial y^2} & \frac{\partial^2 f}{\partial y \partial z} \\ \frac{\partial^2 f}{\partial z \partial x} & \frac{\partial^2 f}{\partial z \partial y} & \frac{\partial^2 f}{\partial z^2} \end{bmatrix},$$

where the first column includes partial derivatives of the first derivative $\frac{\partial f}{\partial x}$, the second column comprises partial derivatives of $\frac{\partial f}{\partial y}$, and the third column includes partial derivatives of $\frac{\partial f}{\partial z}$.

Example 6.9 Find the Hessian matrix of the function $f(x, y) = x^2 y + 2xy^3$.

Solution

$$\mathcal{H}f(x, y) = \begin{bmatrix} H_{11}(x, y) & H_{12}(x, y) \\ H_{21}(x, y) & H_{22}(x, y) \end{bmatrix} = \begin{bmatrix} \frac{\partial^2 f}{\partial x^2} & \frac{\partial^2 f}{\partial x \partial y} \\ \frac{\partial^2 f}{\partial y \partial x} & \frac{\partial^2 f}{\partial y^2} \end{bmatrix} = \begin{bmatrix} 2y & 2x + 6y^2 \\ 2x + 6y^2 & 12xy \end{bmatrix}.$$

Exercises

6.7 Find the Hessian matrix of the following functions:
(1) $f(x, y) = e^x y^2 + e^y x^2$.
(2) $f(x, y, z) = x^3 y^2 z - 2xyz^3$.

6.8 Which of the following statements about the Jacobian matrix and the Hessian matrix is correct?

(1) The Hessian matrix is always non-square, while the Jacobian matrix is always square.
(2) The Jacobian matrix describes the rate of change of each output variable with respect to the input variables. In contrast, the Hessian matrix helps to understand the curvature of the function at a specific point.
(3) The Hessian matrix is always a diagonal matrix for any multivariable function.
(4) The Jacobian matrix is the matrix of second-order partial derivatives of a function.

6.2 Applications of Partial Derivatives

We now come to discuss issues regarding the maxima and minima of multiple variable functions, as indicated earlier.

6.2.1 Local Maxima and Minima

Section 5.3 of Chap. 5 shows how to find the critical points for a function of one variable. This section describes how to find maxima and minima for a function with two variables, $f(x, y)$. As before, these are local maxima and minima.

Necessary condition—suppose $f(x, y)$ has a relative extreme value at the point (x_0, y_0) and the partial derivatives of $f(x, y)$ at (x_0, y_0) exist. Thus, we have

$$\frac{\partial f}{\partial x}\bigg|_{\substack{x=x_0\\y=y_0}} = 0, \text{ and } \frac{\partial f}{\partial y}\bigg|_{\substack{x=x_0\\y=y_0}} = 0.$$

Sufficient condition—suppose the first partial derivatives and the second partial derivatives of $f(x, y)$ at (x_0, y_0) exist, and $\frac{\partial f}{\partial x}\big|_{\substack{x=x_0\\y=y_0}} = 0$, $\frac{\partial f}{\partial x}\big|_{\substack{x=x_0\\y=y_0}} = 0$.

Set $A = \frac{\partial^2 f}{\partial x^2}\big|_{\substack{x=x_0\\y=y_0}}$, $B = \frac{\partial^2 f}{\partial x \partial y}\big|_{\substack{x=x_0\\y=y_0}}$, and $C = \frac{\partial^2 f}{\partial y^2}\big|_{\substack{x=x_0\\y=y_0}}$.

Then, the Hessian matrix $\mathcal{H}f(x, y)$ is $\mathcal{H}f(x, y) = \begin{bmatrix} A & B \\ B & C \end{bmatrix}$.

Then,

- The function has a local maximum value at (x_0, y_0) if $AC - B^2 > 0$ (i.e., $\det(\mathcal{H}f(x, y)) > 0$) and $A < 0$.
- The function has a local minimum value at (x_0, y_0) if $AC - B^2 > 0$ (i.e., $\det(\mathcal{H}f(x, y)) > 0$) and $A > 0$.
- The function does not have an extreme value at (x_0, y_0) if $AC - B^2 < 0$ (i.e., $\det(\mathcal{H}f(x, y)) < 0$);
- More investigation is needed if $AC - B^2 = 0$ (i.e., $\det(\mathcal{H}f(x, y)) = 0$).

Remark 6.4 If $AC - B^2 > 0$ and $A < 0$, then necessarily $C < 0$, and if $AC - B^2 > 0$ and $A > 0$, then necessarily $C > 0$. So the above conditions for local maximums and minimums could have been written equivalently using $C < 0$ for a maximum and $C > 0$ for a minimum. ♦

6.2 Applications of Partial Derivatives

The general procedure of finding local maxima and minima of a function of two variables $f(x, y)$ is

- Step 1: Find critical values (x, y) by solving simultaneous equations:

$$\frac{\partial f(x, y)}{\partial x} = 0 \text{ and } \frac{\partial f(x, y)}{\partial y} = 0.$$

- Step 2: Write the mathematical expression of A, B, and C for the given function.
- Step 3: Evaluate A, B, and C using each pair of critical values.
- Step 4: Check the sign of $AC - B^2$ and decide whether the function has a relative extreme at the corresponding pair of critical values in terms of sufficient conditions.

Example 6.10 Find the critical points of the following function:

$$f(x, y) = x^3 - y^3 + 3x^2 + 3y^2 - 9x.$$

State whether the critical points are local maxima or minima.

Solution

- Step 1: Find critical values by solving simultaneous equations obtained from the first partial derivatives:

$$\begin{cases} \frac{\partial f(x,y)}{\partial x} = 3x^2 + 6x - 9 = 0 \\ \frac{\partial f(x,y)}{\partial y} = -3y^2 + 6y = 0 \end{cases} \Rightarrow \begin{cases} (x+3)(x-1) = 0 \\ y(y-2) = 0 \end{cases} \Rightarrow \begin{cases} x_1 = -3 \\ x_2 = 1 \\ y_1 = 0 \\ y_2 = 2. \end{cases}$$

Therefore, the function has four critical points at $(-3, 0)$, $(-3, 2)$, $(1, 0)$, and $(1, 2)$.

- Step 2: Write the mathematical expression of A, B, and C for the given function:

$$A = \frac{\partial^2 f(x, y)}{\partial x^2} = 6x + 6, \quad B = \frac{\partial^2 f(x, y)}{\partial x \partial y} = 0,$$

$$C = \frac{\partial^2 f(x, y)}{\partial y^2} = -6y + 6.$$

(continued)

Example 6.10 (continued)
- Step 3: Evaluate A, B, and C using each pair of critical values:
  - At point $(-3, 0)$, $A = -12$, $B = 0$, and $C = 6$.
  - At point $(-3, 2)$, $A = -12$, $B = 0$, and $C = -6$.
  - At point $(1, 0)$, $A = 12$, $B = 0$, and $C = 6$.
  - At point $(1, 2)$, $A = 12$, $B = 0$, and $C = -6$.
- Step 4: Check the sign of $AC - B^2$, and decide whether the function has a relative extreme at the corresponding pair of critical values in terms of sufficient conditions:
  - At point $(-3, 0)$, $AC - B^2 = (-12) \times 6 - 0 < 0$.
    Therefore, the function does not have an extreme value at $(-3, 0)$.
  - At point $(-3, 2)$, $AC - B^2 = (-12) \times (-6) - 0 > 0$, and $A < 0$.
    Therefore, $f(x, y)$ has a local maximum value of 31 at $(-3, 2)$.
  - At point $(1, 0)$, $AC - B^2 = 12 \times 6 - 0 > 0$, and $A > 0$.
    Therefore, $f(x, y)$ has a local minimum value of -5 at $(1, 0)$.
  - At point $(1, 2)$, $AC - B^2 = 12 \times (-6) - 0 < 0$.
    Therefore, the function does not have an extreme value at $(1, 2)$.

Example 6.11 Find the critical points of the following function:

$$f(x, y) = 2x^2 + 3y^2 + 3xy + 3x + y.$$

State whether the critical points are local maxima or minima.

Solution

- Step 1: Find critical values by solving simultaneous equations obtained from the first partial derivatives:
  $\frac{\partial f(x,y)}{\partial x} = 4x + 3y + 3 = 0$,
  $\frac{\partial f(x,y)}{\partial y} = 6y + 3x + 1 = 0$.
  Multiply the first equation by 2 and subtracting the second gives
  $5x + 5 = 0$, that is, $x = -1$. Substituting back in either equation gives $y = \frac{1}{3}$.
  So the function has only one critical point at $(-1, \frac{1}{3})$.
- Step 2: Write the mathematical expression of A, B, and C for the given function:

$$A = \frac{\partial^2 f(x, y)}{\partial x^2} = 4, \ B = \frac{\partial^2 f(x, y)}{\partial x \partial y} = 3, \ C = \frac{\partial^2 f(x, y)}{\partial y^2} = 6.$$

(continued)

Example 6.11 (continued)
- Step 3: Evaluate A, B, and C for the critical value.
  Nothing to do here.
- Step 4: Check the sign of $AC - B^2$, and decide whether the function has a relative extreme at the critical value in terms of sufficient conditions.
  At point $(-1, \frac{1}{3})$, $AC - B^2 = 15 > 0$, and $A > 0$.
  Therefore, $f(x, y)$ has a local minimum value of $\frac{-4}{3}$ at $(-1, \frac{1}{3})$.

Exercise

6.9 Find the critical points of the following functions and state whether the critical points are local maxima or minima:
(1) $f(x, y) = 2x^3 + 2y^3 - 3x^2 + 3y^2 - 12x - 12y$.
(2) $f(x, y) = 6 - x^3 - 4xy - 2y^2 - x$.
(3) $f(x, y) = x^2 + y^2 + (x + y + 1)^2$.
(4) $f(x, y) = 2x^3 + 2y^3 + 3y^2 - 9x^2 - 36y + 4$.

6.2.2 Method of Lagrange Multipliers for Maxima and Minima

So far, we have discussed how to find the local maxima and minima of a function with one or two variables. The only condition we have considered is that these functions are defined within their domain. However, in many real-world applications, it is possible to meet problems with other constraints, such as all the solutions have to be on a plane or line. Converting a constraint problem to a non-constrained problem is not always easy.

In this section, we introduce the method of Lagrange multipliers for local maxima and minima created to deal with such constraint problems. This method was proposed by Joseph-Louis Lagrange, an Italian mathematician and astronomer, later naturalised French. Lagrange found that the relative extreme of a function under a constraint is obtained when the gradient of the original function is parallel to the gradient of the constraint condition function.

Suppose we wanted to find the local maxima and minima of a function $z = f(x, y)$ where x and y need to satisfy the constraint $g(x, y) = 0$. The Lagrange multiplier method operates using the following steps:

- First, it constructs a new function, that is,

$$F(x, y) = f(x, y) + \lambda g(x, y),$$

where λ is a constant. This function $F(x, y)$[1] is called the Lagrangian.
- Then, it calculates the first partial derivative of the function $F(x, y)$ with respect to x and y and sets them to zero. Together with the constraint, these form a set of simultaneous equations given by

$$\begin{cases} \frac{\partial f(x,y)}{\partial x} + \lambda \frac{\partial g(x,y)}{\partial x} = 0, \\ \frac{\partial f(x,y)}{\partial y} + \lambda \frac{\partial g(x,y)}{\partial y} = 0, \\ g(x, y) = 0. \end{cases}$$

The solutions of (x, y) from these simultaneous equations are the points at which the function may have a relative extreme.

This method can be extended to functions with more than two variables and with more than one constraint condition.

Example 6.12 Minimise the function $f(x, y) = 4x^2 + y^2$, subject to the constraint $g(x, y) = x + y - 2 = 0$.

Figure 6.2 shows the function and the constraint condition function. As it shows, the plane of the constraint condition function intersects the function not at the bottom of the surface. Therefore, the minimum value of the function is not at the bottom of the surface anymore; rather, it is at the lowest point where the plane intersects the surface.

Solution

- First, construct a new function, that is,

$$F(x, y) = f(x, y) + \lambda g(x, y) = 4x^2 + y^2 + \lambda(x + y - 2).$$

- Then, calculate the first partial derivative of the function $F(x, y)$ with respect to x and y and set them both to zero. Together with the constraint, these form a set of simultaneous equations given by

$$\begin{cases} \frac{\partial f(x,y)}{\partial x} + \lambda \frac{\partial g(x,y)}{\partial x} = 8x + \lambda = 0, \\ \frac{\partial f(x,y)}{\partial y} + \lambda \frac{\partial g(x,y)}{\partial y} = 2y + \lambda = 0, \\ g(x, y) = x + y - 2 = 0. \end{cases} \Rightarrow \begin{cases} x = 0.4 \\ y = 1.6 \\ \lambda = -3.2. \end{cases}$$

Substitute $x = 0.4$ and $y = 1.6$ into the function, and then we have $f(x = 0.4, y = 1.6) = 4 \times 0.4^2 + 1.6^2 = 3.2$. The relative extreme of the function

(continued)

---

[1] Or sometimes written as $F(x, y) = f(x, y) - \lambda g(x, y)$. The sign in front of λ is arbitrary. What matters is that we need to be consistent throughout the derivation for a specific task.

Example 6.12 (continued)

is 3.2 obtained at the point (0.4, 1.6) subject to the constraint $g(x, y) = x + y - 2$.

- From Fig. 6.2, this is clearly a minimum, and Fig. 6.3 shows this as well. Figure 6.3 shows the contour lines of the function. The second innermost contour line tells us that all the points along this ellipse give a function value of 3.2. The dash-dotted line is the constraint condition. As can be seen, the dash-dotted line is a tangent of the contour line with a function value of 3.2, and it touches this contour line at the point ($x = 0.4, y = 1.6$). The arrow shows the gradient direction at this point. It points in the opposite direction of the maximum rate of decrease of the function and is perpendicular to the tangent.

  The gradient of f, $\mathrm{grad} f = [8x, 2y]$, at the point ($x = 0.4, y = 1.6$) can be calculated as $\mathrm{grad} f = [8 \times 0.4, 2 \times 1.6] = [3.2, 3.2]$.

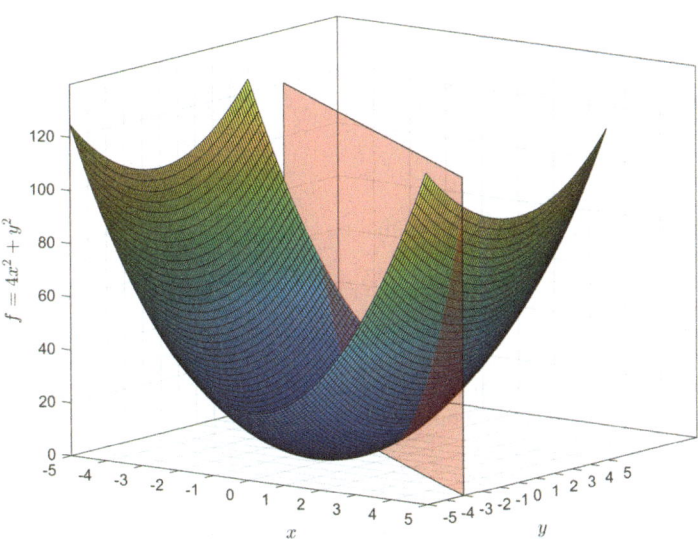

Fig. 6.2 The visualisation of the graph of the function $f(x, y) = 4x^2 + y^2$ together with the constraint $g(x, y) = x + y - 2 = 0$, examined in Example 6.12

Fig. 6.3 Example 6.12: the contour lines and the gradient at the extreme for the function $f(x, y) = 4x^2 + y^2$, and the constraint $g(x, y) = x + y - 2 = 0$

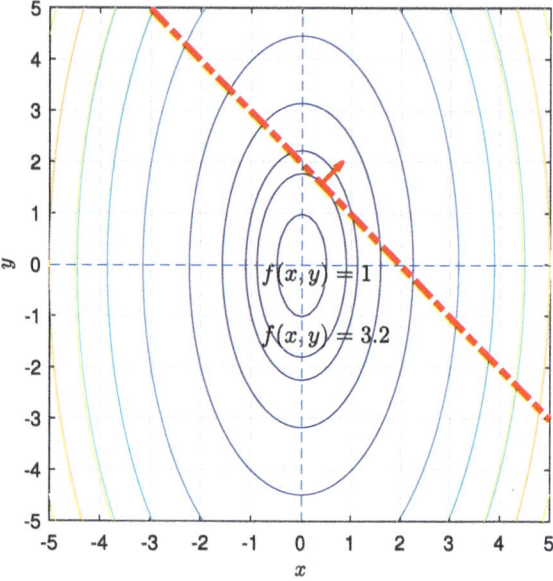

Example 6.13 Maximise the function $f(x, y) = xy$, subject to the constraint $g(x, y) = 2x + y - 1 = 0$.

Solution

- First, construct a new function, that is,

$$F(x, y) = f(x, y) + \lambda g(x, y) = xy + \lambda(2x + y - 1).$$

- Then, calculate the first partial derivative of the function $F(x, y)$ with respect to x and y and set them both to zero. Together with the constraint, these form a set of simultaneous equations given by

$$\begin{cases} \frac{\partial f(x,y)}{\partial x} + \lambda \frac{\partial g(x,y)}{\partial x} = y + 2\lambda = 0, \\ \frac{\partial f(x,y)}{\partial y} + \lambda \frac{\partial g(x,y)}{\partial y} = x + \lambda = 0, \\ g(x, y) = 2x + y - 1 = 0. \end{cases} \Rightarrow \begin{cases} x = \frac{1}{4} \\ y = \frac{1}{2} \\ \lambda = -\frac{1}{4}. \end{cases}$$

Substitute $x = \frac{1}{4}$ and $y = \frac{1}{2}$ into the function, and then we have $f(x = \frac{1}{4}, y = \frac{1}{2}) = \frac{1}{8} = 0.125$. The relative extreme of the function is $\frac{1}{8}$ obtained at the point $(\frac{1}{4}, \frac{1}{2})$ subject to the constraint $g(x, y) = 2x + y - 1$.

(continued)

Example 6.13 (continued)
- To check if it is a local maximum, try points just either side of $(0.25, 0.5)$ on the line $2x + y - 1 = 0$
$f(0.249, 0.502) = 0.124998 < 0.125 = \frac{1}{8}$ and
$f(0.251, 0.498) = 0.124998 < 0.125 = \frac{1}{8}$, so it is a local maximum.

Remark 6.5 To see whether the function has an extreme at the critical values and which sort of extreme it is, we often need to make a judgment in terms of the nature of the problem we are solving. ♦

Exercise

6.10 Applying the Lagrange multiplier method to the following functions:
(1) Minimise the function $f(x, y) = x^2 + y^2 + 1$ subject to the constraint $g(x, y) = x - y + 1 = 0$.
(2) Minimise the function $f(x, y) = x^3 + y^3$ subject to the constraint $g(x, y) = x + y - 1 = 0$.

6.2.3 Gradient Descent Algorithm

This section introduces how to find a minimum value of a function by using the gradient descent algorithm. Earlier, we have seen in Fig. 5.7 in Chap. 5 that if a function monotonically decreases, then the sign of its first derivative is negative; otherwise, it is positive.

The general idea of a gradient descent algorithm is to update the values of variables of a function iteratively to minimise the function. This is used extensively in neural networks. Figure 6.4 displays two functions: one on the left in black has a minimum value at $x = x_0$, and the other one on the right in brown has a maximum value at $x = x_1$. Let us have a look at the one with a minimum value first. When its x value is on the left-hand side of x_0, x needs to move along the positive direction of the x-axis to reach x_0. The moving direction is opposite to the sign of its derivative, which is negative. When its x value is on the right-hand side of x_0, x needs to move along the negative direction of the x-axis to reach x_0. Again, the moving direction is opposite to the sign of its derivative, which is positive. Therefore, the gradient descent algorithm works as follows:

- Step 1: Initialise a value for x, denoted as x^{old}; calculate the function value using x^{old}.

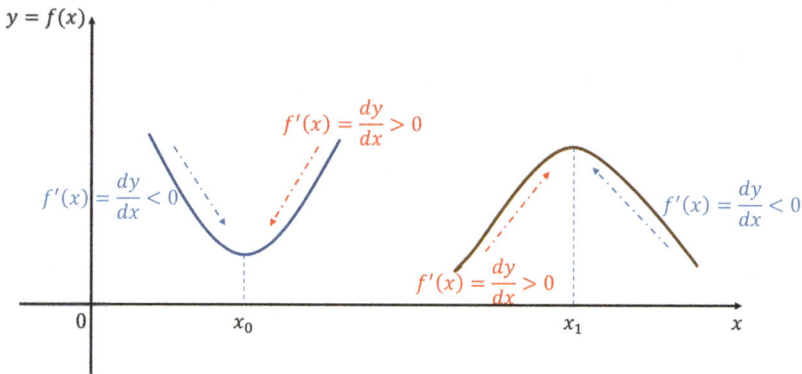

Fig. 6.4 An illustration of the relationship between the signs of gradients and the extreme values

- Step 2: Update x value by moving x along the x-axis with the direction opposite to the corresponding derivative sign. That is,

$$x^{new} = x^{old} - \epsilon \frac{d}{dx} f(x) \Big|_{x=x^{old}}, \tag{6.6}$$

where ϵ is the learning rate, a positive scalar determining the size of the step x moves, and the derivative of the function is evaluated using x^{old}.
- Step 3: Calculate the function value using x^{new}.
- Step 4: Assign $x^{old} = x^{new}$.
- Step 5: Repeat Steps 2–4 until the function reaches its minimum or the iterations satisfy some pre-set criterion.

This can be easily extended to functions with more than one variable. For example, for a function with two variables $f(x, y)$, one can update variable values as follows:

$$x^{new} = x^{old} - \epsilon \frac{\partial f(x, y)}{\partial x}\Big|_{\substack{x=x^{old}\\y=y^{old}}}, \text{ and } y^{new} = y^{old} - \epsilon \frac{\partial f(x, y)}{\partial y}\Big|_{\substack{x=x^{old}\\y=y^{old}}}. \tag{6.7}$$

Similarly, the gradient ascent algorithm updates the values of variables and the function by moving x in the direction matching the sign of the corresponding derivative (see the function on the right in Fig. 6.4). For a function with two variables $f(x, y)$, one can update variable values as follows:

$$x^{new} = x^{old} + \epsilon \frac{\partial f(x, y)}{\partial x}\Big|_{\substack{x=x^{old}\\y=y^{old}}}, \text{ and } y^{new} = y^{old} + \epsilon \frac{\partial f(x, y)}{\partial y}\Big|_{\substack{x=x^{old}\\y=y^{old}}}. \tag{6.8}$$

Remark 6.6 When applying a gradient-based algorithm, the initial value of a variable should be chosen carefully. In addition, the step size ϵ should be small to

6.2 Applications of Partial Derivatives

avoid going past the local minimum or the maximum. However, very small values will take longer to calculate. ♦

Example 6.14 Let $f(x_1, x_2) = 4x_1^2 + x_2^2$. Perform one iteration of the gradient descent algorithm. The initial values are $x_1^{old} = 3$, and $x_2^{old} = 2$. Set the learning rate to $\epsilon = 0.001$.

Solution

- Substitute $x_1^{old} = 3$, and $x_2^{old} = 2$ to $f(x_1, x_2) = 4x_1^2 + x_2^2$, and we have $4 \times 3^2 + 2^2 = 40$.
- Compute partial derivatives of the function:

$$\frac{\partial f}{\partial x_1} = 8x_1, \quad \frac{\partial f}{\partial x_2} = 2x_2.$$

- Update values for x_1 and x_2: substitute initial values to Eq. (6.7), and we have

$$x_1^{new} = x_1^{old} - \epsilon \frac{\partial f}{\partial x_1}\bigg|_{\substack{x_1=x_1^{old} \\ x_2=x_2^{old}}} = 3 - 0.001 \times (8 \times 3) = 2.976,$$

$$x_2^{new} = x_2^{old} - \epsilon \frac{\partial f}{\partial x_2}\bigg|_{\substack{x_1=x_1^{old} \\ x_2=x_2^{old}}} = 2 - 0.001 \times (2 \times 2) = 1.996.$$

Substitute $x_1 = 2.976$, and $x_2 = 1.996$ into $f(x_1, x_2) = 4x_1^2 + x_2^2$, which gives $4 \times 2.976^2 + 1.996^2 = 39.41$. Note that this is smaller than the initial value of 40, so we are moving (slowly) toward the minimum. (If we had set the learning rate, ϵ, to 0.01, we would get $x_1 = 2.76$ and $x_2 = 1.96$, giving $f(x_1, x_2) = f(2.76, 1.96) = 34.31$, which is a faster descent—though it runs the risk of jumping right past the minimum.)
- Assign $x_1^{old} = 2.976$ and $x_2^{old} = 1.996$, to complete the first iteration.

Remark 6.7 The gradient descent algorithm is really a process that requires a computer that is programmed specifically to do this task. Indeed, work in neural networks invariably does access a computer to do all the hard iterative calculations. So, we will not attempt further iterations and examples in this book. Hopefully, the basic idea of the iterative method is clear. ♦

Exercise

6.11 Which of the following statements is correct?

(1) The gradient descent algorithm follows the direction of the gradient to maximise a function, whereas the gradient ascent algorithm follows the gradient to minimise a function.
(2) Gradient descent and gradient ascent algorithms are used for unconstrained optimisation, while Lagrange multipliers are used for constrained optimisation problems.
(3) Gradient descent is used to maximise functions, while Lagrange multipliers are used to minimise functions.
(4) Lagrange multipliers iteratively adjust the variables to minimise or maximise the objective function.

6.3 Double Integrals

In this book, we consider double integrals as an example of multiple definite integrals.

A definite integral of a function of one variable gives the area "under" the curve between two x value limits of integration, which define the "bottom" boundary of the area. In the same way, a double definite integral of a function of two variables gives the volume "under" the surface, where the limits of integration define the area, or region, on the $x - y$ plane that gives the "bottom" boundary of the volume. Intuitively integrating by one variable gives the area, and integrating these "areas" in the other direction gives the volume.

The easiest cases are where the bounding region in the $x - y$ plane is a rectangle. We start with a really easy one.

Example 6.15 Find

$$\int_{x=0}^{1} \int_{y=0}^{2} 6 \, dx \, dy.$$

Solution The boundary of the region is the rectangle from $x = 0$ to $x = 1$ and from $y = 0$ to $y = 2$. The "top" of the volume is the surface $f(x, y) = 6$.

(continued)

Example 6.15 (continued)
Integrating over y first and then x,

$$\int_{x=0}^{1}\int_{y=0}^{2} 6\,dx\,dy = \int_{x=0}^{1}\left(\int_{y=0}^{2} 6\,dy\right)dx$$

$$= \int_{x=0}^{1}\left[6y\right]\Big|_{y=0}^{2} dx = \int_{x=0}^{1} 12\,dx = \left[12x\right]\Big|_{x=0}^{1} = 12.$$

In fact, $f(x, y) = 6$ is a horizontal plane at height 6. So, the volume is a cuboid with base 1 by 2 and height 6. This has volume 12, so the integration has "worked".

Not surprisingly, you get the same result if you integrate in the other order:

$$\int_{y=0}^{2}\int_{x=0}^{1} 6\,dx\,dy = \int_{y=0}^{2}\left(\int_{x=0}^{1} 6\,dx\right)dy$$

$$= \int_{y=0}^{2}\left[6x\right]\Big|_{x=0}^{1} dy = \int_{y=0}^{2} 6\,dx = \left[6y\right]\Big|_{y=0}^{2} = 12.$$

It is generally true for rectangular regions that you can integrate in either order since it is the same volume, and it does not matter which "areas" are summed in the second integration. However, it is sometimes easier to define and integrate the region in the $x - y$ plane in one direction first rather than the other, especially when we have non-rectangular regions.

However, we will do another, more complicated, rectangular region example first. For these examples when integrating over one variable, we treat the other independent variable as a constant.

Example 6.16 Find

$$\int_{x=1}^{2}\int_{y=0}^{2} x + y\,dx\,dy.$$

Solution The boundary of the region is another rectangle from $x = 1$ to $x = 2$ and from $y = 0$ to $y = 2$. The "top" face of the volume is now the

(continued)

Example 6.16 (continued)
surface $f(x, y) = x + y$:

$$\int_{x=1}^{2} \int_{y=0}^{2} x + y \, dx \, dy = \int_{x=1}^{2} \left(\int_{y=0}^{2} x + y \, dy \right) dx.$$

Since we are integrating over y first, we treat x as a constant. So we continue:

$$\int_{x=1}^{2} \left[xy + \frac{y^2}{2} \right]_{y=0}^{2} dx = \int_{x=1}^{2} 2x + 2 \, dx = \left[x^2 + 2x \right]_{x=1}^{2} = 5.$$

Again, you can check if you get the same result if you integrate over x and y in the other order.

Exercise

6.12 Calculate the following integrals that also have rectangular regions:

(1) $\int_{x=0}^{2} \int_{y=0}^{1} xy \, dx \, dy$.
(2) $\int_{x=-1}^{1} \int_{y=-2}^{2} 2x^2 + 3y^2 + 1 \, dx \, dy$.
(3) $\int_{x=0}^{2} \int_{y=0}^{\frac{\pi}{2}} x \cos y \, dx \, dy$.

More generally, regions over which the integration takes place are not rectangles. Sometimes, one or more of the boundaries is a curve or a sloping line. Then, one of the limits is expressed in terms of a variable since it is bounded by a curve or sloping line. So, we need to look at the limits (or endpoints) of integration. If there is an integration variable in the limits of integration, we must perform the integral with the variable limit first. As before, when integrating over one variable, we treat the other independent variable as a constant.

For our first example, we will integrate over the region that is a triangle bounded by the x-axis, the line $x = 1$, and the line $y = x$ (see Fig. 6.5). If we integrate over y first, then the "areas" are between the x-axis, that is, $y = 0$ and the line $y = x$. So, the top y limit is x. We then integrate over x from $x = 0$ to $x = 1$.

6.3 Double Integrals

Fig. 6.5 This figure shows the region (shaded) used in Example 6.17, over which the double integration is performed to form the base of the volume

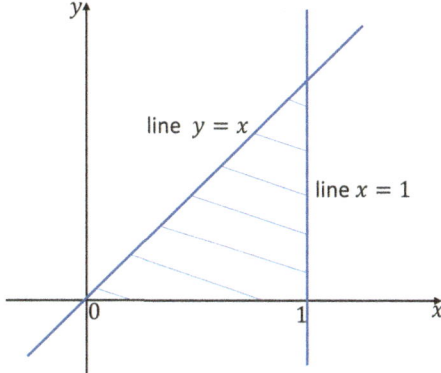

Example 6.17 Perform the following integral:

$$\int_{x=0}^{1} \int_{y=0}^{x} 1 + x^2 + y^2 \, dx \, dy.$$

Solution The volume defined is that of a triangular-based shape bounded at the "top" by the surface $f(x, y) = 1 + x^2 + y^2$. We must integrate over y first since the upper limit of the integral over y is a variable limit (x). When integrating over y, we treat the independent variable x as a constant:

$$\int_{x=0}^{1} \int_{y=0}^{x} 1 + x^2 + y^2 \, dx \, dy = \int_{x=0}^{1} \left(\int_{y=0}^{x} 1 + x^2 + y^2 \, dy \right) dx$$

$$= \int_{x=0}^{1} \left[y + x^2 y + \frac{y^3}{3} \right]_{y=0}^{x} dx$$

$$= \int_{x=0}^{1} x + x^3 + \frac{x^3}{3} - 0 \, dx = \int_{x=0}^{1} x + \frac{4x^3}{3} \, dx = \left[\frac{x^2}{2} + \frac{x^4}{3} \right]_{x=0}^{1} = \frac{1}{2} + \frac{1}{3} = \frac{5}{6}.$$

Integrating over this particular region can be thought of another way. We could integrate over x first. This gives "areas" from the line $y = x$ to the line $x = 1$. So, the x limits are y and 1. We then integrate y from $y = 0$ to $y = 1$. This gives the integral:

$$\int_{y=0}^{1} \int_{x=y}^{1} 1 + x^2 + y^2 \, dx \, dy.$$

Now, you have to integrate over x first. We set this as an exercise (see below).

It should be noted again that sometimes, it is easier to define the region in the $x - y$ plane in one direction first rather than the other when we have non-rectangular regions, so the order of integration cannot be reversed easily by changing the limits.

Example 6.18 Perform the following integral:

$$\int_{x=1}^{2} \int_{y=0}^{2x} xy\,dx\,dy.$$

Solution This is an integration over the region bounded by the x-axis, the lines $x = 1$ and $x = 2$, and the line $y = 2x$. It is a trapezium. The "top" boundary of the volume is the surface $f(x, y) = xy$.

We must integrate over y first since the upper limit of the integral over y is a variable limit (x). When integrating over y, we treat the independent variable x as a constant:

$$\int_{x=1}^{2} \int_{y=0}^{2x} xy\,dx\,dy = \int_{x=1}^{2} \left(\int_{y=0}^{2x} xy\,dy \right) dx = \int_{x=1}^{2} \left[\frac{xy^2}{2} \right]_{y=0}^{2x} dx$$

$$= \int_{x=1}^{2} [2x^3 - 0]\,dx = \int_{x=1}^{2} 2x^3\,dx = \left[\frac{x^4}{2} \right]_{x=1}^{2} = \frac{1}{2}(2^4 - 1^4) = \frac{15}{2}.$$

Now, consider a boundary for our integration region to be a semicircle above the x-axis, centred at the origin of radius 1. This circle has equation $x^2 + y^2 = 1$. We can form limits of an integral as y goes from $y = 0$ to $y = \sqrt{1 - x^2}$, and then x goes from -1 to 1. This gives us the next example, where the "top" boundary of the volume is the surface: $f(x, y) = x^2 y$.

Example 6.19 Perform the following integral:

$$\int_{x=-1}^{1} \int_{y=0}^{\sqrt{1-x^2}} x^2 y\,dx\,dy.$$

Solution

$$\int_{x=-1}^{1} \int_{y=0}^{\sqrt{1-x^2}} x^2 y\,dx\,dy = \int_{x=-1}^{1} \left(\int_{y=0}^{\sqrt{1-x^2}} x^2 y\,dy \right) dx$$

(continued)

Example 6.19 (continued)

$$= \int_{x=-1}^{1} \left[\frac{x^2 y^2}{2} \right]_{y=0}^{\sqrt{1-x^2}} dx$$

$$= \int_{x=-1}^{1} \frac{x^2(1-x^2)}{2} dx = \int_{x=-1}^{1} \frac{x^2 - x^4}{2} dx = \left[\frac{x^3}{6} - \frac{x^5}{10} \right]_{x=-1}^{1} = \frac{2}{15}.$$

Exercise

6.13 Calculate the following integrals:

(1) $\int_{x=0}^{y} \int_{y=0}^{1} xy \, dx \, dy$.
(2) $\int_{y=0}^{1} \int_{x=y}^{1} 1 + x^2 + y^2 \, dx \, dy$.
(3) $\int_{x=1}^{2} \int_{y=0}^{2x} 2x^2 y + 3xy^2 \, dx \, dy$.
(4) $\int_{x=0}^{1} \int_{y=0}^{1-x^2} 3x^2 y^2 \, dx \, dy$.

6.3.1 Integration of Double Integrals Using Polar Coordinates

Some double integrals can be expressed in a simpler form if we can transform the area from rectangular Cartesian coordinates (x, y) to polar coordinates (r, θ). As we saw in Example 6.19 above, the limits for integration in some examples can be really complicated when expressed in Cartesian coordinates (e.g., $y = \sqrt{1 - x^2}$) and potentially lead to some difficult integrals. However, the region in Example 6.19, which is a half-disc centred on the origin of unit radius, can be easily expressed in polar coordinates. It is

$$0 \le r \le 1 \text{ and } 0 \le \theta \le \pi.$$

So, the limits of integration would be simple.

To convert to polar coordinates, we need the following relationships between Cartesian and polar coordinates:

$$\begin{cases} x = r \cos \theta, \\ y = r \sin \theta. \end{cases}$$

We also need to replace the $dxdy$. In brief, $dxdy$ essentially represents a small area in the plane, denoted as dA. The Jacobian matrix J describes the change in coordinates from (x, y) to (r, θ), that is, $dA = dxdy = |J|drd\theta$, where the determinant of the Jacobian matrix represents the scaling factor by which areas are scaled during the transformation.

The Jacobian matrix for this transformation is

$$J = \begin{bmatrix} \frac{\partial x}{\partial r} & \frac{\partial x}{\partial \theta} \\ \frac{\partial y}{\partial r} & \frac{\partial y}{\partial \theta} \end{bmatrix} = \begin{bmatrix} \cos\theta & -r\sin\theta \\ \sin\theta & r\cos\theta \end{bmatrix}.$$

The determinant of this Jacobian matrix is r. Therefore, the area element in polar coordinates is $rdrd\theta$, which accounts for the fact that the segments of a circle increase in size as you move further from the centre.

So, the transformation formula is shown as follows:

$$\iint_D f(x, y)\, dx\, dy = \iint_D f(r\cos\theta, r\sin\theta) r\, dr\, d\theta. \tag{6.9}$$

Let us redo Example 6.19 from the previous section by converting to polar coordinates in Example 6.20.

Example 6.20 Converting Example 6.19 to polar coordinates,

$$\int_{x=-1}^{1} \int_{y=0}^{\sqrt{1-x^2}} x^2 y\, dx\, dy \text{ becomes } \int_{\theta=0}^{\pi} \int_{r=0}^{1} (r\cos\theta)^2 (r\sin\theta) r\, dr\, d\theta.$$

Solution

$$\int_{\theta=0}^{\pi} \int_{r=0}^{1} (r\cos\theta)^2 (r\sin\theta) r\, dr\, d\theta = \int_{\theta=0}^{\pi} \left(\int_{r=0}^{1} r^4 \cos^2\theta \sin\theta\, dr \right) d\theta$$

$$= \int_{\theta=0}^{\pi} \left[\frac{r^5}{5} \right]_{r=0}^{1} \cos^2\theta \sin\theta\, d\theta = \frac{1}{5} \int_{\theta=0}^{\pi} \cos^2\theta \sin\theta\, d\theta$$

$$= \frac{1}{5} \left[\frac{-\cos^3\theta}{3} \right]_{\theta=0}^{\pi} = \frac{2}{15}.$$

When integrating with respect to θ, we have used integration by substitution with $u = \cos\theta$, so that $du = -\sin\theta d\theta$. We have obtained the same answer as before for this example.

6.3 Double Integrals

Now, to do an example that is impossible without converting to polar coordinates.

Example 6.21 Perform the following integral:

$$\iint_D e^{-x^2-y^2}\, dx\, dy,$$

where D is a closed circular area with the origin as the centre and a as the radius.

Solution If not expressed in polar coordinates, the integral would be

$$\int_{x=-a}^{a} \int_{y=-\sqrt{a^2-x^2}}^{\sqrt{a^2-x^2}} e^{-x^2-y^2}\, dx\, dy.$$

This is impossible to integrate non-numerically, so we express D in the polar coordinate system as follows:

$$0 \le r \le a,\ 0 \le \theta \le 2\pi.$$

Applying Eq. (6.9), we have

$$\iint_D e^{-x^2-y^2}\, dx\, dy = \iint_D e^{-r^2} r\, dr\, d\theta$$

$$= \int_{\theta=0}^{2\pi} \left[\int_{r=0}^{a} e^{-r^2} r\, dr \right] d\theta$$

$$= \int_{\theta=0}^{2\pi} \left[-\frac{1}{2} e^{-r^2} \right]_{r=0}^{a} d\theta$$

$$= \frac{1}{2}(1 - e^{-a^2}) \int_{\theta=0}^{2\pi} d\theta$$

$$= \pi(1 - e^{-a^2}).$$

In the first equation line of the above, we have used the general equation of a circle centred at $(0, 0)$, that is, $x^2 + y^2 = r^2$. In the second line, we have applied integration by substitution. That is, we set $u = r^2$, and then we have $du = 2r\, dr$.

Exercise

6.14 Convert the following to polar coordinates and hence evaluate the following:

(1) $\iint_D e^{x^2+y^2} dxdy$, where D is a closed circle area with the origin as the centre and 2 as the radius.
(2) $\iint_D xy\,dxdy$, where D is the area in the first quadrant between the circles with radius 1 and 3 centred at the origin, that is, a quarter of a ring around the origin.
(3) $\iint_D \sin(x^2+y^2)dxdy$, where D is a closed circle area with the origin as centre and one as the radius.

Chapter 7
Algorithms 1: Principal Component Analysis

This chapter and the next two chapters, (8 and 9), represent the culmination of a lot of mathematics theory, specifically linear algebra and calculus. The material has been divided into three chapters to indicate the separate nature of each topic since each chapter revisits and completes one of the case studies introduced in Chap. 1.

Hence, these three chapters aim to show how we can apply the knowledge introduced in previous chapters to formulate three widely used algorithms in the Data Science field: principal component analysis, simple linear regression, and simple two-layer neural networks trained by gradient descent.

This chapter deals with principal component analysis. In Chap. 4, we have described the basic idea of principal component analysis (PCA). This chapter will further help us understand the relationship between eigenvalues produced in the PCA analysis and variances among the data projected in the PCA space.

7.1 Revisit Principal Component Analysis

In Sect. 4.2 of Chap. 4, you learned how to find the principal components for a set of data points. The aim was to find the directions with the most variance in the data. If all you want to do is to find principal components for data, then the work in Chap. 4 is all you need, and this new section in this chapter is unnecessary for you. However, this is a book giving the maths behind the algorithms, so we will now explain why defining principal components as eigenvectors of the covariance matrix of data $\mathbf{X}$, called Σ, gives the directions of most variance. To do this, we need to find the maximum value of the variance with various constraints, such as the direction being a unit vector. Finding maximum values with a constraint means we will appeal to the Lagrange multipliers method for maxima and minima as given in Sect. 6.2.2 of Chap. 6. Before going through the details, we need further knowledge regarding vectors, matrices, and their differentiation, as given in the next subsection.

7.2 Preliminary Knowledge

Apart from the basic maths knowledge introduced in Sect. 3.3.11.1 of Chap. 3 and Sect. 4.2.1 of Chap. 4, we need a bit more to deal with the reasons that using eigenvectors and eigenvalues for principal component analysis does what we require. After giving each result, we will illustrate that they are true with one or more examples.

Suppose **X** is a $n \times d$ matrix and **u** is a $d \times 1$ vector.

- The variance of **Xu** is given by

$$var(\mathbf{Xu}) = \mathbf{u}^T cov(\mathbf{X})\mathbf{u}. \tag{7.1}$$

If we denote the matrix $cov(\mathbf{X})$ as Σ, then we have

$$var(\mathbf{Xu}) = \mathbf{u}^T \Sigma \mathbf{u}. \tag{7.2}$$

This result looks at the variance of a matrix multiplied by a vector. This is important since it is about projecting the data onto a vector—see Sect. 4.2.2 of Chap. 4.

Example 7.1 Let **X** be a 5×2 matrix of data and **u** a 2×1 vector given by

$$\mathbf{X} = \begin{bmatrix} 3 & 2 \\ 4 & 3 \\ 2 & 1 \\ 2 & 2 \\ 4 & 2 \end{bmatrix} \text{ and } \mathbf{u} = \begin{bmatrix} 2 \\ 1 \end{bmatrix}, \text{ so } \mathbf{Xu} \text{ is } \begin{bmatrix} 8 \\ 11 \\ 5 \\ 6 \\ 10 \end{bmatrix}.$$

We will show that $var(\mathbf{Xu})$ is the same as $\mathbf{u}^T cov(\mathbf{X})\mathbf{u}$ for this example.

Xu is a set of numbers and has a mean of 8 and a variance of 6.5. So $var(\mathbf{Xu}) = 6.5$.

Now for our data, the mean of the first column is 3, and the mean of the second column is 2. Remember that the covariance of two sets of numbers has been defined in Sect. 4.2.1 of Chap. 4. For example, for our data,

$$cov(\mathbf{x}_1, \mathbf{x}_2) = \frac{\sum_{i=1}^{5} (x_{i,1} - \bar{x}_1)(x_{i,2} - \bar{x}_2)}{5 - 1}$$

$$= \frac{0 + 1 \times 1 + (-1) \times (-1) + 0 + 0}{4}$$

$$= \frac{2}{4}.$$

(continued)

7.2 Preliminary Knowledge

Example 7.1 (continued)
So, the covariance matrix is a 2×2 matrix:

$$\Sigma = cov(\mathbf{X}) = \begin{bmatrix} cov(x_1, x_1) & cov(x_1, x_2) \\ cov(x_2, x_1) & cov(x_2, x_2) \end{bmatrix} = \frac{1}{4}\begin{bmatrix} 4 & 2 \\ 2 & 2 \end{bmatrix}.$$

Hence, $\mathbf{u}^T cov(\mathbf{X})\mathbf{u}$ is

$$\frac{1}{4}\begin{bmatrix} 2 & 1 \end{bmatrix}\begin{bmatrix} 4 & 2 \\ 2 & 2 \end{bmatrix}\begin{bmatrix} 2 \\ 1 \end{bmatrix} = \frac{1}{4}\begin{bmatrix} 10 & 6 \end{bmatrix}\begin{bmatrix} 2 \\ 1 \end{bmatrix} = 6.5.$$

So, both are 6.5. Hence, we have demonstrated that $var(\mathbf{Xu}) = \mathbf{u}^T cov(\mathbf{X})\mathbf{u} = \mathbf{u}^T \Sigma \mathbf{u}$ as required.

Example 7.2 Let us try a more complicated example. So, let $\mathbf{X}$ be a 6×3 matrix of data and $\mathbf{u}$ a 3×1 vector given by

$$\mathbf{X} = \begin{bmatrix} 2 & 3 & 4 \\ 1 & 0 & 1 \\ 4 & 3 & 2 \\ 1 & 2 & 2 \\ 2 & 2 & 1 \\ 2 & 2 & 2 \end{bmatrix} \text{ and } \mathbf{u} = \begin{bmatrix} 2 \\ 3 \\ 1 \end{bmatrix},$$

then $\mathbf{Xu}$ is $\begin{bmatrix} 17 \\ 3 \\ 19 \\ 10 \\ 11 \\ 12 \end{bmatrix}$ with a mean of 12 and a variance of 32. So, $var(\mathbf{Xu}) = 32$.

Now for our data, the mean of all three columns is 2. The covariance matrix is a 3×3 matrix:

$$\Sigma = cov(\mathbf{X}) = \begin{bmatrix} cov(x_1, x_1) & cov(x_1, x_2) & cov(x_1, x_3) \\ cov(x_2, x_1) & cov(x_2, x_2) & cov(x_2, x_3) \\ cov(x_3, x_1) & cov(x_3, x_2) & cov(x_3, x_3) \end{bmatrix} = \frac{1}{5}\begin{bmatrix} 6 & 4 & 1 \\ 4 & 6 & 4 \\ 1 & 4 & 6 \end{bmatrix}.$$

(continued)

Example 7.2 (continued)

So, $\mathbf{u}^T cov(\mathbf{X})\mathbf{u}$ is

$$\frac{1}{5}\begin{bmatrix}2 & 3 & 1\end{bmatrix}\begin{bmatrix}6 & 4 & 1 \\ 4 & 6 & 4 \\ 1 & 4 & 6\end{bmatrix}\begin{bmatrix}2 \\ 3 \\ 1\end{bmatrix} = \frac{1}{5}\begin{bmatrix}25 & 30 & 20\end{bmatrix}\begin{bmatrix}2 \\ 3 \\ 1\end{bmatrix} = 32.$$

Hence, $var(\mathbf{Xu}) = \mathbf{u}^T cov(\mathbf{X})\mathbf{u} = \mathbf{u}^T \Sigma \mathbf{u}$ as required since both are 32.

Exercise

7.1 Show that $var(\mathbf{Xu}) = \mathbf{u}^T cov(\mathbf{X})\mathbf{u} = \mathbf{u}^T \Sigma \mathbf{u}$ for

(1)

$$\mathbf{X} = \begin{bmatrix}3 & 4 \\ 4 & 2 \\ 2 & 3 \\ 2 & 2 \\ 4 & 4\end{bmatrix} \text{ and } \mathbf{u} = \begin{bmatrix}1 \\ -1\end{bmatrix}.$$

(2)

$$\mathbf{X} = \begin{bmatrix}2 & 5 & 4 \\ 4 & 2 & 3 \\ 2 & 2 & 2 \\ 3 & 2 & 5 \\ 5 & 4 & 2 \\ 2 & 3 & 2\end{bmatrix} \text{ and } \mathbf{u} = \begin{bmatrix}2 \\ 3 \\ 1\end{bmatrix}.$$

Now, suppose $\mathbf{A}$ is a $d \times d$ symmetric matrix, $\mathbf{x}$ is a $d \times 1$ vector with $\mathbf{x}^T = (x_1, x_2, \cdots, x_d)$, and α is a scalar.

- Let the scalar α be defined by

$$\alpha = \mathbf{x}^T \mathbf{A} \mathbf{x},$$

7.2 Preliminary Knowledge

where $\mathbf{A}$ does not depend on $\mathbf{x}$, and then in general, we have

$$\frac{\partial \alpha}{\partial \mathbf{x}} = (\mathbf{A} + \mathbf{A}^T)\mathbf{x}. \tag{7.3}$$

Since $\mathbf{A}$ is also symmetric, then $\mathbf{A}^T = \mathbf{A}$ so that

$$\frac{\partial \alpha}{\partial \mathbf{x}} = 2\mathbf{A}\mathbf{x}. \tag{7.4}$$

The product $\mathbf{x}^T \mathbf{A} \mathbf{x}$ is just a scalar value (i.e., not a vector or matrix), which is an equation in $x_1, x_2, \cdots, x_d$. So, $\frac{\partial \alpha}{\partial \mathbf{x}}$ is the gradient of the function as a vector as in Sect. 6.1.4 of Chap. 6. Hence,

$$\frac{\partial \alpha}{\partial \mathbf{x}} = \text{grad}\alpha = \left[\frac{\partial \alpha}{\partial x_1}, \frac{\partial \alpha}{\partial x_2}, \cdots, \frac{\partial \alpha}{\partial x_d} \right]^T.$$

Of course, we are only interested in symmetric matrices since $cov(\mathbf{X})$ is always symmetric. However, let us start with a non-symmetric example (see Example 7.3).

Example 7.3 Let $d = 2$, $\mathbf{x}^T = (x_1, x_2)$, and $\mathbf{A} = \begin{bmatrix} a & b \\ c & d \end{bmatrix}$.

Then, $\alpha = \mathbf{x}^T \mathbf{A} \mathbf{x} = \begin{bmatrix} ax_1 + cx_2 & bx_1 + dx_2 \end{bmatrix} \begin{bmatrix} x_1 \\ x_2 \end{bmatrix} = ax_1^2 + cx_1x_2 + bx_1x_2 + dx_2^2$.

So, α is a scalar function of x_1 and x_2. Hence,

$$\frac{\partial \alpha}{\partial \mathbf{x}} = \begin{bmatrix} \frac{\partial \alpha}{\partial x_1} \\ \frac{\partial \alpha}{\partial x_2} \end{bmatrix} = \begin{bmatrix} 2ax_1 + cx_2 + bx_2 \\ cx_1 + bx_1 + 2dx_2 \end{bmatrix}.$$

Also,

$$(\mathbf{A} + \mathbf{A}^T)\mathbf{x} = \begin{bmatrix} 2a & b+c \\ b+c & 2d \end{bmatrix} \begin{bmatrix} x_1 \\ x_2 \end{bmatrix} = \begin{bmatrix} 2ax_1 + cx_2 + bx_2 \\ cx_1 + bx_1 + 2dx_2 \end{bmatrix}.$$

So, $\frac{\partial \alpha}{\partial \mathbf{x}}$ is the same as $(\mathbf{A} + \mathbf{A}^T)\mathbf{x}$ and the result is as required.

Now, let us see a symmetric example (see Example 7.4).

Example 7.4 Let $\mathbf{A} = \begin{bmatrix} a & b \\ b & d \end{bmatrix}$. So, $\mathbf{A}$ is the same as in Example 7.3 except that $c = b$.

Then, looking at the result in Example 7.3 and putting $c = b$, we have

$$\alpha = \mathbf{x}^T \mathbf{A} \mathbf{x} = ax_1^2 + 2bx_1 x_2 + dx_2^2,$$

and so

$$\frac{\partial \alpha}{\partial \mathbf{x}} = \begin{bmatrix} \frac{\partial \alpha}{\partial x_1} \\ \frac{\partial \alpha}{\partial x_2} \end{bmatrix} = \begin{bmatrix} 2ax_1 + 2bx_2 \\ 2bx_1 + 2dx_2 \end{bmatrix} = 2 \begin{bmatrix} a & b \\ b & d \end{bmatrix} \begin{bmatrix} x_1 \\ x_2 \end{bmatrix} = 2\mathbf{A}\mathbf{x}.$$

Hence, $\frac{\partial \alpha}{\partial \mathbf{x}} = 2\mathbf{A}\mathbf{x}$ as required.

Exercise

7.2 Let $d = 3$, $\mathbf{x}^T = (x_1,\ x_2,\ x_3)$, and $\mathbf{A} = \begin{bmatrix} a & b & c \\ b & d & e \\ c & e & f \end{bmatrix}$, which is symmetric. If $\alpha = \mathbf{x}^T \mathbf{A} \mathbf{x}$, show that $\frac{\partial \alpha}{\partial \mathbf{x}} = 2\mathbf{A}\mathbf{x}$.

Now, suppose $\mathbf{x}$ and $\mathbf{a}$ are $d \times 1$ vectors and β and γ are scalars.

- Let the scalar β be defined by

$$\beta = \mathbf{x}^T \mathbf{x},$$

then we have:

$$\frac{\partial \beta}{\partial \mathbf{x}} = 2\mathbf{x}. \tag{7.5}$$

- Let the scalar γ be defined by

$$\gamma = \mathbf{x}^T \mathbf{a} = \mathbf{a}^T \mathbf{x},$$

then we have:

$$\frac{\partial \gamma}{\partial \mathbf{x}} = \mathbf{a}. \tag{7.6}$$

Example 7.5 Let $d = 2$. So, we have $\mathbf{x}^T = (x_1, x_2)$ and $\mathbf{a}^T = (a_1, a_2)$.
Then, $\beta = \mathbf{x}^T \mathbf{x} = \begin{bmatrix} x_1 & x_2 \end{bmatrix} \begin{bmatrix} x_1 \\ x_2 \end{bmatrix} = x_1^2 + x_2^2$,

and so $\frac{\partial \beta}{\partial \mathbf{x}} = \begin{bmatrix} \frac{\partial \beta}{\partial x_1} \\ \frac{\partial \beta}{\partial x_2} \end{bmatrix} = \begin{bmatrix} 2x_1 \\ 2x_2 \end{bmatrix} = 2\mathbf{x}$.

Also $\gamma = \mathbf{x}^T \mathbf{a} = \mathbf{a}^T \mathbf{x} = x_1 a_1 + x_2 a_2$,

and so $\frac{\partial \gamma}{\partial \mathbf{x}} = \begin{bmatrix} \frac{\partial \gamma}{\partial x_1} \\ \frac{\partial \gamma}{\partial x_2} \end{bmatrix} = \begin{bmatrix} a_1 \\ a_2 \end{bmatrix} = \mathbf{a}$.

Exercise

7.3 When $d = 3$, $\mathbf{x}^T = (x_1, x_2, x_3)$ and $\mathbf{a}^T = (a_1, a_2, a_3)$. If $\beta = \mathbf{x}^T \mathbf{x}$ and $\gamma = \mathbf{x}^T \mathbf{a}$, show that $\frac{\partial \beta}{\partial \mathbf{x}} = 2\mathbf{x}$ and $\frac{\partial \gamma}{\partial \mathbf{x}} = \mathbf{a}$.

7.3 Problem Setting

Recall that in Sect. 4.2.2 of Chap. 4, we have claimed that if the first principal component of the data $\mathbf{X}$ is the eigenvector $\mathbf{u}_1$ of the covariance matrix of data $\mathbf{X}$, then the projection of the data onto $\mathbf{u}_1$ is such that:

- $\mathbf{X}\mathbf{u}_1$ has the largest variance,
- where this is subject to the normalising constraint $\mathbf{u}_1^T \mathbf{u}_1 = 1$.

We can re-express these using mathematical equations and prove that the direction of the first principal component of the data is the direction with the largest eigenvalue:

- First, the variance of $\mathbf{X}\mathbf{u}_1$ can be written as $var(\mathbf{X}\mathbf{u}_1)$, which is equal to $\mathbf{u}_1^T \Sigma \mathbf{u}_1$ according to Eqs. (7.1) and (7.2), where $\Sigma = cov(\mathbf{X})$ is a symmetric matrix. Hence, $\mathbf{u}_1^T \Sigma \mathbf{u}_1$ is the objective function we wish to maximise.
- We want to maximise the objective function subject to the constraint that $\mathbf{u}_1^T \mathbf{u}_1 = 1$. Hence, this is an optimisation problem with constraints.

Applying the Lagrange multiplier method (see Sect. 6.2.2 of Chap. 6), the new objective function is shown as follows:

$$F_1 = \mathbf{u}_1^T \Sigma \mathbf{u}_1 - \lambda_1 (\mathbf{u}_1^T \mathbf{u}_1 - 1), \qquad (7.7)$$

where λ_1 is a Lagrange multiplier. The task has been converted to maximising F_1 with respect to both $\mathbf{u}_1$ and λ_1.

Similarly, we can set up an objective function for all other principal components. For example, for the second principal component ($\mathbf{u}_2$), we need to maximise $var(\mathbf{X}\mathbf{u}_2) = \mathbf{u}_2^T \Sigma \mathbf{u}_2$ subject to certain constraints. However, this time, we have not only $\mathbf{u}_2^T \mathbf{u}_2 = 1$ but also $\mathbf{u}_2^T \mathbf{u}_1 = 0$, since we are looking for a coordinate system, where axes are perpendicular to each other. Again, applying the Lagrange multiplier method, the new objective function is shown as follows:

$$F_2 = \mathbf{u}_2^T \Sigma \mathbf{u}_2 - \lambda_2(\mathbf{u}_2^T \mathbf{u}_2 - 1) - \rho \mathbf{u}_2^T \mathbf{u}_1, \tag{7.8}$$

where λ_2 and ρ are Lagrange multipliers.

We now have two objective functions, F_1 and F_2, that we can maximise to get the first and second largest variances.

7.4 The Formulation of Principal Component Analysis

Let us maximise each of the functions in turn.

7.4.1 The First Principal Component

To find the maximum of F_1, we calculate the partial derivative $\frac{\partial F_1}{\partial \mathbf{u}_1}$ from Eq. (7.7):

$$\frac{\partial F_1}{\partial \mathbf{u}_1} = 2\Sigma \mathbf{u}_1 - 2\lambda_1 \mathbf{u}_1, \tag{7.9}$$

where we have applied Eqs. (7.4) and (7.5).

Setting the partial derivative (7.9) to zero gives us the following:

$$2\Sigma \mathbf{u}_1 - 2\lambda_1 \mathbf{u}_1 = 0.$$

That is,

$$\Sigma \mathbf{u}_1 = \lambda_1 \mathbf{u}_1. \tag{7.10}$$

As can be seen, Eq. (7.10) coincides with the definition of eigendecomposition (see Definition 4.1 of Chap. 4) since Σ is a square matrix. This tells us that the solution for the first principal component $\mathbf{u}_1$, an eigenvector satisfying Eq. (7.10), points to the direction of maximum variance. Hence, the direction of the first principal component of the data is the direction with the largest variance, as claimed. So how big is this variance?

7.4 The Formulation of Principal Component Analysis 193

Since $var(\mathbf{X}\mathbf{u}_1) = \mathbf{u}_1^T \Sigma \mathbf{u}_1$, if we substitute $\Sigma \mathbf{u}_1 = \lambda_1 \mathbf{u}_1$ to the variance of $\mathbf{X}\mathbf{u}_1$ and consider the constraint condition of $\mathbf{u}_1^T \mathbf{u}_1 = 1$, we can obtain

$$var(\mathbf{X}\mathbf{u}_1) = \mathbf{u}_1^T \Sigma \mathbf{u}_1 = \mathbf{u}_1^T \lambda_1 \mathbf{u}_1 = \lambda_1 \mathbf{u}_1^T \mathbf{u}_1 = \lambda_1.$$

This says that the variance of data projections along the first principal component equals the eigenvalue λ_1 of the first principal component. Since the direction of the first principal component $\mathbf{u}_1$ captures the largest variation in the data projections, which is proved to be λ_1, we can say that the first principal component has the largest variance among all principal components.

7.4.2 The Second Principal Component

We now consider the second principal component and calculate the partial derivative $\frac{\partial F_2}{\partial \mathbf{u}_2}$ from Eq. (7.8). By applying Eqs. (7.4), (7.5) and (7.6), we have

$$\frac{\partial F_2}{\partial \mathbf{u}_2} = 2\Sigma \mathbf{u}_2 - 2\lambda_2 \mathbf{u}_2 - \rho \mathbf{u}_1. \tag{7.11}$$

Setting the partial derivative (7.11) to zero gives us the following:

$$2\Sigma \mathbf{u}_2 - 2\lambda_2 \mathbf{u}_2 - \rho \mathbf{u}_1 = 0. \tag{7.12}$$

Multiplying $\mathbf{u}_1^T$ from the left side on both sides of Eq. (7.12), we have

$$\mathbf{u}_1^T 2\Sigma \mathbf{u}_2 - \mathbf{u}_1^T 2\lambda_2 \mathbf{u}_2 - \mathbf{u}_1^T \rho \mathbf{u}_1 = 0.$$

If we take scalars, including Lagrange multipliers in front of vectors, we obtain

$$2\mathbf{u}_1^T \Sigma \mathbf{u}_2 - 2\lambda_2 \mathbf{u}_1^T \mathbf{u}_2 - \rho \mathbf{u}_1^T \mathbf{u}_1 = 0. \tag{7.13}$$

Since $\mathbf{u}_1^T \mathbf{u}_2 = 0$ and $\mathbf{u}_1^T \mathbf{u}_1 = 1$, from the equation above, we have

$$\rho = 2\mathbf{u}_1^T \Sigma \mathbf{u}_2. \tag{7.14}$$

We will now show that, in fact, $\rho = 0$. First, multiplying $\mathbf{u}_2^T$ from the left side on both sides of Eq. (7.10), it gives us the following:

$$\mathbf{u}_2^T \Sigma \mathbf{u}_1 = \mathbf{u}_2^T \lambda_1 \mathbf{u}_1 = \lambda_1 \mathbf{u}_2^T \mathbf{u}_1 = 0. \tag{7.15}$$

However, from the matrix transpose rule (see Sect. 3.3.11.1 of Chap. 3), we have $(\mathbf{u}_2^T \Sigma \mathbf{u}_1)^T = \mathbf{u}_1^T \Sigma \mathbf{u}_2$. Note that the covariance matrix Σ is symmetrical, and the transpose of a symmetrical matrix is itself. Therefore, we also now have $\mathbf{u}_1^T \Sigma \mathbf{u}_2 = 0$ since from Eq. (7.15) $\mathbf{u}_2^T \Sigma \mathbf{u}_1 = 0$. Together with Eq. (7.14), we obtain $\rho = 0$. Further, substituting $\rho = 0$ into Eq. (7.12), gives us

$$2\Sigma \mathbf{u}_2 - 2\lambda_2 \mathbf{u}_2 = 0 \rightarrow \Sigma \mathbf{u}_2 = \lambda_2 \mathbf{u}_2. \qquad (7.16)$$

Hence, we have shown that the solution for the second principal component $\mathbf{u}_2$, given by Eq. (7.16), gives us the direction of the second most maximum variance.

Furthermore, since $var(\mathbf{X}\mathbf{u}_2) = \mathbf{u}_2^T \Sigma \mathbf{u}_2$, if we substitute $\Sigma \mathbf{u}_2 = \lambda_2 \mathbf{u}_2$ to the variance of $\mathbf{X}\mathbf{u}_2$ and consider the constraint condition of $\mathbf{u}_2^T \mathbf{u}_2 = 1$, we can obtain

$$var(\mathbf{X}\mathbf{u}_2) = \mathbf{u}_2^T \Sigma \mathbf{u}_2 = \mathbf{u}_2^T \lambda_2 \mathbf{u}_2 = \lambda_2 \mathbf{u}_2^T \mathbf{u}_2 = \lambda_2.$$

It says that the variance of data projections along the second principal component equals the eigenvalue of the second principal component (see Eq. (7.16)).

7.4.3 Data Normalisation

Let us complete Examples 7.1 and 7.2 given earlier in Sect. 7.2 of this chapter. We will find the principal components three times in Example 7.1 to illustrate some important facts about the process.

> **Example 7.6** Example 7.1 continued—part 1
> First, we take the data as given and found in Example 7.1. We have
>
> $$\mathbf{X} = \begin{bmatrix} 3 & 2 \\ 4 & 3 \\ 2 & 1 \\ 2 & 2 \\ 4 & 2 \end{bmatrix},$$
>
> and the covariance matrix is
>
> $$\Sigma = cov(\mathbf{X}) = \frac{1}{4}\begin{bmatrix} 4 & 2 \\ 2 & 2 \end{bmatrix} = \begin{bmatrix} 1 & \frac{1}{2} \\ \frac{1}{2} & \frac{1}{2} \end{bmatrix},$$
>
> where the total variance in the two features (columns) is 1.5, 1 for the first column and $\frac{1}{2}$ for the second column.
>
> (continued)

7.4 The Formulation of Principal Component Analysis

Example 7.6 (continued)

To find the direction of maximum variance, we have proved that it is in the direction $\mathbf{u}_1$ of the first eigenvector when using principal component analysis on $cov(\mathbf{X})$ and has the value given by the largest eigenvalue, namely, the eigenvalue λ_1. Similarly, the second most maximum variance is in the direction $\mathbf{u}_2$ of the second eigenvector and has the value given by the second largest eigenvalue, λ_2, found using principal component analysis on $cov(\mathbf{X})$. So, we need to carry out a principal component analysis.

The characteristic polynomial of Σ is found via the following:

$$\Sigma - \lambda \mathbf{I} = \begin{bmatrix} 1 - \lambda & \frac{1}{2} \\ \frac{1}{2} & \frac{1}{2} - \lambda \end{bmatrix},$$

$$|\Sigma - \lambda \mathbf{I}| = (1 - \lambda)(\frac{1}{2} - \lambda) - \frac{1}{4} = \frac{1}{4}(4\lambda^2 - 6\lambda + 1).$$

The eigenvalues are obtained by solving

$$4\lambda^2 - 6\lambda + 1 = 0,$$

which gives $\lambda_1 = \frac{1}{4}(3 + \sqrt{5})$ and $\lambda_2 = \frac{1}{4}(3 - \sqrt{5})$ as the eigenvalues of Σ. So, the largest eigenvalue is $\frac{1}{4}(3 + \sqrt{5}) = 1.31$, capturing about 87.3% of the total variation (which is 1.5), and the second largest is $\frac{1}{4}(3 - \sqrt{5}) = 0.19$, capturing about 12.7% of the total variation.

Find $\mathbf{u}_1$ by using $\lambda_1 = \frac{1}{4}(3 + \sqrt{5})$ in $\Sigma - \lambda \mathbf{I}$ and solving $(\Sigma - \lambda \mathbf{I})\mathbf{u} = \mathbf{0}$:

$$\frac{1}{4} \begin{bmatrix} 1 - \sqrt{5} & 2 \\ 2 & -1 - \sqrt{5} \end{bmatrix} \begin{bmatrix} u_1 \\ u_2 \end{bmatrix} = \begin{bmatrix} 0 \\ 0 \end{bmatrix}.$$

That is, :

$$\begin{cases} (1 - \sqrt{5})u_1 + 2u_2 = 0 \\ 2u_1 - (1 + \sqrt{5})u_2 = 0. \end{cases}$$

The solution to the above simultaneous equations is $u_1 = 1 + \sqrt{5}$ and $u_2 = 2$.

So, the direction of maximum variance is the first eigenvector $\mathbf{u}_1 = \begin{bmatrix} 1 + \sqrt{5} \\ 2 \end{bmatrix}$, which has unit vector $\hat{\mathbf{u}}_1 = \begin{bmatrix} 0.85 \\ 0.53 \end{bmatrix}$, at approximately 32 degrees to the x-axis.

(continued)

Example 7.6 (continued)

A similar calculation for $\lambda_2 = \frac{1}{4}(3 - \sqrt{5})$ gives $u_1 = 1 - \sqrt{5}$ and $u_2 = 2$. So, the direction with the second largest variance is the second eigenvector $\mathbf{u}_2 = \begin{bmatrix} 1 - \sqrt{5} \\ 2 \end{bmatrix}$, which has unit vector $\hat{\mathbf{u}}_2 = \begin{bmatrix} -0.53 \\ 0.85 \end{bmatrix}$.

The original data with the two principal component directions $\mathbf{u}_1$ and $\mathbf{u}_2$ are shown in Fig. 7.1a, and the data as projected onto the two principal component directions are shown in Fig. 7.1b.

Example 7.7 Example 7.1 continued—part 2

Next, we will take the dataset $\mathbf{X}$ and make it zero mean. That is, each column has a zero mean. This is done by subtracting the mean of the column from each item, giving

$$\mathbf{X} = \begin{bmatrix} 0 & 0 \\ 1 & 1 \\ -1 & -1 \\ -1 & 0 \\ 1 & 0 \end{bmatrix}.$$

Now, if we calculate the covariance matrix, we get

$$\Sigma = cov(\mathbf{X}) = \frac{1}{4}\begin{bmatrix} 4 & 2 \\ 2 & 2 \end{bmatrix}.$$

This is the same covariance matrix as shown in Example 7.1, which is not surprising since covariance is calculated by taking each item and subtracting the mean.

Hence, the solution to this is identical to the previous calculation. It is usual to subtract the mean because it gives smaller numbers and has axes at the centre of the picture. It is conventional to do this and will be expected, so it is always done.

The original data that has been made zero mean with the two principal component directions $\mathbf{u}_1$ and $\mathbf{u}_2$ are shown in Fig. 7.2a, and the data as projected onto the two principal component directions are shown in Fig. 7.2b.

7.4 The Formulation of Principal Component Analysis

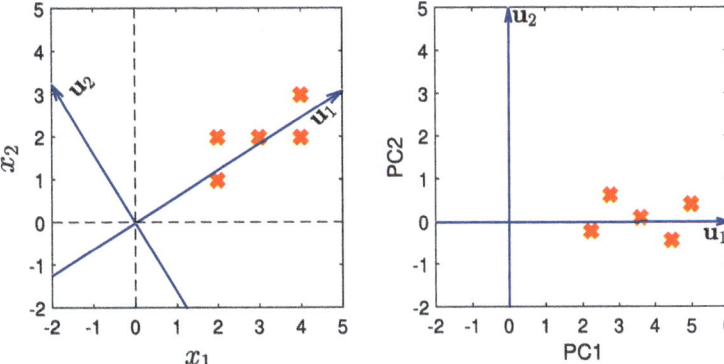

Fig. 7.1 The left panel illustrates the original data along with the two principal component directions, u_1 and u_2; the right displays the data projected onto these two principal component directions

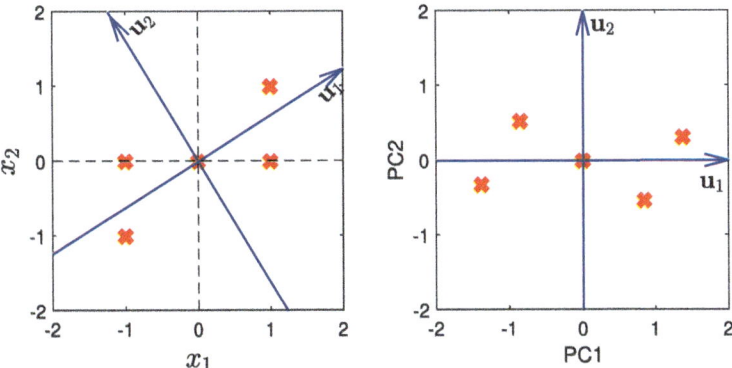

Fig. 7.2 The left panel illustrates the original, zero-mean, data along with the two principal component directions, u_1 and u_2; the right panel displays the data projected onto these two principal component directions

Example 7.8 Example 7.1 continued—part 3
Now, we also normalise each column by dividing by the standard deviation of each column. This means that the total variance of each column will be 1.

(continued)

Example 7.8 (continued)
This gives the following form to the data:

$$\mathbf{X} = \begin{bmatrix} 0 & 0 \\ 1 & \sqrt{2} \\ -1 & -\sqrt{2} \\ -1 & 0 \\ 1 & 0 \end{bmatrix}.$$

Calculating the covariance matrix, we get

$$\Sigma = cov(\mathbf{X}) = \begin{bmatrix} 1 & \frac{1}{\sqrt{2}} \\ \frac{1}{\sqrt{2}} & 1 \end{bmatrix},$$

where the total variance in the two features is 2.

The characteristic polynomial of Σ is obtained via the following:

$$\Sigma - \lambda \mathbf{I} = \begin{bmatrix} 1-\lambda & \frac{1}{\sqrt{2}} \\ \frac{1}{\sqrt{2}} & 1-\lambda \end{bmatrix},$$

$$|\Sigma - \lambda \mathbf{I}| = (1-\lambda)(1-\lambda) - \frac{1}{2} = \frac{1}{2}(2\lambda^2 - 4\lambda + 1).$$

The eigenvalues are obtained by solving

$$2\lambda^2 - 4\lambda + 1 = 0,$$

which gives the largest eigenvalue as $\lambda_1 = \frac{2+\sqrt{2}}{2} = 1.71$, capturing about 85.4% of the total variation, and second largest eigenvalue as $\lambda_2 = \frac{2-\sqrt{2}}{2} = 0.29$, capturing about 14.6% of the total variation.

Finally, we use λ_1 to find the direction of the maximum variance, $\mathbf{u}_1$, by solving

$$\begin{bmatrix} -\frac{1}{\sqrt{2}} & \frac{1}{\sqrt{2}} \\ \frac{1}{\sqrt{2}} & -\frac{1}{\sqrt{2}} \end{bmatrix} \begin{bmatrix} u_1 \\ u_2 \end{bmatrix} = \begin{bmatrix} 0 \\ 0 \end{bmatrix}.$$

That is,

$$\begin{cases} -\frac{1}{\sqrt{2}}u_1 + \frac{1}{\sqrt{2}}u_2 = 0 \\ \frac{1}{\sqrt{2}}u_1 - \frac{1}{\sqrt{2}}u_2 = 0. \end{cases}$$

(continued)

7.4 The Formulation of Principal Component Analysis

Example 7.8 (continued)
The solution to the above simultaneous equations is $u_1 = u_2$. Therefore, any non-zero vector satisfying the condition $u_1 = u_2$ is a solution to the eigenvector. For instance, $u_1 = 1$ and $u_2 = 1$. So, the direction of maximum variance is the first eigenvector $\mathbf{u}_1 = \begin{bmatrix} 1 \\ 1 \end{bmatrix}$, which has unit vector $\hat{\mathbf{u}}_1 = \begin{bmatrix} \frac{1}{\sqrt{2}} \\ \frac{1}{\sqrt{2}} \end{bmatrix}$, which is at 45 degrees to the x-axis.

A similar calculation for λ_2 gives $u_1 = -u_2$. Again, any non-zero vector satisfying the condition $u1 = -u2$ is a solution to the eigenvector. For instance, $u_1 = -1$ and $u_2 = 1$. So, the direction with the second largest variance is the second eigenvector $\mathbf{u}_2 = \begin{bmatrix} -1 \\ 1 \end{bmatrix}$, which has unit vector $\hat{\mathbf{u}}_2 = \begin{bmatrix} -\frac{1}{\sqrt{2}} \\ \frac{1}{\sqrt{2}} \end{bmatrix}$.

These values and directions are different from the original ones and show that normalisation does have an effect. In this case, it has increased the importance of the second feature of $\mathbf{X}$ so that the first eigenvector is rotated round to 45 degrees from the x-axis from 32 degrees as before. The original data that has been made zero mean and normalised with the two principal component directions $\mathbf{u}_1$ and $\mathbf{u}_2$ are shown in Fig. 7.3a, and the data as projected onto the two principal component directions are shown in Fig. 7.3b.

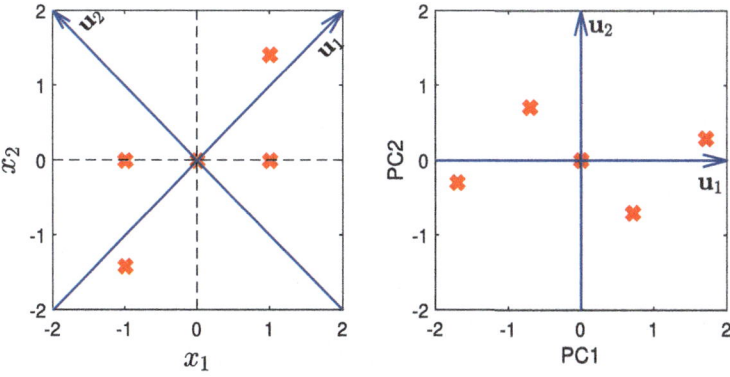

Fig. 7.3 The left panel illustrates the original, zero-mean, normalised, data along with the two principal component directions, $\mathbf{u}_1$ and $\mathbf{u}_2$; the right panel displays the data projected onto these two principal component directions

Example 7.9 Example 7.2 continued
Again starting with the raw data, we have found that for our data,

$$X = \begin{bmatrix} 2 & 3 & 4 \\ 1 & 0 & 1 \\ 4 & 3 & 2 \\ 1 & 2 & 2 \\ 2 & 2 & 1 \\ 2 & 2 & 2 \end{bmatrix},$$

the covariance matrix is

$$\Sigma = cov(X) = \frac{1}{5}\begin{bmatrix} 6 & 4 & 1 \\ 4 & 6 & 4 \\ 1 & 4 & 6 \end{bmatrix}.$$

Now, to find the eigenvalues and eigenvectors for this matrix involves solving a cubic equation and solving three simultaneous equations. As in all realistic exercises, this is done with the aid of suitable programs on a computer. So, for the sake of completeness, this has been done and gives the three eigenvalues in descending order:

$$\lambda_1 = 2.44, \lambda_2 = 1, \lambda_3 = 0.16.$$

And it gives the three unit eigenvectors corresponding to these eigenvalues as

$$u_1 = \begin{bmatrix} 0.52 \\ 0.67 \\ 0.52 \end{bmatrix} \quad u_2 = \begin{bmatrix} -0.71 \\ 0 \\ 0.71 \end{bmatrix} \quad u_3 = \begin{bmatrix} 0.48 \\ -0.74 \\ 0.48 \end{bmatrix}.$$

So, u_1 is the direction of most variance, u_2 is the direction of the second most variance, and u_3 is the direction of least variance.

If we were to plot the data projected onto the first and second principal components, then we would capture $2.44 + 1 = 3.44$ out of the total of $2.44 + 1 + 0.16 = 3.6$ variance, that is, 95.6% of the total.

Now, we convert each column to have a zero mean and a unit variance. After making each column zero mean, each column has a variance of $\frac{6}{5} \neq 1$. So, we divide each element by the standard deviation, which is $\frac{\sqrt{6}}{\sqrt{5}}$, and the

(continued)

Example 7.9 (continued)
data becomes

$$X = \begin{bmatrix} 0 & \frac{\sqrt{5}}{\sqrt{6}} & \frac{2\sqrt{5}}{\sqrt{6}} \\ -\frac{\sqrt{5}}{\sqrt{6}} & -\frac{2\sqrt{5}}{\sqrt{6}} & -\frac{\sqrt{5}}{\sqrt{6}} \\ \frac{2\sqrt{5}}{\sqrt{6}} & \frac{\sqrt{5}}{\sqrt{6}} & 0 \\ -\frac{\sqrt{5}}{\sqrt{6}} & 0 & 0 \\ 0 & 0 & -\frac{\sqrt{5}}{\sqrt{6}} \\ 0 & 0 & 0 \end{bmatrix},$$

and the covariance matrix is

$$\Sigma = cov(X) = \begin{bmatrix} 1 & \frac{2}{3} & \frac{1}{6} \\ \frac{2}{3} & 1 & \frac{2}{3} \\ \frac{1}{6} & \frac{2}{3} & 1 \end{bmatrix}.$$

So again, solving using a computer, we get

$$\lambda_1 = 2.03, \lambda_2 = 0.83, \lambda_3 = 0.14.$$

And it gives the three unit eigenvectors corresponding to these eigenvalues as

$$\mathbf{u}_1 = \begin{bmatrix} 0.52 \\ 0.67 \\ 0.52 \end{bmatrix} \mathbf{u}_2 = \begin{bmatrix} -0.71 \\ 0 \\ 0.71 \end{bmatrix} \mathbf{u}_3 = \begin{bmatrix} -0.48 \\ 0.74 \\ -0.48 \end{bmatrix}.$$

Figure 7.4 shows the normalised data and the eigenvectors in the data space. Figure 7.5 presents the directions of the principal components (eigenvectors) in the PCA space. Both figures show $\mathbf{u}_1$ in red, $\mathbf{u}_2$ in green, and $\mathbf{u}_3$ in blue.

Notice that the three eigenvectors, the three principal components, are the same as previously. This is because the original data was chosen so that each column had the same variance. Therefore, dividing by the standard deviation meant dividing all the values by the same amount. Not surprisingly, this had no effect on the principal component directions. The eigenvalues, or variances, are different but are all, in fact, the original ones divided by the variance (which was $\frac{6}{5} = 1.2$). For real problems, having the same variance to start with will not be the case! So, dividing by the standard deviation is important.

(continued)

Example 7.9 (continued)

Again, if we were to plot the data projected onto the first and second principal components, then we would capture $2.03 + 0.83 = 2.86$ out of the total of $2.03 + 0.83 + 0.14 = 3$ variance, that is, 95.3% of the total.

Figure 7.6 shows projections of the normalised data in the PCA space, from left to right, displaying PC1 against PC2, PC2 against PC3, and PC1 against PC3, respectively. We can see that the largest range among projections along each principal component axis decreases from PC1 to PC3.

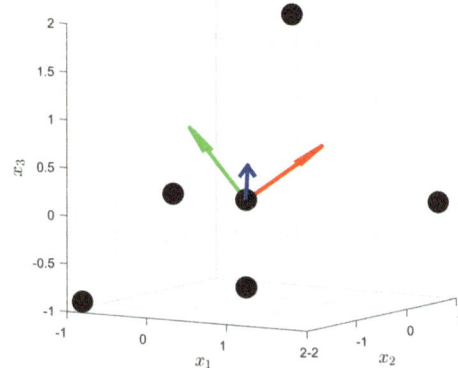

Fig. 7.4 The normalised data and the three eigenvector directions ($\mathbf{u}_1$ in red, $\mathbf{u}_2$ in green, and $\mathbf{u}_3$ in blue) in the data space

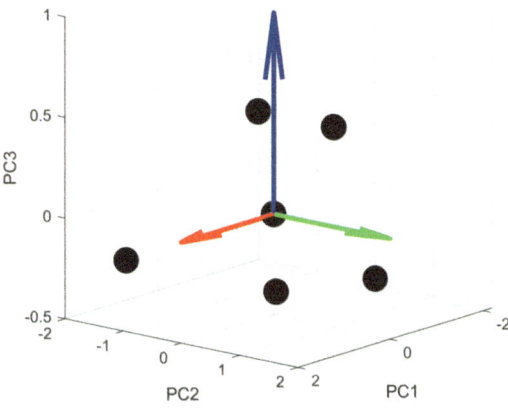

Fig. 7.5 The PCA space represented by the eigenvectors ($\mathbf{u}_1$ in red, $\mathbf{u}_2$ in green, and $\mathbf{u}_3$ in blue)

7.4 The Formulation of Principal Component Analysis

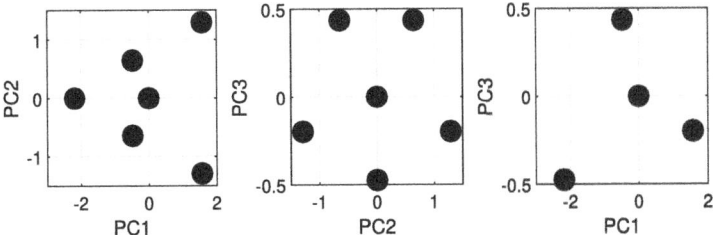

Fig. 7.6 Projections of the normalised data in the PCA space: (from left to right) PC1 versus PC2, PC2 versus PC3, and PC1 versus PC3

Remark 7.1 As mentioned in Example 7.8, one can find more than one non-zero eigenvector for a corresponding eigenvalue. That is, there may be more than one solution for the required unit vector or principal component as long as the condition solved from the simultaneous equations is satisfied. For example, for a condition $u_1 = u_2$, the principal component may be $\hat{\mathbf{u}}_1 = \begin{bmatrix} \frac{1}{\sqrt{2}} \\ \frac{1}{\sqrt{2}} \end{bmatrix}$, or $\hat{\mathbf{u}}_1 = \begin{bmatrix} -\frac{1}{\sqrt{2}} \\ -\frac{1}{\sqrt{2}} \end{bmatrix}$. Both vectors lie on the same line but point in opposite directions. It does not affect visualising the structure in the data when projecting the same data in these two directions, though one visualisation plot may seem to be a flip of the other one. ♦

Remark 7.2 As mentioned in Example 7.7, it is expected that you should centre the data by making each feature have a zero mean. It centres the picture and which makes it easier to interpret. It is also expected that you should normalise the data by making each feature (column) have a standard deviation, or variance, of 1, as mentioned in Example 7.8. This is so that one feature does not dominate the calculation just because it has much larger values. As we have demonstrated here for really simple data, these two tasks are often not really needed, but do not get mislead—for real data, they are important tasks to perform. ♦

Exercise

7.4 Find the principal components for the following data with and without normalisation (having both zero means and unit standard deviations):

(1)

$$\mathbf{X} = \begin{bmatrix} 3 & 4 \\ 4 & 2 \\ 2 & 3 \\ 2 & 2 \\ 4 & 4 \end{bmatrix}.$$

(continued)

(2)
$$X = \begin{bmatrix} 2 & 2 \\ 1 & 2 \\ 4 & 3 \\ 1 & 0 \\ 2 & 3 \\ 2 & 2 \end{bmatrix}.$$

(3) If you are feeling brave, try this larger one. In fact, it is not too difficult since *four* of the *nine* values in the covariance matrix are zero, and the ones on the main diagonal are all the same. This makes getting the first eigenvalue easy, and the other two are found by factorising a quadratic equation. Hence, it is possible to do it by hand.

$$X = \begin{bmatrix} 2 & 5 & 4 \\ 4 & 2 & 3 \\ 2 & 2 & 2 \\ 3 & 2 & 5 \\ 5 & 4 & 2 \\ 2 & 3 & 2 \end{bmatrix}.$$

7.5 Case Study 2 from Chap. 1: Continued

We are now ready to answer those five questions asked in Sect. 1.3.2 of Chap. 1.

1. What are those principal components (PC) axes?
   Principal component axes are the eigenvectors computed via eigendecomposition on the data covariance matrix.
2. What is the relationship between those PCs and the original four features in the dataset?
   Recall in Sect. 4.2.3 of Chap. 4 that we can obtain positions of data projections along the first PC axis using the following equation:

$$\text{projected\_data} = \mathbf{X}_{n \times d} \mathbf{u}_{d \times 1} = u_{11}\mathbf{x}_{,1} + u_{21}\mathbf{x}_{,2} + \cdots + u_{d1}\mathbf{x}_{,d},$$

where $\mathbf{x}_{,i}$ is the ith column of $\mathbf{X}_{n \times d}$. A more general expression to each principal component is

$$\text{pc} = \mathbf{X}\mathbf{u}_i = u_{1i}\mathbf{x}_{,1} + u_{2i}\mathbf{x}_{,2} + \cdots + u_{di}\mathbf{x}_{,d}, \text{ where } i = 1, \cdots, d.$$

It shows that each PC is a linear combination of all d features, weighted by the element in the corresponding eigenvector. Note that the number of elements of each eigenvector (d) is determined by the number of features included in the data covariance matrix, that is, the number of columns in $\mathbf{X}_{n \times d}$, which is also d.

3. Why is it necessary to report the variance percentage value?

Now, we know that the principle behind the PCA analysis is to find the direction that can capture the most significant variance among the data projections in the PCA space. When doing feature extraction using PCA, reporting how much percentage of the total variance has been captured by each PC will help us to decide on how many features to use. Note that each feature extracted via PCA is a linear combination of all the original features. When visualising the data using PCA, reporting how much percentage of the total variance has been captured, especially by the first two PCs, will give us a sense of whether this linear data visualisation method is a suitable way to visualise the data.

4. How is the variance percentage value calculated?

This has been shown in Sect. 4.2.3 of Chap. 4: the amount of information contained in the ith principal component is calculated as $\frac{\lambda_i}{\sum \lambda_j}$. However, it should be clear now why the eigenvalue is used when calculating the variance percentage.

5. How is the position of each data in the coordinate plane determined?

This is similar to point (2). In practice, first, we remove the mean value from each feature in the dataset. Then, we substitute the corresponding data values and eigenvector elements into the following equation to obtain the coordinate value in the principal component space:

$$\text{projected\_data} = u_1 x_1 + u_2 x_2 + \cdots + u_d x_d.$$

Chapter 8
Algorithms 2: Linear Regression

This is the second of three chapters that aim to show how we can apply the knowledge introduced in previous chapters to formulate three widely used algorithms in the Data Science field. This chapter applies the least-squares technique for formulating a simple linear algorithm. This algorithm aims to find a linear relationship between variables that will enable us to estimate the new value of one variable, called the dependent variable, given new values for one or more independent variables.

8.1 Simple Linear Regression Algorithm

Linear regression is a technique that statisticians use to describe the relationship between a dependent variable, also called a regressor, and one or more independent variables, also called predictors. Figure 8.1 shows the first three chemicals (represented by crosses) displayed in Table 1.2 of Chap. 1. In this example, we want to estimate enhancement ratios in terms of the molecular weights of chemicals. The molecular weight is an independent variable, and the enhancement ratio is the dependent variable. The linear regression algorithm aims to fit a linear line among the data. However, many straight lines can be fit, for example, the three dashed lines shown in Fig. 8.1. Which one shall we use? Since we want to use the linear line to estimate values for the dependent variable, we need to find the one that can provide the estimations as accurately as possible.

Let us start with a simple form with only one independent variable, illustrated in Fig. 8.2. Since we are in two dimensions, then this is a simple straight line, and it can be mathematically expressed as follows:

$$f_\mathbf{a}(x) = a_0 + a_1 x, \tag{8.1}$$

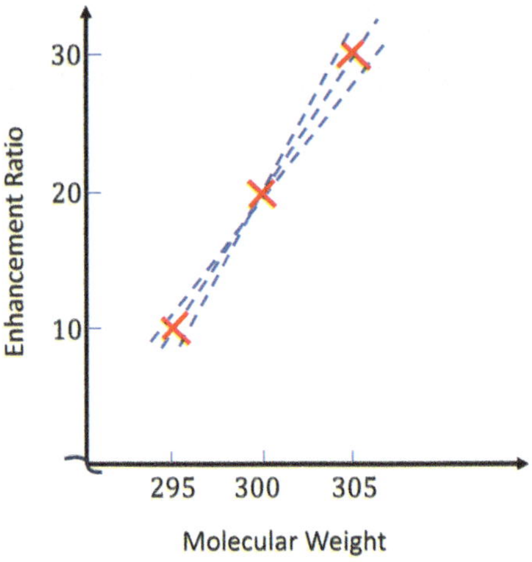

Fig. 8.1 Three possible linear regression lines illustrating the relationship between molecular weights and enhancement ratios. The three cross markers represent the available chemical compounds

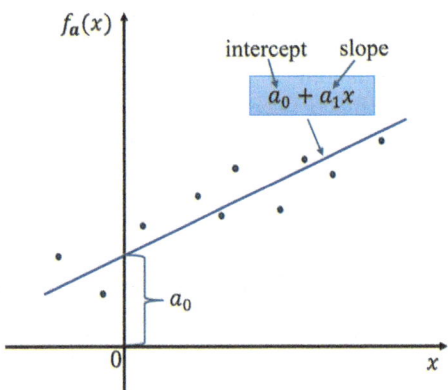

Fig. 8.2 An illustration of a linear regression line

where a_0 is the intercept, a_1 is the slope of the line, and x is the variable for which we have observations, the independent variable. It is similar to the one shown in Sect. 2.3.2.1 of Chap. 2. However, here we use a subscript letter **a** to indicate that $\mathbf{a} = \{a_0, a_1\}$ is the parameter set we need to estimate.

8.2 Least-Squares Estimation

How can we find suitable values for a_0 and a_1 in Eq. (8.1) from the given data? Ideally, we want to select values for a_0 and a_1, such as the line can pass through all the given data points. However, it is impossible in most real-world applications since data points do not exist on the same line. Therefore, all we can do is to choose values for a_0 and a_1 such that the differences between estimations from the fitted line $f(x)$ and actual measurements are as small as possible. This is another optimisation

8.2 Least-Squares Estimation

problem. That is, we want to minimise the sum of differences over all the given data with respect to a_0 and a_1.

Now, the question is how we construct the objective function involving the sum of differences. The differences will be expressed in terms of how far the observation of the dependent variable is from what the straight line calculates for that corresponding value of the independent variable. Suppose there are N data points $(x_i, y_i), i = 1, \ldots, N$, where x is the independent variable and y is the dependent variable. For any data point (x_i, y_i), the actual value of the dependent variable is y_i, and the value given by the line, the estimated value of y_i, is given by $a_0 + a_1 x_i$. The possible objective functions will be in terms of sums of the differences between these two values, that is, the sums of $y_i - (a_0 + a_1 x_i)$ in some form. Three possible objective functions are

$$Q1(a_0, a_1) = \frac{1}{N} \sum_{i=1}^{N} (y_i - (a_0 + a_1 x_i)), \qquad (8.2)$$

$$Q2(a_0, a_1) = \frac{1}{N} \sum_{i=1}^{N} |y_i - (a_0 + a_1 x_i)|, \qquad (8.3)$$

and

$$Q(a_0, a_1) = \frac{1}{N} \sum_{i=1}^{N} (y_i - (a_0 + a_1 x_i))^2. \qquad (8.4)$$

The problem with $Q1$ is that the differences or errors are signed values, and positive values can cancel with negative values when adding up all errors. The problem with $Q2$ is that the absolute value function is not differentiable (recall Example 5.7 in Chap. 5). Thus, it is not convenient for further analysis. The problem with Q is that differences are not weighted equally. That is, large differences are given more weight than smaller differences. There are problems with all the above three objective functions. However, Q is not so bad compared with the other two as differences cannot be cancelled and it is differentiable. Therefore, we minimise Q with respect to a_0 and a_1 to find the line that best fits the data. This is called the least-squares method. By convention, Q is divided by 2 as shown as follows:

$$Q(a_0, a_1) = \frac{1}{2N} \sum_{i=1}^{N} (y_i - (a_0 + a_1 x_i))^2. \qquad (8.5)$$

8.2.1 Deriving the Estimates Using the Least-Squares Objective Function

Finding the estimates of the parameters of the regression models means we need to minimise Eq. (8.5). We set the partial derivatives of Q with respect to a_0 and a_1 equal to zero:

$$\frac{\partial Q}{\partial a_1} = \sum_{i=1}^{N}(y_i - (a_0 + a_1 x_i))(-x_i) = 0, \qquad (8.6)$$

$$\frac{\partial Q}{\partial a_0} = \sum_{i=1}^{N}(y_i - (a_0 + a_1 x_i))(-1) = 0. \qquad (8.7)$$

Note that to calculate the above partial derivatives, we have applied the addition and chain rules for differentiation shown in Sect. 5.2.2 of Chap. 5. This does not look very easy due to the use of the $\sum$ sign. However, it just means many terms like the first term, $(y_1 - (a_0 + a_1 x_1))^2$. Each can be differentiated by using the chain rule for the squared part, giving two parts u^2 and $u = y_1 - (a_0 + a_1 x_1)$. After differentiating, all the parts would then be just summed up again. In fact, in the next part of the text, remember that it is just lots of terms conveniently summed together.

To obtain their mathematical expressions, we can rewrite Eqs. (8.6) and (8.7), respectively, as follows:

$$\left(\sum_{i=1}^{N} x_i^2\right) a_1 + \left(\sum_{i=1}^{N} x_i\right) a_0 = \sum_{i=1}^{N} x_i y_i,$$

$$\left(\sum_{i=1}^{N} x_i\right) a_1 + \left(\sum_{i=1}^{N} 1\right) a_0 = \sum_{i=1}^{N} y_i.$$

Further, we can write the above two equations in a matrix equation as was implied in Sect. 3.3.5 of Chap. 3:

$$\begin{bmatrix} \sum_{i=1}^{N} x_i^2 & \sum_{i=1}^{N} x_i \\ \sum_{i=1}^{N} x_i & \sum_{i=1}^{N} 1 \end{bmatrix} \begin{bmatrix} a_1 \\ a_0 \end{bmatrix} = \begin{bmatrix} \sum_{i=1}^{N} x_i y_i \\ \sum_{i=1}^{N} y_i \end{bmatrix}. \qquad (8.8)$$

Now, to solve simultaneous equations of the form $\mathbf{Aa} = \mathbf{x}$, where $\mathbf{A}$ is a matrix, and $\mathbf{a}$ and $\mathbf{x}$ are vectors, we need to multiply the left-hand side of both sides of the equation by the inverse matrix to $\mathbf{A}$, namely, $\mathbf{A}^{-1}$, giving $\mathbf{a} = \mathbf{A}^{-1}\mathbf{x}$.

8.2 Least-Squares Estimation

So, to obtain estimates for a_0 and a_1, we need to compute the inverse of

$$\begin{bmatrix} \sum_{i=1}^{N} x_i^2 & \sum_{i=1}^{N} x_i \\ \sum_{i=1}^{N} x_i & \sum_{i=1}^{N} 1 \end{bmatrix}.$$

To do so, we need to determine whether its inverse exists. Let us denote this matrix as $\mathbf{A}$, and the mean of x as $\bar{x}$, which is $\bar{x} = \frac{1}{N} \sum_{i=1}^{N} x_i$.

To show that the inverse exists, we must show that the determinant of $\mathbf{A}$ is non-zero. A standard way to do this is to show that the determinant is the square of something or the sum of lots of squares where none, or not all, of the square terms could be zero. This works because all the non-zero square terms are positive, so they cannot cancel with any negative terms to give an overall total of zero. This is what we will do in the following calculation.

Note that in this calculation, we use another mathematical trick: adding and subtracting the same thing in an expression so that we can reorganise the expression into a convenient form. We will do this in the fourth line of the following (Eq. 8.9) by adding and subtracting $\bar{x}^2$.

The determinant of $\mathbf{A}$ is computed as follows:

$$\begin{aligned} \det \mathbf{A} &= \sum_{i=1}^{N} x_i^2 \cdot \sum_{i=1}^{N} 1 - \sum_{i=1}^{N} x_i \cdot \sum_{i=1}^{N} x_i \\ &= N \sum_{i=1}^{N} x_i^2 - (N\bar{x})^2 \\ &= N^2 \left(\frac{1}{N} \sum_{i=1}^{N} x_i^2 - \bar{x}^2 \right) \\ &= N^2 \left(\frac{1}{N} \sum_{i=1}^{N} x_i^2 - 2\bar{x}^2 + \bar{x}^2 \right) \\ &= N^2 \left(\frac{1}{N} \sum_{i=1}^{N} x_i^2 - 2\bar{x} \frac{1}{N} \sum_{i=1}^{N} x_i + \frac{1}{N} \sum_{i=1}^{N} \bar{x}^2 \right) \\ &= N \sum_{i=1}^{N} (x_i - \bar{x})^2. \end{aligned} \quad (8.9)$$

We have applied $\bar{x} = \frac{1}{N} \sum_{i=1}^{N} x_i$ and $\bar{x}^2 = \frac{1}{N} N\bar{x}^2 = \frac{1}{N} \sum_{i=1}^{N} \bar{x}^2$ in the second last line in Eq. (8.9).

Equation (8.9) shows that as long as all x_i are not equal, which would make $x_i = \bar{x}$ for all i, the determinant of $\mathbf{A}$ will not be zero and the inverse of $\mathbf{A}$ exists.

Let us do that again with $N = 2$ and expand so that we do not have the awkward looking $\sum$ signs.

If $N = 2$, then $A = \begin{bmatrix} x_1^2 + x_2^2 & x_1 + x_2 \\ x_1 + x_2 & 2 \end{bmatrix}$ and $\bar{x} = \frac{1}{2}(x_1 + x_2)$.

$$\begin{aligned}
\det A &= 2(x_1^2 + x_2^2) - (x_1 + x_2)^2 \\
&= 2(x_1^2 + x_2^2) - (2\bar{x})^2 \\
&= 2[(x_1^2 + x_2^2) - 2(\bar{x})^2] \\
&= 2[(x_1^2 + x_2^2) - 4(\bar{x})^2 + 2(\bar{x})^2] \\
&= 2[(x_1^2 + x_2^2) - 4\bar{x}\frac{1}{2}(x_1 + x_2) + ((\bar{x})^2 + (\bar{x})^2)] \\
&= 2[[x_1^2 - 2x_1\bar{x} + (\bar{x})^2] + [x_2^2 - 2x_2\bar{x} + (\bar{x})^2]] \\
&= 2[(x_1 - \bar{x})^2 + (x_2 - \bar{x})^2]
\end{aligned}$$

as required.

Since the inverse exists, we can calculate a_0 and a_1 from Eq. (8.8). That is, we can multiply the inverse of the matrix from the left side on both sides of Eq. (8.8). We have

$$\begin{bmatrix} a_1 \\ a_0 \end{bmatrix} = \begin{bmatrix} \sum_{i=1}^N x_i^2 & \sum_{i=1}^N x_i \\ \sum_{i=1}^N x_i & \sum_{i=1}^N 1 \end{bmatrix}^{-1} \begin{bmatrix} \sum_{i=1}^N x_i y_i \\ \sum_{i=1}^N y_i \end{bmatrix}.$$

After calculating the inverse of $\mathbf{A}$, we have the following:

$$\begin{bmatrix} a_1 \\ a_0 \end{bmatrix} = \frac{1}{N \sum_{i=1}^N (x_i - \bar{x})^2} \begin{bmatrix} \sum_{i=1}^N 1 & -\sum_{i=1}^N x_i \\ -\sum_{i=1}^N x_i & \sum_{i=1}^N x_i^2 \end{bmatrix} \begin{bmatrix} \sum_{i=1}^N x_i y_i \\ \sum_{i=1}^N y_i \end{bmatrix}.$$

Therefore,

$$a_1 = \frac{1}{N \sum_{i=1}^N (x_i - \bar{x})^2} \left[N \sum_{i=1}^N x_i y_i - \sum_{i=1}^N x_i \sum_{i=1}^N y_i \right], \tag{8.10}$$

and

$$a_0 = \frac{1}{N \sum_{i=1}^N (x_i - \bar{x})^2} \left[\sum_{i=1}^N x_i^2 \sum_{i=1}^N y_i - \sum_{i=1}^N x_i \sum_{i=1}^N x_i y_i \right]. \tag{8.11}$$

8.2 Least-Squares Estimation

Remark 8.1 Alternatively, a_1 and a_0 can be rewritten in a more concise way as follows:

$$a_1 = \frac{\sum_{i=1}^{N}(x_i - \bar{x})(y_i - \bar{y})}{\sum_{i=1}^{N}(x_i - \bar{x})^2}, \tag{8.12}$$

and

$$a_0 = \bar{y} - a_1\bar{x}, \tag{8.13}$$

where $\bar{x}$ and $\bar{y}$ are the mean value of x and y, respectively.

We have mainly ignored proofs in this book. However, readers are encouraged to do the two proofs that the values of a_1 in Eqs. (8.10) and (8.12) are the same and that the values of a_0 in Eqs. (8.11) and (8.13) are the same as an exercise by themselves. The tricks that may be used in the proof include $\bar{x} = \frac{1}{N}\sum x_i$, $N\bar{x} = \sum x_i$, and $\sum \bar{x} = \sum x_i$. You will find it easier to do the proof for just the $N = 2$ case. ♦

To illustrate how these formulae work and give you some examples to try, we now do some examples and exercises for really small values of N. It should be noted that for any realistic values of N, this would be calculated using a computer program, as was the case for real principal component analysis problems in the previous chapter. The values that satisfy the two simultaneous equations, (8.10) and (8.11), or (8.12) and (8.13), are the least-squares *estimates* for a_1 and a_0 and are denoted as $\hat{a}_1$ and $\hat{a}_0$, respectively.

Example 8.1 Find the regression line when we have just two points, so $N = 2$, where the two points are (2, 2) and (4, 3).

Solution The average of the independent variable is $\bar{x} = \frac{2+4}{2} = 3$, and the average of the dependent variable is $\bar{y} = \frac{2+3}{2} = 2.5$.

Using Eq. (8.10), we get

$$\hat{a}_1 = \frac{1}{2(1+1)}(2 \times (4+12) - 6 \times 5) = \frac{1}{2}.$$

Using Eq. (8.11), we get

$$\hat{a}_0 = \frac{1}{2(1+1)}(20 \times 5 - 6 \times (4+12)) = 1.$$

(continued)

Example 8.1 (continued)
Alternately,
using Eq. (8.12), we get

$$\hat{a}_1 = \frac{(-1)(-\frac{1}{2}) + (1)(\frac{1}{2})}{(-1)^2 + (1)^2} = \frac{1}{2}.$$

Using Eq. (8.13), we get

$$\hat{a}_0 = \frac{5}{2} - \frac{1}{2} \times 3 = 1.$$

So either way, we get the same answers, that is, $\hat{a}_1 = \frac{1}{2}$ and $\hat{a}_0 = 1$.

Of course, with just two different points, you get a unique line that goes through both points. This is illustrated in Fig. 8.3.

Example 8.2 Find the regression line with three points, so that $N = 3$. The points are $(1, 2)$, $(2, 4)$, and $(3, 3)$.

Solution The average of the independent variable is $\bar{x} = \frac{6}{3} = 2$, and the average of the dependent variable is $\bar{y} = \frac{9}{3} = 3$.

Again, do it using both sets of equations to show that you get the same answer:

Using Eq. (8.10), we get

$$\hat{a}_1 = \frac{1}{3(1+0+1)}(3 \times (2+8+9) - 6 \times 9) = \frac{3}{6} = \frac{1}{2}.$$

Using Eq. (8.11), we get

$$\hat{a}_0 = \frac{1}{3(1+0+1)}(14 \times 9 - 6 \times (2+8+9)) = \frac{12}{6} = 2.$$

Alternately,
using Eq. (8.12), we get

$$\hat{a}_1 = \frac{(-1)(-1) + (0)(1) + (1)(0)}{(-1)^2 + 0 + (1)^2} = \frac{1}{2}.$$

(continued)

8.2 Least-Squares Estimation

Example 8.2 (continued)
Using Eq. (8.13), we get

$$\hat{a}_0 = 3 - \frac{1}{2} \times 2 = 2.$$

So either way, we get the same answers, that is, $\hat{a}_1 = \frac{1}{2}$ and $\hat{a}_0 = 2$. This is illustrated in Fig. 8.4.

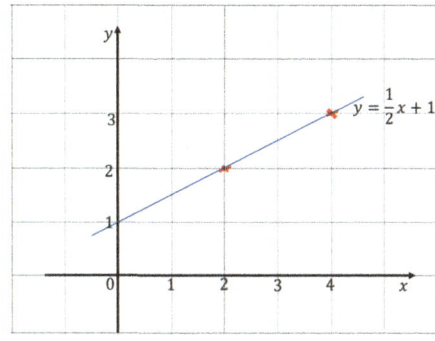

Fig. 8.3 The regression line for Example 8.1

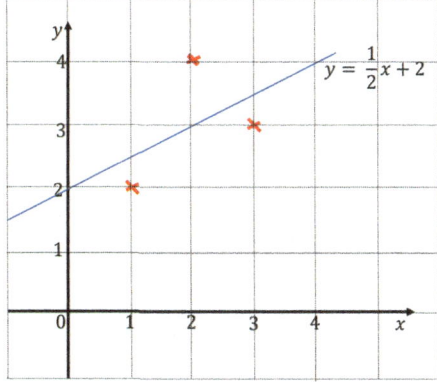

Fig. 8.4 The regression line for Example 8.2.

Exercise

8.1 Find the regression line for the following:

(1) $N = 2$. Points $(2, 3)$ and $(3, 5)$.
(2) $N = 3$. Points $(1, 3), (2, 2)$, and $(3, 4)$.
(3) $N = 3$. Points $(1, 4), (2, 2)$, and $(3, 1.5)$.
(4) $N = 3$. Points $(1, 1), (3, 4)$, and $(5, 4)$.

8.3 Linear Regression with Multiple Variables

So far, we have considered using just one independent variable to estimate the dependent variable relationship. We now consider having d independent variables. With one independent variable and one dependent variable, we are working in two dimensions, and the linear regression line is just a straight line in two dimensions, namely, $f(x) = a_0 + a_1 x$. With two independent variables, our regression "line" is a plane in three dimensions, namely, $f(x) = a_0 + a_1 x_1 + a_2 x_2$. This generalises for d independent variables to

$$f_\mathbf{a}(\mathbf{x}) = a_0 + a_1 x_1 + a_2 x_2 + \cdots + a_d x_d = \sum_{j=0}^{d} a_j x_j.$$

Note that we define x_0 as 1 to give a convenient summation.

So, for each of the N points $\mathbf{x}_i = (x_{i1}, x_{i2}, \cdots, x_{id})$, this gives the linear regression model as follows:

$$f_\mathbf{a}(\mathbf{x}_i) = a_0 + a_1 x_{i1} + a_2 x_{i2} + \cdots + a_d x_{id} = \sum_{j=0}^{d} a_j x_{ij}. \qquad (8.14)$$

The objective function is then given by the following:

$$Q = \sum_{i=1}^{N} (y_i - \sum_{j=0}^{d} a_j x_{ij})^2. \qquad (8.15)$$

So, for $d = 1$, we have

$$Q = \sum_{i=1}^{N} (y_i - (a_0 + a_1 x_{i1}))^2,$$

8.3 Linear Regression with Multiple Variables

which is the same as previously, apart from the $\frac{1}{2N}$ factor, which is a constant, and so does not affect things.

And for $d = 2$, we have

$$Q = \sum_{i=1}^{N}(y_i - (a_0 + a_1 x_{i1} + a_2 x_{i2}))^2.$$

This is just the sum of the squares of the differences between the real value y_i and the corresponding point on the plane, as expected.

If we write the data as a matrix $\mathbf{X}$ of size of $N \times (d+1)$, where $\mathbf{x}_0$ is a column vector including N ones, $\mathbf{x}_0 = \begin{bmatrix} 1 \\ 1 \\ \vdots \\ 1 \end{bmatrix}_{N \times 1}$ and $\mathbf{a} = \begin{bmatrix} a_0 \\ a_1 \\ \vdots \\ a_d \end{bmatrix}$, then Eq. (8.14) can be rewritten as $f_{\mathbf{a}}(\mathbf{x}_i) = \mathbf{X}\mathbf{a}$, and Eq. (8.15) can be rewritten as

$$Q = (\mathbf{y} - \mathbf{X}\mathbf{a})^T (\mathbf{y} - \mathbf{X}\mathbf{a}).$$

Example 8.3 To illustrate the above, consider $d = 1$ and $N = 2$. This has points (x_{11}, y_1) and (x_{21}, y_2). Then, $\mathbf{X}$, $\mathbf{a}$, and $\mathbf{y}$ are

$$\mathbf{X} = \begin{bmatrix} 1 & x_{11} \\ 1 & x_{21} \end{bmatrix}, \; \mathbf{a} = \begin{bmatrix} a_0 \\ a_1 \end{bmatrix}, \; \mathbf{y} = \begin{bmatrix} y_1 \\ y_2 \end{bmatrix}.$$

So, $\mathbf{X}\mathbf{a}$ is $\begin{bmatrix} a_0 + a_1 x_{11} \\ a_0 + a_1 x_{21} \end{bmatrix}$, where each row is $f_{\mathbf{a}}(\mathbf{x}_i)$.

Hence, we have

$$(\mathbf{y} - \mathbf{X}\mathbf{a}) = \begin{bmatrix} y_1 - (a_0 + a_1 x_{11}) \\ y_2 - (a_0 + a_1 x_{21}) \end{bmatrix},$$

and

$$(\mathbf{y} - \mathbf{X}\mathbf{a})^T = \begin{bmatrix} y_1 - (a_0 + a_1 x_{11}), & y_2 - (a_0 + a_1 x_{21}) \end{bmatrix},$$

respectively.

(continued)

Example 8.3 (continued)

Therefore, $(\mathbf{y} - \mathbf{X}\mathbf{a})^T(\mathbf{y} - \mathbf{X}\mathbf{a})$ is

$$(y_1 - (a_0 + a_1 x_{11}))^2 + (y_2 - (a_0 + a_1 x_{21}))^2 = \sum_{i=1}^{2}(y_i - (a_0 + a_1 x_{i1}))^2.$$

So, $(\mathbf{y} - \mathbf{X}\mathbf{a})^T(\mathbf{y} - \mathbf{X}\mathbf{a}) = \sum_{i=1}^{2}(y_i - (a_0 + a_1 x_{i1}))^2 = Q$ as required.

To obtain a formula for $\mathbf{a}$, we need to find the partial derivative of Q with respect to $\mathbf{a}$. That is,

$$\begin{aligned}\frac{\partial Q}{\partial \mathbf{a}} &= \frac{\partial (\mathbf{y} - \mathbf{X}\mathbf{a})^T(\mathbf{y} - \mathbf{X}\mathbf{a})}{\partial \mathbf{a}} \\ &= \frac{\partial (\mathbf{y}^T - \mathbf{a}^T\mathbf{X}^T)(\mathbf{y} - \mathbf{X}\mathbf{a})}{\partial \mathbf{a}} \\ &= \frac{\partial (\mathbf{y}^T\mathbf{y} - \mathbf{a}^T\mathbf{X}^T\mathbf{y} - \mathbf{y}^T\mathbf{X}\mathbf{a} + \mathbf{a}^T\mathbf{X}^T\mathbf{X}\mathbf{a})}{\partial \mathbf{a}} \\ &= \frac{\partial (\mathbf{y}^T\mathbf{y} - 2\mathbf{a}^T\mathbf{X}^T\mathbf{y} + \mathbf{a}^T\mathbf{X}^T\mathbf{X}\mathbf{a})}{\partial \mathbf{a}} \\ &= -2\mathbf{X}^T\mathbf{y} + 2\mathbf{X}^T\mathbf{X}\mathbf{a} \\ &= -2\mathbf{X}^T(\mathbf{y} - \mathbf{X}\mathbf{a}).\end{aligned} \qquad (8.16)$$

From the third line to the fourth line in Eq. (8.16), we have used the property that the transpose of a scalar is still the scalar itself, that is, the product of $\mathbf{a}^T\mathbf{X}^T\mathbf{y}$ is a scalar (by checking the size of each factor (see Sect. 3.3.3 of Chap. 3), and $\mathbf{a}^T\mathbf{X}^T\mathbf{y} = (\mathbf{a}^T\mathbf{X}^T\mathbf{y})^T = \mathbf{y}^T\mathbf{X}\mathbf{a}$. Going from the fourth line to the fifth line, we have differentiated using the two results, Eqs. (7.6) and (7.4), from Chap. 7.

Setting the partial derivative to zero, that is, $\mathbf{X}^T(\mathbf{y} - \mathbf{X}\mathbf{a}) = 0$, we obtain

$$\mathbf{X}^T\mathbf{X}\mathbf{a} = \mathbf{X}^T\mathbf{y}.$$

If the inverse of $\mathbf{X}^T\mathbf{X}$ exists, then multiplying the inverse from the left side of both sides of the above equation gives

$$\mathbf{a} = (\mathbf{X}^T\mathbf{X})^{-1}\mathbf{X}^T\mathbf{y}. \qquad (8.17)$$

Equation (8.17) is called the normal equation. Applying the normal equation is the method to solve for $\mathbf{a}$ analytically.

8.3 Linear Regression with Multiple Variables

Example 8.4 Let us illustrate the partial differentiation result by returning to our Example 8.3. We have already shown that

$$Q = (\mathbf{y} - \mathbf{Xa})^T (\mathbf{y} - \mathbf{Xa}) = (y_1 - (a_0 + a_1 x_{11}))^2 + (y_2 - (a_0 + a_1 x_{21}))^2,$$

which is a scalar. We can use the formula for gradient shown in Sect. 6.1.4 of Chap. 6. So,

$$\frac{\partial Q}{\partial \mathbf{a}} = \begin{bmatrix} \frac{\partial Q}{\partial a_0} \\ \frac{\partial Q}{\partial a_1} \end{bmatrix} = \begin{bmatrix} -2(y_1 - (a_0 + a_1 x_{11})) - 2(y_2 - (a_0 + a_1 x_{21})) \\ -2(y_1 - (a_0 + a_1 x_{11}))x_{11} - 2(y_2 - (a_0 + a_1 x_{21}))x_{21} \end{bmatrix}.$$

This has again used the chain rule to substitute for the squared bits in brackets. If we now substitute $\mathbf{X}$, $\mathbf{a}$, and $\mathbf{y}$ into Eq. (8.16) for computing $\frac{\partial Q}{\partial \mathbf{a}}$, we can see that

$$-2\mathbf{X}^T (\mathbf{y} - \mathbf{Xa}) = -2 \begin{bmatrix} 1 & 1 \\ x_{11} & x_{21} \end{bmatrix} \begin{bmatrix} y_1 - (a_0 + a_1 x_{11}) \\ y_2 - (a_0 + a_1 x_{21}) \end{bmatrix}.$$

Multiplying out the matrices, we get

$$\begin{bmatrix} -2(y_1 - (a_0 + a_1 x_{11})) - 2(y_2 - (a_0 + a_1 x_{21})) \\ -2(y_1 - (a_0 + a_1 x_{11}))x_{11} - 2(y_2 - (a_0 + a_1 x_{21}))x_{21} \end{bmatrix},$$

which is the result we got for $\frac{\partial Q}{\partial \mathbf{a}}$ before as required.

Before we look at the real way to find these solutions using a gradient descent algorithm and a computer program, let us look at a couple of simple examples that can be done by hand to illustrate this result. We will start by re-doing Example 8.2.

Example 8.5 Example 8.2 revisited
Remember, this example had $N = 3$ and was for $d = 1$ since we just had one independent variable. The points were $(1, 2)$, $(2, 4)$, and $(3, 3)$.
We are going to use the new formula for finding $\mathbf{a}$, namely,

$$\mathbf{a} = (\mathbf{X}^T \mathbf{X})^{-1} \mathbf{X}^T \mathbf{y}.$$

(continued)

Example 8.5 (continued)

Solution First, $\mathbf{X} = \begin{bmatrix} 1 & 1 \\ 1 & 2 \\ 1 & 3 \end{bmatrix}$ and $\mathbf{y} = \begin{bmatrix} 2 \\ 4 \\ 3 \end{bmatrix}$, and we wish to find $\mathbf{a} = \begin{bmatrix} a_0 \\ a_1 \end{bmatrix}$.

So, $\mathbf{X}^T\mathbf{X} = \begin{bmatrix} 1 & 1 & 1 \\ 1 & 2 & 3 \end{bmatrix} \begin{bmatrix} 1 & 1 \\ 1 & 2 \\ 1 & 3 \end{bmatrix} = \begin{bmatrix} 3 & 6 \\ 6 & 14 \end{bmatrix}$ and $(\mathbf{X}^T\mathbf{X})^{-1} = \frac{1}{6} \begin{bmatrix} 14 & -6 \\ -6 & 3 \end{bmatrix}$.

Also $\mathbf{X}^T\mathbf{y} = \begin{bmatrix} 1 & 1 & 1 \\ 1 & 2 & 3 \end{bmatrix} \begin{bmatrix} 2 \\ 4 \\ 3 \end{bmatrix} = \begin{bmatrix} 9 \\ 19 \end{bmatrix}$.

So, $\mathbf{a} = \begin{bmatrix} a_0 \\ a_1 \end{bmatrix} = \frac{1}{6} \begin{bmatrix} 14 & -6 \\ -6 & 3 \end{bmatrix} \begin{bmatrix} 9 \\ 19 \end{bmatrix} = \begin{bmatrix} 2 \\ \frac{1}{2} \end{bmatrix}$.

Hence, $a_0 = 2$ and $a_1 = \frac{1}{2}$ as before.

Example 8.6 We will now do a $d = 2$ example, which is an example with two independent variables, x_1 and x_2, and one dependent variable y. We will do one with $N = 3$. The points are listed in Table 8.1. Apply Eq. (8.17) to find $\mathbf{a}$.

Solution First, $\mathbf{X} = \begin{bmatrix} 1 & 1 & 2 \\ 1 & 2 & 1 \\ 1 & 2 & 2 \end{bmatrix}$ and $\mathbf{y} = \begin{bmatrix} 1 \\ 2 \\ -2 \end{bmatrix}$, and we wish to find $\mathbf{a} = \begin{bmatrix} a_0 \\ a_1 \\ a_2 \end{bmatrix}$.

So,

$$\mathbf{X}^T\mathbf{X} = \begin{bmatrix} 1 & 1 & 1 \\ 1 & 2 & 2 \\ 2 & 1 & 2 \end{bmatrix} \begin{bmatrix} 1 & 1 & 2 \\ 1 & 2 & 1 \\ 1 & 2 & 2 \end{bmatrix} = \begin{bmatrix} 3 & 5 & 5 \\ 5 & 9 & 8 \\ 5 & 8 & 9 \end{bmatrix},$$

and

$$(\mathbf{X}^T\mathbf{X})^{-1} = \begin{bmatrix} 17 & -5 & -5 \\ -5 & 2 & 1 \\ -5 & 1 & 2 \end{bmatrix}.$$

(You can check that last part by showing that $\mathbf{X}^T\mathbf{X}(\mathbf{X}^T\mathbf{X})^{-1} = \mathbf{I}$.)

(continued)

Example 8.6 (continued)

Also $\mathbf{X}^T\mathbf{y} = \begin{bmatrix} 1 & 1 & 1 \\ 1 & 2 & 2 \\ 2 & 1 & 2 \end{bmatrix} \begin{bmatrix} 1 \\ 2 \\ -2 \end{bmatrix} = \begin{bmatrix} 1 \\ 1 \\ 0 \end{bmatrix}$.

So, $\mathbf{a} = \begin{bmatrix} a_0 \\ a_1 \\ a_2 \end{bmatrix} = \begin{bmatrix} 17 & -5 & -5 \\ -5 & 2 & 1 \\ -5 & 1 & 2 \end{bmatrix} \begin{bmatrix} 1 \\ 1 \\ 0 \end{bmatrix} = \begin{bmatrix} 12 \\ -3 \\ -4 \end{bmatrix}$.

Hence, $a_0 = 12$, $a_1 = -3$, and $a_2 = -4$.

Exercise

8.2 These are the same as the last two examples in Exercise 8.1. You should now calculate the answer using the new method of this section and check you get the same answer:

(1) $d = 1$, $N = 3$. Points (1, 4), (2, 2), and (3, 1.5).
(2) $d = 1$, $N = 3$. Points (1, 1), (3, 4), and (5, 4).

8.4 Numerical Computation: Case Study 1 from Chap. 1—Continued

The normal equation provides a nice way to find the parameters of linear regression models. Recall $\mathbf{X}^T\mathbf{X}$ is a $(d+1) \times (d+1)$ matrix. When d is large, computing the inverse of $\mathbf{X}^T\mathbf{X}$ can be very slow. If $\mathbf{X}^T\mathbf{X}$ is non-invertible, we cannot use the normal equation. Alternatively, we can apply the gradient descent algorithm described in Sect. 6.2.3 of Chap. 6 to obtain estimates.

Table 8.1 Three data points with two independent variables, x_1 and x_2, and one dependent variable, y

x_1	x_2	y
1	2	1
2	1	2
2	2	-2

Table 8.2 The original data and the scaled data

MW: raw_X	Enhancement ratio: y	scaledX $= \frac{\text{raw\_}X-\min}{\max-\min}$
295	10	0
305	30	1
300	20	0.5

Table 8.3 The initial values

a_0	a_1	scaled$X$	y	$y_{\text{pred}} = a_0 + a_1 \times$ scaledX	error $= \frac{1}{2}(y - y_{\text{pred}})^2$
5	1	0	10	5	12.5
		1	30	6	288
		0.5	20	5.5	105.125
Total error: 405.625					

In Sect. 1.3.1 of Chap. 1, we have

$$\text{raw\_}X = \begin{bmatrix} 295 \\ 305 \\ 300 \end{bmatrix} \text{ and } \mathbf{y} = \begin{bmatrix} 10 \\ 30 \\ 20 \end{bmatrix}.$$

We will use this example, where $d = 1$ and $N = 3$, to show how the gradient descent algorithm can be used to estimate a_0 and a_1 of the linear regression model. We keep four decimal places when it is not divisible.

- Step 1: Data normalisation/scaling. We scale raw_X using $\frac{\text{raw\_}X-\min}{\max-\min}$, where $\min$ and $\max$ denote the minimum and maximum values of $\text{raw\_}X$. We keep the values of the target variable y unchanged. Note that there are many different normalisation methods. In this example, we simply rescale data between 0 and 1 (Table 8.2).
- Step 2: To fit a line $y_{\text{pred}} = a_0 + a_1 \times$ scaledX, we initialise random values for a_0 and a_1 and calculate the error given by $\frac{1}{2}(y - y_{\text{pred}})^2$ for each data point. This error is the same as Q given in Eq. (8.15) when $d = 1$. For example, we start with the random initial values of $a_0 = 5$ and $a_1 = 1$ (Table 8.3).
- Step 3: Calculate the partial derivative with respect to a_1 and a_0, respectively. The error for each data point is given by

$$\text{error} = \frac{1}{2}(y - y_{\text{pred}})^2 = \frac{1}{2}(y - (a_0 + a_1 \times \text{scaledX}))^2.$$

From Eqs. (8.6) and (8.7), we have the following:

$$\frac{\partial \text{error}}{\partial a_1} = -(y - (a_0 + a_1 \times \text{scaledX})) \times \text{scaledX} = -(y - y_{\text{pred}}) \times \text{scaledX},$$

8.4 Numerical Computation: Case Study 1 from Chap. 1—Continued

Table 8.4 Results of the partial derivative with respect to a_1 and a_0

a_0	a_1	scaledX	y	y_{pred}	error	$\frac{\partial \text{error}}{\partial a_0}$	$\frac{\partial \text{error}}{\partial a_1}$
5	1	0	10	5	12.5	−5	0
		1	30	6	288	−24	−24
		0.5	20	5.5	105.125	−14.5	−7.25
Total: 405.625						−43.5	−31.25

Table 8.5 Results after the first iteration

a_0	a_1	scaledX	y	y_{pred}	error	$\frac{\partial \text{error}}{\partial a_0}$	$\frac{\partial \text{error}}{\partial a_1}$
5.435	1.3125	0	10	5.435	10.4196	−4.565	0
		1	30	6.7475	270.3394	−23.2525	−23.2525
		0.5	20	6.0913	96.7267	−13.9088	−6.9544
Total: 377.486						−41.7263	−30.2069

and

$$\frac{\partial \text{error}}{\partial a_0} = -(y - (a_0 + a_1 \times \text{scaledX})) = -(y - y_{\text{pred}}).$$

Results are shown in the last two columns of Table 8.4.
- Step 4: Set the learning rate, for example, $\epsilon = 0.01$. Update the estimates by applying Eq. (6.7) in Chap. 6 with the corresponding total partial derivatives as follows:

$$a_0^{new} = 5 - 0.01 \times (-43.5) = 5.435,$$

and

$$a_1^{new} = 1 - 0.01 \times (-31.25) = 1.3125.$$

- Step 5: Use the updated value for a_0 and a_1 for computing predictions. Then, calculate new total errors and total partial derivatives.

We can see that the total error has decreased from 405.625 in Table 8.4 to 377.485 in Table 8.5 after the first iteration using the gradient descent algorithm. As we know, the gradient descent algorithm is an iterative procedure. We have shown the first iteration in this example. In practice, more iterations are needed so that the total errors can be minimised and converge to a value as small as possible.

Of course, this example is here to illustrate how the gradient descent algorithm works. But because it is a really simple example with $d = 1$ and $N = 3$, we can find a_0 and a_1 directly using Eqs. (8.13) and (8.12) on scaledX and y to get $a_0 = 10$ and $a_1 = 20$. The gradient descent values can be seen to be heading in the right direction!

Remark 8.2 Data normalisation is an important step in data pre-processing. In this example, we have scaled original molecular weights using the minimum and maximum values among the available molecular weight values. We have not scaled y, the target values. If we scale the target values, the estimated values from the fitted regression line need to be transformed back to the original target value space. For example, if the scaling is done simply by removing the mean value of targets, then the mean value must be added to the estimated value to obtain the final prediction. ♦

Exercise

8.3 Do the first iteration of the gradient descent algorithm for the following example:

- $d = 1$, $N = 3$. Points $(1, 1)$, $(3, 4)$, and $(5, 4)$. Start with estimates for a_0 and a_1 using $a_0 = 1$ and $a_1 = 1$. Suppose the learning rate is 0.01.

Since the numbers are small, do not bother to scale the values of x (it makes the calculation easier too). Note that it is the same example as the last exercise in Exercise 8.2—so you know the real answer!

8.5 Some Useful Results

These are properties and formulae that apply once you have found the "regression line" of best fit and indicate how good your results are.

8.5.1 Residuals

A residual is defined as $e_i = y_i - \tilde{y}_i$, where $\tilde{y}_i$ is the estimate of the ith of the N points. We have the following useful properties:

- The sum of the residuals is zero.
- The sum of observed target values equals the sum of the estimated values.

If the regression line is found by calculation, then these properties are correct. We may not find the exact solution for iterative procedures, but one that is close enough that these two properties are very close to being correct. They give an indication of how close you have come to the exact solution.

8.5 Some Useful Results

Example 8.7 Example 8.1 revisited.
Here, we just had two points and found the unique answer, giving a line that goes through both points. So, the residuals are both zero, and the observed target and the estimated values are the same. Thus, both properties are correct.

Example 8.8 Example 8.2 revisited.
Here, we had three points, $(1, 2)$, $(2, 4)$, and $(3, 3)$, and found that $a_0 = \frac{1}{2}$ and $a_1 = 2$.
The estimated points on the line $y = 2 + \frac{1}{2}x$ are the following:
For $x = 1$, $\tilde{y}_1 = \frac{5}{2}$; for $x = 2$, $\tilde{y}_2 = 3$; and for $x = 3$, $\tilde{y}_3 = \frac{7}{2}$.
The sum of the residuals is therefore $(2 - \frac{5}{2}) + (4 - 3) + (3 - \frac{7}{2}) = 0$.
The sum of targets is $2+4+3 = 9$, and the sum of estimates is $\frac{5}{2}+3+\frac{7}{2} = 9$ as required.

8.5.2 The Coefficient of Determination

The coefficient of determination, denoted as R^2, is defined as follows:

$$R^2 = 1 - \frac{\sum_i^N (y_i - \tilde{y}_i)^2}{\sum_i^N (y_i - \bar{y})^2},$$

where $\tilde{y}_i$ is the estimate of the ith of the N points and $\bar{y}$ is the mean value of the dependent variable.

The closer the value of R^2 is to 1, the better the fit. We have ignored the proof; however, the coefficient of determination can be interpreted as the square of Pearson's correlation coefficient (see Sect. 4.2.1 in Chap. 4) between the observed target values y_i and the estimated values $\tilde{y}_i$.

Example 8.9 Let us do Examples 8.1 and 8.2 again.
For Example 8.1: all of $y_i = \bar{y}$, so $R^2 = 1$.
For Example 8.2: $\bar{y} = 3$ and

$$R^2 = 1 - \frac{(-\frac{1}{2})^2 + (1)^2 + (-\frac{1}{2})^2}{(1)^2 + (1)^2 + 0} = \frac{1}{4}.$$

Exercise

8.4 These data points are the same as Exercise 8.2. In each case, find
(a) the sum of the residuals, (b) the sum of the targets, (c) the sum of the estimates, and (d) R^2:

(1) $d = 1, N = 3$. Points $(1, 4)$, $(2, 2)$, and $(3, 1.5)$.
(2) $d = 1, N = 3$. Points $(1, 1)$, $(3, 4)$, and $(5, 4)$.

Chapter 9
Algorithms 3: Neural Networks

This is the third of three chapters that aim to show how we can apply the knowledge introduced in previous chapters to formulate three widely used algorithms in the Data Science field. This chapter considers neural networks. Neural networks are a huge topic, and there are many textbooks dedicated to describing all the different types and giving the details of how they work. There are unsupervised and supervised neural networks dedicated to different tasks. We are only going to consider one type of supervised neural network: the single-layered and multilayered perceptrons trained using back-propagation of errors. Knowledge of this type of network is a good entry point to lots of other networks. This chapter introduces the basic idea of input data being passed through the network in a forward direction and the error being propagated backwards through the network with the network weights being updated using a gradient descent algorithm of the type described in Sect. 6.2.3 of Chap. 6.

9.1 Training a Neural Network by Gradient Descent

We are leading up to describing the training of a two-layer neural network (NN) using a gradient descent algorithm. This algorithm uses the gradient of the error between the outputs of the neural network and the desired target values. It then adjusts the weights in a neural network by considering the error relative to each weight by looking backwards through the neural network in a method known as back-propagation. This is Case Study 3 in Chap. 1. We will do this by first illustrating the principles on a simple one-layer network.

© The Author(s), under exclusive license to Springer Nature Singapore Pte Ltd. 2025
Y. Sun, R. Adams, *A Mathematical Introduction to Data Science*,
https://doi.org/10.1007/978-981-96-5639-4_9

9.2 A Simple One-Layer Neural Network

This section explains the training principle behind artificial neural networks using a simple example of a one-layer neural network with just two inputs and two outputs. It is a very limited network but will illustrate how the inputs are fed forwards and the errors fed backwards.

Figure 9.1 shows the architecture of the one-layer neural network used in this example, where we consider only one input (training) example **x**, which has two attributes or features x_1 and x_2. Squares in Fig. 9.1 represent the two input features, forming the neural network's input layer. Suppose each input example has two targets, denoted as t_1 and t_2. y_1 and y_2 are the outputs or predictions of the neural network for the given **x**. We follow the notations used in [5] for weights. That is, we denote each weight as w_{ji}, where j is the jth output unit and i is the ith node of the input layer. For example, w_{21} denotes the weight going from the first input feature x_1 to output unit 2 in the output layer. The training of this neural network aims to adjust weight values to reduce the error, that is, the difference between the targets and predictions.

In general, the input values are fed to the output nodes as a weighted sum formed as a dot product of the input vector **x** and the weight vector **w**, that is, **x**·**w**. Expressed in full for our two-node example, we get

$$a_1 = w_{11}x_1 + w_{12}x_2,$$

and

$$a_2 = w_{21}x_1 + w_{22}x_2.$$

The output node can transform this to give an output by using an activation function, denoted as g, which for these two units can be written as follows:

$$y_1 = g(a_1),$$

and

$$y_2 = g(a_2).$$

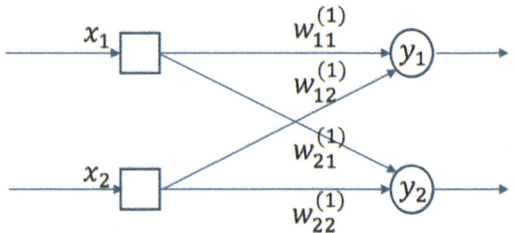

Fig. 9.1 A simple one-layer neural network

9.2 A Simple One-Layer Neural Network

We are going to consider a linear activation function and then a logistic sigmoid activation function for g.

Remark 9.1 We can represent the complete operation of finding the inputs to the first layer of the simple neural network by using the multiplication operation of an input vector and a matrix of weights.

Let $\mathbf{w}_1 = \begin{bmatrix} w_{11} \\ w_{12} \end{bmatrix}$, $\mathbf{w}_2 = \begin{bmatrix} w_{21} \\ w_{22} \end{bmatrix}$, and $\mathbf{W} = \begin{bmatrix} \mathbf{w}_1^T \\ \mathbf{w}_2^T \end{bmatrix}$. Then, we have

$$\begin{bmatrix} a_1 \\ a_2 \end{bmatrix} = \begin{bmatrix} \mathbf{w}_1^T \\ \mathbf{w}_2^T \end{bmatrix} \begin{bmatrix} x_1 \\ x_2 \end{bmatrix}.$$

♦

9.2.1 Linear Activation Function

Figure 9.2 illustrates a linear activation function with $y = g(x) = x$. Note that this activation function is differentiable and $g'(x) = 1$.

Applying this linear activation function to our simple one-layer neural network example, we have $y_1 = g(a_1) = a_1$ and $y_2 = g(a_2) = a_2$. With targets t_1 and t_2, we get the error (E) as

$$E = \frac{1}{2} \sum_{j=1}^{2} (t_j - y_j)^2 = \frac{1}{2} \sum_{j=1}^{2} (t_j - a_j)^2 = \frac{1}{2} \sum_{j=1}^{2} \left(t_j - \left(\sum_{i=1}^{2} w_{ji} x_i \right) \right)^2. \quad (9.1)$$

This error is associated with just one training example, the simplest training method to explain. So, we are going to update the weights for each input training example.

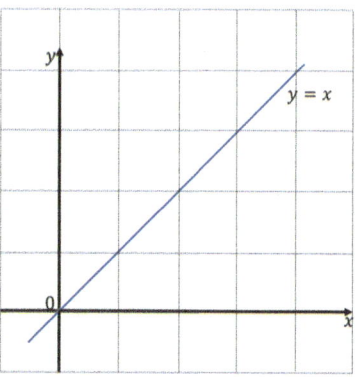

Fig. 9.2 Linear activation function

More will be said about this in Sect. 9.5.2. Expanding the equation for E out, we have

$$E = \frac{1}{2}\left(\left(t_1 - (w_{11}x_1 + w_{12}x_2)\right)^2 + \left(t_2 - (w_{21}x_1 + w_{22}x_2)\right)^2\right).$$

We will update the weights by back-propagation of the error. The basic idea is to apply the gradient descent algorithm (see Sect. 6.2.3 of Chap. 6). That is,

$$w_{ji} \leftarrow w_{ji} - \epsilon \frac{\partial E}{\partial w_{ji}}. \tag{9.2}$$

So, we need to differentiate Eq. (9.1). Looking at the formula for E, we can see it contains two terms, so $E = \frac{1}{2}(E_1 + E_2)$. Each term needs the use of the differentiation of composite functions. Looking at the first term, we have $E_1 = u^2$, where $u = t_1 - (w_{11}x_1 + w_{12}x_2)$. Let us differentiate with respect to w_{11}:

$$\frac{\partial E_1}{\partial w_{11}} = \frac{\partial E_1}{\partial u}\frac{\partial u}{\partial w_{11}} = 2u(-x_1) = -2(t_1 - (w_{11}x_1 + w_{12}x_2))x_1 = -2(t_1 - y_1)x_1.$$

When we work out $\frac{\partial E_2}{\partial w_{11}}$, we get 0 since E_2 has no w_{11} in it, and all other variables are treated as constants when differentiating with respect to w_{11}.

So, adding the two results together and dividing by two, we get

$$\frac{\partial E}{\partial w_{11}} = -(t_1 - y_1)x_1.$$

Repeating this for the other weights, we get

$$\frac{\partial E}{\partial w_{12}} = -(t_1 - y_1)x_2,$$

$$\frac{\partial E}{\partial w_{21}} = -(t_2 - y_2)x_1,$$

and

$$\frac{\partial E}{\partial w_{22}} = -(t_2 - y_2)x_2.$$

Or in general:

$$\frac{\partial E}{\partial w_{ji}} = -(t_j - y_j)x_i. \tag{9.3}$$

9.2 A Simple One-Layer Neural Network

Armed with this result, we can then update the weights using Eq. (9.2) as the following Example illustrates.

Example 9.1 We will show how the gradient descent algorithm updates weights for a really simple example and just one iteration. Assume the initial input vector is $x_1 = x_2 = 1$, and the target vector is $t_1 = 0.5$ and $t_2 = 0$. Also assume that $w_{11} = w_{12} = 0.5$, $w_{21} = w_{22} = 0.25$, and finally $\epsilon = 0.1$. Then,

$$y_1 = a_1 = w_{11}x_1 + w_{12}x_2 = 1,$$

and

$$y_2 = a_2 = w_{21}x_1 + w_{22}x_2 = 0.5.$$

So,

$$E = \frac{1}{2}((t_1 - y_1)^2 + (t_2 - y_2)^2) = \frac{1}{4} = 0.25.$$

Also,

$$\frac{\partial E}{\partial w_{11}} = -(t_1 - y_1)x_1 = 0.5,$$

$$\frac{\partial E}{\partial w_{12}} = -(t_1 - y_1)x_2 = 0.5,$$

$$\frac{\partial E}{\partial w_{21}} = -(t_2 - y_2)x_1 = 0.5,$$

and

$$\frac{\partial E}{\partial w_{22}} = -(t_2 - y_2)x_2 = 0.5.$$

Using Eq. (9.2), we get the new values for the weights as

$$w_{11} = 0.5 - (0.1)(0.5) = 0.45,$$

$$w_{12} = 0.5 - (0.1)(0.5) = 0.45,$$

$$w_{21} = 0.25 - (0.1)(0.5) = 0.2,$$

(continued)

Example 9.1 (continued)
and

$$w_{22} = 0.25 - (0.1)(0.5) = 0.2.$$

So after one iteration, the weights have changed. The new $y_1 = 0.9$ and the new $y_2 = 0.4$. The new error is

$$E = \frac{1}{2}((0.5 - 0.9)^2 + (0 - 0.4)^2) = 0.16.$$

Hence, the error was reduced after one iteration.

Remark 9.2 Of course, realistically, applying the gradient descent algorithm requires many iterations to reduce the error to zero and would be done using an appropriate computer program.

◆

Exercise

9.1 Here is one for you to try. You will probably need a calculator! Do one iteration for this one-layer neural network with a linear activation function. The initial input vector is $x_1 = 1$ and $x_2 = 0$; the target is $t_1 = 0.25$ and $t_2 = 0.5$.
Initially, $w_{11} = w_{12} = 0.5$, $w_{21} = w_{22} = 0.25$ and finally $\epsilon = 0.1$.

9.2.2 Logistic Sigmoid Activation Function

Figure 9.3 illustrates a logistic sigmoid activation function with $y = g(x) = \sigma(x) = \frac{1}{1+e^{-x}}$. Note that this activation function is differentiable.

Applying the sigmoid activation function to our simple one-layer neural network example, we have $y_1 = g(a_1)$ and $y_2 = g(a_2)$, where $g(a_j) = \sigma(a_j) = \frac{1}{1+e^{-a_j}}$ and $j = 1, 2$. With targets t_1 and t_2, we get the error as

$$E = \frac{1}{2}\sum_{j=1}^{2}(t_j - y_j)^2 = \frac{1}{2}\sum_{j=1}^{2}(t_j - g(a_j))^2 = \frac{1}{2}\sum_{j=1}^{2}\left(t_j - g\left(\sum_{i=1}^{2}w_{ji}x_i\right)\right)^2. \tag{9.4}$$

9.2 A Simple One-Layer Neural Network

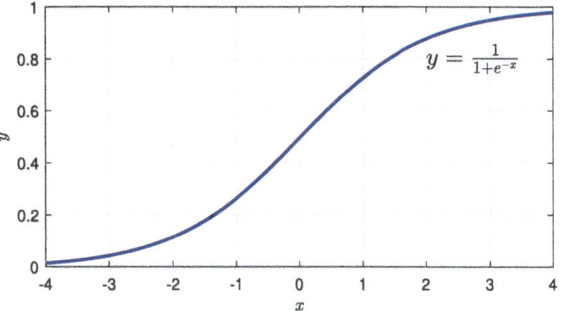

Fig. 9.3 Logistic sigmoid activation function

Again, we are updating the weights for each input training example. Expanding the equation for E out, we have

$$E = \frac{1}{2}\left(\left(t_1 - g(w_{11}x_1 + w_{12}x_2)\right)^2 + \left(t_2 - g(w_{21}x_1 + w_{22}x_2)\right)^2\right).$$

We need to differentiate Eq. (9.4) to update the weights using Eq. (9.2). The only difference between Eqs. (9.4) and (9.1) is the addition of the function $g(a_j)$. So, when we come to differentiate E after dealing with the square term using u, as before, we need to find $\frac{dg}{da_j} = g'(a_j)$ before we can get inside the composite function g and differentiate the expression for a_j in terms of weights and input values.

Hence, if we consider $E = \frac{1}{2}(E_1 + E_2)$, then in terms of composite functions, we have $E_1 = u^2$, where $u = t_1 - g(a_1)$ and $a_1 = w_{11}x_1 + w_{12}x_2$.

Now to differentiate with respect to w_{12}, we can use

$$\frac{\partial E_1}{\partial w_{12}} = \frac{\partial E_1}{\partial u} \frac{\partial u}{\partial a_1} \frac{\partial a_1}{\partial w_{12}},$$

where $\frac{\partial u}{\partial a_1} = -\frac{\partial g(a_1)}{\partial a_1} = -g'(a_1)$. Again differentiating E_2 gives 0 since E_2 does not contain w_{12}. So,

$$\frac{\partial E}{\partial w_{12}} = u(-g'(a_1))x_2 = -(t_1 - g(w_{11}x_1 + w_{12}x_2))g'(a_1)x_2 = -(t_1 - y_1)g'(a_1)x_2.$$

In general, we get the following:

$$\begin{aligned}\frac{\partial E}{\partial w_{ji}} &= u(-g'(a_j))x_i \\ &= -(t_j - g(w_{j1}x_1 + w_{j2}x_2))g'(a_j)x_i \\ &= -(t_j - y_j)g'(a_j)x_i.\end{aligned} \quad (9.5)$$

Luckily, we already have found the derivative of the sigmoid function from Example 5.13 of Sect. 5.2.2 of Chap. 5, and we know that $g'(x) = \sigma(x)(1 - \sigma(x))$. So, $g'(a_j) = \sigma(a_j)(1 - \sigma(a_j))$. In particular, we have

$$g'(a_1) = \sigma(a_1)(1 - \sigma(a_1)),$$

and

$$g'(a_2) = \sigma(a_2)(1 - \sigma(a_2)).$$

So finally,

$$\frac{\partial E}{\partial w_{12}} = -(t_1 - y_1)g'(a_1)x_2 = -(t_1 - y_1)\sigma(a_1)(1 - \sigma(a_1))x_2,$$

and in general,

$$\frac{\partial E}{\partial w_{ji}} = -(t_j - y_j)g'(a_j)x_i = -(t_j - y_j)\sigma(a_j)(1 - \sigma(a_j))x_i.$$

If we collect all the parts relating to j together and define

$$\delta_j = (t_j - y_j)\sigma(a_j)(1 - \sigma(a_j)), \tag{9.6}$$

then we can express the final result as

$$\frac{\partial E}{\partial w_{ji}} = -\delta_j x_i. \tag{9.7}$$

Readers will see why it is useful to define δ_j in this way in Sect. 9.4.

Remark 9.3 After defining Eq. (9.5), we have put in the logistic sigmoid for $g(x)$. However, other functions can be used to give different algorithms. The hyperbolic tangent activation function could be slotted in where now $g(x) = \frac{e^x - e^{-x}}{e^x + e^{-x}}$. In fact, if we put in $g(x) = x$, that is, the linear activation function, then since $g'(x) = 1$, we get all the same results as in the previous section on the linear activation function, namely, Eq. (9.3).

♦

9.2 A Simple One-Layer Neural Network

Example 9.2 To illustrate the logistic sigmoid activation function, $g(x) = \sigma(x) = \frac{1}{1+e^{-x}}$, we will do one iteration as we did in Example 9.1. In fact, if we take the same start values and targets as Example 9.1, we have $x_1 = x_2 = 1$, $t_1 = 0.5$, and $t_2 = 0$. $w_{11} = w_{12} = 0.5$, $w_{21} = w_{22} = 0.25$, and $\epsilon = 0.1$. We keep three decimal places in the following calculation.
First, since

$$a_1 = w_{11}x_1 + w_{12}x_2 = 1,$$

and

$$a_2 = w_{21}x_1 + w_{22}x_2 = 0.5,$$

we have

$$y_1 = \sigma(1) = 0.731,$$

and

$$y_2 = \sigma(0.5) = 0.622.$$

So,

$$E = \frac{1}{2}((t_1 - y_1)^2 + (t_2 - y_2)^2) = 0.220,$$

Then, using Eq. (9.6),

$$\delta_1 = (t_1 - y_1)\sigma(a_1)(1 - \sigma(a_1)) = (-0.231) \times 0.731 \times 0.269 = -0.045,$$

and

$$\delta_2 = (t_2 - y_2)\sigma(a_2)(1 - \sigma(a_2)) = (-0.622) \times 0.622 \times 0.378 = -0.146.$$

Also, using Eq. (9.7),

$$\frac{\partial E}{\partial w_{11}} = -\delta_1 x_1 = 0.045 \times 1 = 0.045,$$

$$\frac{\partial E}{\partial w_{12}} = -\delta_1 x_2 = 0.045 \times 1 = 0.045,$$

(continued)

Example 9.2 (continued)

$$\frac{\partial E}{\partial w_{21}} = -\delta_2 x_1 = 0.146 \times 1 = 0.146,$$

and

$$\frac{\partial E}{\partial w_{22}} = -\delta_2 x_2 = 0.146 \times 1 = 0.146.$$

Finally, we update the weights using Eq. (9.2) to obtain

$$w_{11} = 0.5 - (0.1 \times 0.045) = 0.496,$$

$$w_{12} = 0.5 - (0.1 \times 0.045) = 0.496,$$

$$w_{21} = 0.25 - (0.1 \times 0.146) = 0.235,$$

and

$$w_{22} = 0.25 - (0.1 \times 0.146) = 0.235.$$

So after one iteration, we have new weights and will get a new value for an error of $E = 0.215$.

Exercise

9.2 Here is a harder one for you to try. You will definitely need a calculator! Do one iteration for this one-layer neural network with a logistic sigmoid activation function:
The initial input vector is $x_1 = 1$ and $x_2 = 0$; the target is $t_1 = 0.25$ and $t_2 = 0.5$.
Initially, $w_{11} = w_{12} = 0.5$ and $w_{21} = w_{22} = 0.25$. Finally, $\epsilon = 0.1$.
These are the same values as in Exercise 9.1, but now you have the complication of a logistic sigmoid activation function.

9.3 A Simple Two-Layer Neural Network: Case Study 3 from Chap. 1

We have now set the scene for dealing with Case Study 3 from Chap. 1, a simple example of a two-layer neural network with two hidden units only in each layer. This is a simple two-layer neural network but represents the classic back-propagation model. The mathematics behind the two-layer neural network follows the same pattern as the one-layer neural network. However, it is complicated by having two layers and needing to back-propagate the error to the first-level weights to update them.

Figure 9.4 shows the architecture of a two-layer neural network used in this example, where we consider one input (training) example **x**, which has two attributes or features x_1 and x_2 only. This figure is identical to Fig. 1.9 in Chap. 1. For the reader's convenience, it is repeated here. Notations in Fig. 9.4 are the same as those in Fig. 9.1. The two layers mean there are two layers of adaptive weights. The nodes in between two weight layers are called hidden units. We follow notations used in [5] for weights, where now we have to distinguish between weights in the two layers. That is, we denote each weight as $w_{ji}^{(l)}$, where (l) denotes the lth layer, j the jth hidden unit in the corresponding layer or the jth output, and i the ith node of the immediate layer to the left. For example, $w_{21}^{(1)}$ denotes the weight going from the first input feature, x_1, to hidden unit 2 in the first layer; and $w_{12}^{(2)}$ denotes the weight going from the second hidden unit to the output unit 1 in the second layer. $z_j^{(1)}$ denotes the output of the jth node in the hidden layer. The training of this neural network again aims to adjust weight values to reduce the error, that is, the difference between the targets and predictions.

9.3.1 The Feed-Forward Propagation

Each input in the input layer is connected to hidden units via weights of the first layer. Each hidden unit is a linear combination of the input attributes that are

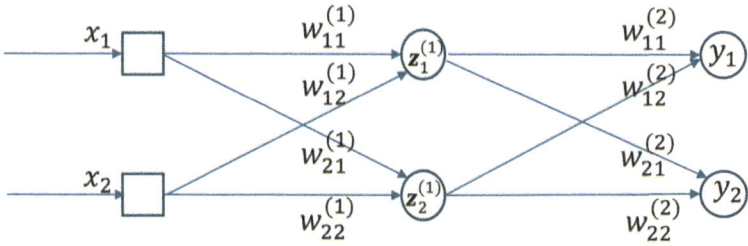

Fig. 9.4 An illustration of the feed-forward of a simple two-layer neural network

transformed by an activation function. The linear combinations of the two input attributes can be written as follows:

$$a_1^{(1)} = w_{11}^{(1)} x_1 + w_{12}^{(1)} x_2,$$

and

$$a_2^{(1)} = w_{21}^{(1)} x_1 + w_{22}^{(1)} x_2.$$

The transformations by a non-linear activation function, denoted as g_1, for these two hidden units can be written as follows:

$$z_1^{(1)} = g_1(a_1^{(1)}),$$

and

$$z_2^{(1)} = g_1(a_2^{(1)}).$$

As can be seen, $z_j^{(1)}$ is the output of a composite function, where the output of a linear function, $a_j^{(1)}$, is the input of a non-linear activation function $g_1(\cdot)$. Here, we use g_1 as the first activation function in case we want to use different activation functions in the different layers.

Similarly, for nodes in the output layer, we have

$$a_1^{(2)} = w_{11}^{(2)} z_1^{(1)} + w_{12}^{(2)} z_2^{(1)},$$

so

$$y_1 = g_2(a_1^{(2)}) = g_2(w_{11}^{(2)} z_1^{(1)} + w_{12}^{(2)} z_2^{(1)}),$$

and

$$a_2^{(2)} = w_{21}^{(2)} z_1^{(1)} + w_{22}^{(2)} z_2^{(1)},$$

so

$$y_2 = g_2(a_2^{(2)}) = g_2(w_{21}^{(2)} z_1^{(1)} + w_{22}^{(2)} z_2^{(1)}).$$

9.3 A Simple Two-Layer Neural Network: Case Study 3 from Chap. 1

So, in general, the outputs can be written as

$$y_k = g_2\left(\sum_{j=1}^{2} w_{kj}^{(2)} z_j^{(1)}\right)$$

$$= g_2\left(\sum_{j=1}^{2} w_{kj}^{(2)} g_1\left(a_j^{(1)}\right)\right) \qquad (9.8)$$

$$= g_2\left(\sum_{j=1}^{2} w_{kj}^{(2)} g_1\left(\sum_{i=1}^{2} w_{ji}^{(1)} x_i\right)\right),$$

where k is the index of the nodes in the second, output, layer, j is the index of the hidden unit of the first layer, and i is the index of the inputs.

In the feed-forward process, the input information is passed as a forward flow through the network. Note that the activation function used in different layers for different hidden units can be the same or different.

9.3.2 The Error Back-Propagation

The weights in any feed-forward network are updated by applying the back-propagation algorithm. The basic idea is to apply the gradient descent algorithm (see Sect. 6.2.3 of Chap. 6). In general, this is written as

$$w_{nm}^{(l)} \leftarrow w_{nm}^{(l)} - \epsilon \frac{\partial E}{\partial w_{nm}^{(l)}},$$

where E denotes the error, that is, the difference between the target values and the neural network outputs, l is the index of the layer, and m and n represent the layers to the left and right, respectively.

We update weights layer by layer in the direction from the output layer to the input layer, which is opposite to the feed-forward propagation. So, in particular, we use

$$w_{kj}^{(2)} \leftarrow w_{kj}^{(2)} - \epsilon \frac{\partial E}{\partial w_{kj}^{(2)}}, \qquad (9.9)$$

to update the weights from the hidden layer to the output layer. And we use

$$w_{ji}^{(1)} \leftarrow w_{ji}^{(1)} - \epsilon \frac{\partial E}{\partial w_{ji}^{(1)}}, \qquad (9.10)$$

to update the weights from the input to the hidden layer.

9.3.2.1 Updating Weights of the Second Layer

Considering Fig. 9.4, let us again suppose the error measure is given again by $\frac{1}{2}((t_1 - y_1)^2 + (t_2 - y_2)^2)$. We update the weights of the second layer, that is, for the weights connecting the hidden nodes to the outputs first.

We compute the error as follows:

$$E = \frac{1}{2}\sum_{k=1}^{2}(t_k - y_k)^2 = \frac{1}{2}\sum_{k=1}^{2}\left(t_k - g_2(a_k^{(2)})\right)^2$$

$$= \frac{1}{2}\sum_{k=1}^{2}\left(t_k - g_2\left(\sum_{j=1}^{2} w_{kj}^{(2)} z_j^{(1)}\right)\right)^2, \quad (9.11)$$

where k in the first summation indicates the index of targets and j in the last summation is the index of hidden units. Again, we will update the weights after each input training example.

If you look at Eq. (9.11) and compare it to Eq. (9.4) used in Sect. 9.2.2, you will see that they are virtually identical. The differences are the naming of indices in the summations, having superscripts that represent the layer, having g_2 instead of g, and having a $z_j^{(1)}$ here instead of the x_i previously. Then, when we compute the partial derivative of E with respect to the second-layer weights using the chain rule, we get a result virtually identical to Eq. (9.5) in Sect. 9.2.2. That is, we get

$$\frac{\partial E}{\partial w_{kj}^{(2)}} = -(t_k - y_k) g_2'(a_k^{(2)}) z_j^{(1)}. \quad (9.12)$$

Note that the sum sign has disappeared in Eq. (9.12), as before, because we are specifying particular values for k and j when calculating the partial derivative. In this case, we treat other weights and other hidden layer units as constants when differentiating with respect to $w_{kj}^{(2)}$. For instance, if $k = 1$ and $j = 2$, we get

$$\frac{\partial E}{\partial w_{12}^{(2)}} = -(t_1 - y_1) g_2'(a_1^{(2)}) z_2^{(1)}.$$

The above equation containing just $z_2^{(1)}$ is because there is only one occurrence of the particular weight, $w_{12}^{(2)}$, in Eq. (9.11) and that is multiplied by $z_2^{(1)}$. This result is again similar to the result we obtained in Sect. 9.2.2.

We can now collect all the terms containing k together in Eq. (9.12) and define

$$\delta_k^{(2)} = (t_k - y_k) g_2'(a_k^{(2)}), \quad (9.13)$$

9.3 A Simple Two-Layer Neural Network: Case Study 3 from Chap. 1

and then Eq. (9.12) can be rewritten as follows:

$$\frac{\partial E}{\partial w_{kj}^{(2)}} = -\delta_k^{(2)} z_j^{(1)}. \tag{9.14}$$

Readers will see why it is useful to define $\delta_k^{(l)}$ in Sect. 9.4.

Suppose the output units have linear activation functions, that is, $g_2(a) = a$, then the derivative of these activation functions is 1. This gives us $\delta_k^{(2)} = t_k - y_k$ and $\frac{\partial E}{\partial w_{kj}^{(2)}} = -(t_k - y_k)z_j^{(1)}$.

Therefore, we can use Eq. (9.9) and update $w_{11}^{(2)}$ as follows:

$$w_{11}^{(2)} \leftarrow w_{11}^{(2)} + \epsilon(t_1 - y_1)z_1^{(1)}.$$

Similarly, we have

$$w_{12}^{(2)} \leftarrow w_{12}^{(2)} + \epsilon(t_1 - y_1)z_2^{(1)},$$

$$w_{21}^{(2)} \leftarrow w_{21}^{(2)} + \epsilon(t_2 - y_2)z_1^{(1)},$$

and

$$w_{22}^{(2)} \leftarrow w_{22}^{(2)} + \epsilon(t_2 - y_2)z_2^{(1)}.$$

This completes the updating of the weights in the second layer.

9.3.2.2 Updating Weights of the First Layer

Now let us consider updating the weights of the first layer, that is, the weights connecting the inputs to the hidden units. To do it, we rewrite Eq. (9.11) by first replacing $z_j^{(1)}$ with $g_1(a_j^{(1)})$ and then replacing each $a_j^{(1)}$ with its summation of products of first layer weights and the input values x_i as follows:

$$E = \frac{1}{2}\sum_{k=1}^{2}\left(t_k - g_2\left(\sum_{j=1}^{2} w_{kj}^{(2)} g_1(a_j^{(1)})\right)\right)^2$$

$$= \frac{1}{2}\sum_{k=1}^{2}\left(t_k - g_2\left(\sum_{j=1}^{2} w_{kj}^{(2)} g_1\left(\sum_{i=1}^{2} w_{ji}^{(1)} x_i\right)\right)\right)^2. \tag{9.15}$$

Updating weights of the first layer is to update $w_{ji}^{(1)}$ shown in Eq. (9.15). To compute the partial derivative of E with respect to $w_{ji}^{(1)}$, we apply the chain rule through several composite functions until we get deep inside the functions to the actual weight we are differentiating with respect to, for example, $w_{21}^{(1)}$. We obtain the following:

$$\frac{\partial E}{\partial w_{ji}^{(1)}} = -\sum_{k=1}^{2}(t_k - y_k)g_2'(a_k^{(2)})w_{kj}^{(2)}g_1'(a_j^{(1)})x_i$$
$$= -\sum_{k=1}^{2}\delta_k^{(2)}w_{kj}^{(2)}g_1'(a_j^{(1)})x_i. \qquad (9.16)$$

Here, we have used the value of $\delta_k^{(2)}$ previously given in Eq. (9.13). Note that the sum signs over j and i have disappeared since we are specifying particular values for them when computing the partial derivatives, as before, and all other weights are treated as constants. There is, however, still a sum of two terms as shown by the summation over k. This is because each $w_{ji}^{(1)}$ appears twice in the error function E. We can see this by looking at Fig. 9.4 and considering, for instance, $w_{21}^{(1)}$. This weight contributes part of the value of $z_2^{(1)}$. But $z_2^{(1)}$ is propagated to both the two output nodes and so contributes to both y_1 and y_2, that is, to both y_k's. Hence, we get $w_{21}^{(1)}$ appearing twice in the final value of E, once for each value of k.

If we collect all the parts of the expression for $\frac{\partial E}{\partial w_{ji}^{(1)}}$ in Eq. (9.16) that don't contain i and define

$$\delta_j^{(1)} = g_1'(a_j^{(1)})\sum_{k=1}^{2}w_{kj}^{(2)}\delta_k^{(2)}, \qquad (9.17)$$

as we did before when defining $\delta_k^{(2)}$, then Eq. (9.16) can be simply rewritten as follows:

$$\frac{\partial E}{\partial w_{ji}^{(1)}} = -\delta_j^{(1)}x_i. \qquad (9.18)$$

We compute $\delta_1^{(1)}$ and $\delta_2^{(1)}$ as follows:

$$\delta_1^{(1)} = g_1'(a_1^{(1)})(w_{11}^{(2)}\delta_1^{(2)} + w_{21}^{(2)}\delta_2^{(2)}), \qquad (9.19)$$

$$\delta_2^{(1)} = g_1'(a_2^{(1)})(w_{12}^{(2)}\delta_1^{(2)} + w_{22}^{(2)}\delta_2^{(2)}), \qquad (9.20)$$

9.3 A Simple Two-Layer Neural Network: Case Study 3 from Chap. 1

respectively. Using Eqs. (9.10) and (9.18), noting that we are subtracting the negative differential, then we can update the first layer weights using

$$w_{ji}^{(1)} \leftarrow w_{ji}^{(1)} + \epsilon \delta_j^{(1)} x_i. \quad (9.21)$$

As an example, suppose the activation function g_1 of hidden units is the sigmoid activation function, and g_2 of output units is the linear function as before. We can then derive the update method for $w_{11}^{(1)}$ by substituting

$$g_1'(a_1^{(1)}) = \sigma(a_1^{(1)})(1 - \sigma(a_1^{(1)})),$$

$$\delta_1^{(2)} = t_1 - y_1,$$

and

$$\delta_2^{(2)} = t_2 - y_2$$

into $\delta_1^{(1)}$ giving the update for $w_{11}^{(1)}$ as follows:

$$w_{11}^{(1)} \leftarrow w_{11}^{(1)} + \epsilon \delta_1^{(1)} x_1 = w_{11}^{(1)} + \epsilon \sigma(a_1^{(1)})(1 - \sigma(a_1^{(1)}))(w_{11}^{(2)}(t_1 - y_1) + w_{21}^{(2)}(t_2 - y_2))x_1.$$

The other weights are dealt with similarly.

Example 9.3 Again, we will do one iteration of a very simple example. Assume the initial input vector is $x_1 = x_2 = 1$ and that $w_{11}^{(1)} = w_{12}^{(1)} = 0.5$ and $w_{21}^{(1)} = w_{22}^{(1)} = 0.25$. Also assume $w_{11}^{(2)} = w_{12}^{(2)} = 0.25$, $w_{21}^{(2)} = w_{22}^{(2)} = 0.5$, and the target is $t_1 = t_2 = 0.5$. As in the text, we will have g_1 as a sigmoid function $g_1(x) = \sigma(x) = \frac{1}{1+e^{-x}}$ and g_2 as a linear function $g_2(x) = x$. We also use $\epsilon = 0.1$ again.

Feeding the input values forwards through the network, we have

$$a_1^{(1)} = w_{11}^{(1)} x_1 + w_{12}^{(1)} x_2 = 1,$$

and

$$a_2^{(1)} = w_{21}^{(1)} x_1 + w_{22}^{(1)} x_2 = 0.5.$$

So,

$$z_1^{(1)} = \sigma(a_1^{(1)}) = 0.731,$$

(continued)

Example 9.3 (continued)
and

$$z_2^{(1)} = \sigma(a_2^{(1)}) = 0.622.$$

Continuing through to the next layer, we get

$$a_1^{(2)} = w_{11}^{(2)} z_1^{(1)} + w_{12}^{(2)} z_2^{(1)} = 0.338,$$

$$y_1 = a_1^{(2)} = 0.338,$$

$$a_2^{(2)} = w_{21}^{(2)} z_1^{(1)} + w_{22}^{(2)} z_2^{(1)} = 0.677,$$

and

$$y_2 = a_2^{(2)} = 0.677.$$

This gives

$$E = \frac{1}{2} \sum_{k=1}^{2} (t_k - y_k)^2 = 0.0288.$$

Now to feed back on the error gradient, we update weights in the second layer first. We use Eq. (9.13) to calculate the values of $\delta_k^{(2)}$. This is quite straightforward since the activation function here is linear, and so

$$g_2'(a_k^{(2)}) = 1.$$

So,

$$\delta_1^{(2)} = t_1 - y_1 = 0.162,$$

and

$$\delta_2^{(2)} = t_2 - y_2 = -0.177.$$

Then, using Eqs. (9.14) and (9.9), we can calculate new values for the weights. Note that Eq. (9.9) says we subtract the ϵ term, but (9.14) says that the gradient

(continued)

Example 9.3 (continued)
is negative, so we end up adding the update term to the old weight:

$$w_{11}^{(2)} = 0.25 + (0.1 \times 0.162 \times 0.731) = 0.262,$$

$$w_{12}^{(2)} = 0.25 + (0.1 \times 0.162 \times 0.622) = 0.260,$$

$$w_{21}^{(2)} = 0.5 + (0.1 \times -0.177 \times 0.731) = 0.487,$$

$$w_{22}^{(2)} = 0.5 + (0.1 \times -0.177 \times 0.622) = 0.489.$$

Now going back to the first layer, we have a sigmoid activation function. To calculate $\delta_j^{(1)}$, we need $g_1'(a_j^{(1)})$:

$$g_1'(a_1^{(1)}) = \sigma(a_1^{(1)})(1 - \sigma(a_1^{(1)})) = \sigma(1)(1 - \sigma(1)) = 0.197,$$

and

$$g_1'(a_2^{(1)}) = \sigma(a_2^{(1)})(1 - \sigma(a_2^{(1)})) = \sigma(0.5)(1 - \sigma(0.5)) = 0.235.$$

Hence, using Eqs. (9.19) and (9.20), we obtain

$$\delta_1^{(1)} = (0.197)((0.25)(0.162) + (0.5)(-0.177)) = -0.00946,$$

and

$$\delta_2^{(1)} = (0.235)((0.25)(0.162) + (0.5)(-0.177)) = -0.0113.$$

Finally, if we take Eq. (9.21), we can calculate new values for the weights:

$$w_{11}^{(1)} = 0.5 + (0.1 \times -0.00946 \times 1) = 0.499,$$

$$w_{12}^{(1)} = 0.5 + (0.1 \times -0.00946 \times 1) = 0.499,$$

$$w_{21}^{(1)} = 0.25 + (0.1 \times -0.0113 \times 1) = 0.249,$$

and

$$w_{22}^{(1)} = 0.25 + (0.1 \times -0.0113 \times 1) = 0.249.$$

(continued)

> **Example 9.3** (continued)
> So that has updated all the weights using back-propagation of errors. We will finally compute the new error. Now, we have
>
> $$a_1^{(1)} = 0.998 \text{ and } a_2^{(1)} = 0.498.$$
>
> So,
>
> $$z_1^{(1)} = \sigma(a_1^{(1)}) = 0.731 \text{ and } z_2^{(1)} = \sigma(a_2^{(1)}) = 0.622.$$
>
> Continuing through to the next layer, we get
>
> $$a_1^{(2)} = 0.353 \text{ and } a_2^{(2)} = 0.660.$$
>
> So,
>
> $$y_1 = a_1^{(2)} = 0.353 \text{ and } y_2 = a_2^{(2)} = 0.660.$$
>
> This gives
>
> $$E = \frac{1}{2} \sum_{k=1}^{2} (t_k - y_k)^2 = 0.0236$$
>
> Hence, after one iteration, y_1 is closer to t_1 and y_2 is closer to t_2, and the error E has reduced.

Remark 9.4 This sort of calculation would definitely be done using an appropriate computer program. It is dreadfully boring to do by hand! It was only done here so that you can check the use of the methods. It should also be noted that ϵ would normally be a smaller number than used in all these examples.

♦

> **Exercise**
>
> **9.3** Here is a two-layer neural network for you to try. This is easier than Example 9.3 since both layers use a linear activation function. This means that both $g_1(x) = x$ and $g_2(x) = x$. Also, we have $g_1'(a_j^{(1)}) = 1$ and $g_2'(a_k^{(2)}) = 1$. However, you will still definitely need a calculator!

(continued)

> Do one iteration for this two-layer neural network with both layers having a linear activation function:
> The initial input vector is $x_1 = 1$ and $x_2 = 0$; the target is $t_1 = 0.25$ and $t_2 = 0.5$.
> Initially, $w_{11}^{(1)} = w_{12}^{(1)} = 0.5$ and $w_{21}^{(1)} = w_{22}^{(1)} = 0.25$. Also $w_{11}^{(2)} = w_{12}^{(2)} = 0.25$, $w_{21}^{(2)} = w_{22}^{(2)} = 0.5$ and finally $\epsilon = 0.1$.
> This exercise, using linear activation functions, is partly relevant as you will see when you meet the rectified linear activation function in Sect. 9.6.

9.4 The Delta Rule

We may consider $\delta_k^{(l)}$ as a quantity measuring the error passing through each node in different layers. Equation (9.17) says that the error of each node is a weighted linear combination of errors from the layer on the immediate right. That is, $\delta_j^{(1)}$ is a linear sum of $\delta_k^{(2)}$. This is the principle of the back-propagation algorithm. In general, we have the Delta rule as follows:

$$\delta_j^{(l-1)} = g'(a_j^{(l-1)}) \sum_k w_{kj}^{(l)} \delta_k^{(l)}, \qquad (9.22)$$

where $g'(a_j^{(l-1)})$ is the derivative of the relevant activation function.

Figure 9.5 shows that the errors are propagated left from the output layer. That is, the weighted errors of $t_k - y_k$ are saved in $\delta_k^{(2)}$ using Eq. (9.13), which are further propagated left to the hidden units and saved as $\delta_j^{(1)}$ using Eq. (9.17). Therefore, the training of the neural network is iterative. In each iteration, weights are updated from right to left. Then, new errors are calculated using these updated weight values as the input is propagated from left to right.

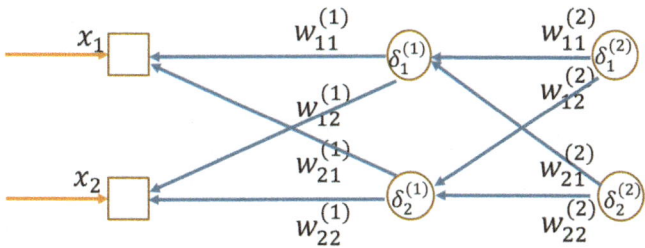

Fig. 9.5 An illustration of the back-propagation algorithm of a simple two-layer neural network

9.5 Implementation Details

9.5.1 Bias

Usually, an extra node with a value of 1 is considered in both the input layer and the hidden layers. The weight connecting from this extra node is called the bias (or threshold), which is denoted as w_0. We have not considered the bias in these simple examples.

To illustrate its use, consider the input to the first unit in a network with an n unit input vector. This would now be

$$w_{10} + x_1 w_{11} + x_2 w_{12} + \cdots + x_n w_{1n},$$

where w_{10} is the bias, instead of

$$x_1 w_{11} + x_2 w_{12} + \cdots + x_n w_{1n}.$$

without bias. So, the output of an activation function would now be

$$g(w_{10} + x_1 w_{11} + x_2 w_{12} + \cdots + x_n w_{1n}),$$

instead of

$$g(x_1 w_{11} + x_2 w_{12} + \cdots + x_n w_{1n}).$$

The net effect of the bias is to move the activation function sideways. As an illustration, let $x = x_1 w_{11} + x_2 w_{12} + \cdots + x_n w_{1n}$ and $w_{10} = 1$ or $w_{10} = -1$, and also suppose $g(x)$ is the logistic activation function. Figure 9.6 illustrates this in *two* dimensions and present the three corresponding function curves with bias w_{10} equal to $-1, 0$ and 1. It shows that involving a constant (1 or -1 in the figure) allows us to shift the function horizontally, left or right, along the x-axis. This means that with a bias, the most variable part of the curve (i.e., the part with the most slope) can be

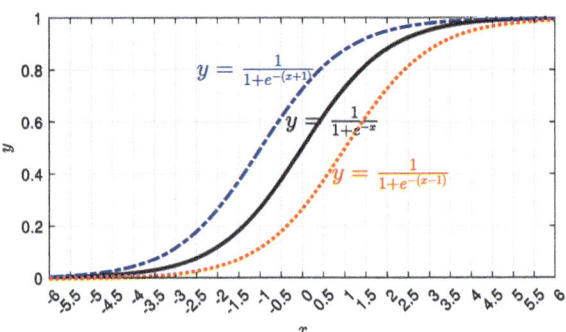

Fig. 9.6 An illustration of shifting a sigmoid function left or right along the x-axis

anywhere on the x-axis rather than always at $x = 0$. Also, with different biases on each unit, this allows each unit to exploit different parts of the activation function curve.

9.5.2 Stochastic Gradient Descent and Batch

9.5.2.1 Stochastic Gradient Descent

The stochastic gradient descent (SGD) calculates the error between the target and the prediction for each training example and then updates the weights immediately. This is the method we have illustrated in our discussion and all the examples. If we have N training examples, usually the SGD method updates weights N times at least.

9.5.2.2 Batch

Here, each weight is updated only once after calculating errors over all the training examples. So, we need to sum all the errors before updating any of the weights. This requires storing the sum of the errors as each training example is used, but it runs more quickly.

9.6 Deep Neural Networks

We have only dealt with one- and two-layer networks, but deeper networks are more useful, and recently, much deeper neural networks have become extremely fashionable. Here, we briefly mention a couple of factors that have made such really deep networks possible.

Improvements in computer technology, especially that of the graphics processing unit (GPU), mean that potentially deeper neural networks could be used without them taking forever to train, especially when they are used to analyse the vast amounts of data that are standard now. Deeper networks increase the ability of the network to learn more complicated relationships between inputs and outputs, so they have become increasingly more desirable. The use of such networks is often called deep learning.

Secondly, a purely linear activation function effectively can only deal with linear relationships, so it is desirable to have a non-linear activation function. However, the common non-linear functions of the logistic sigmoid or the hyperbolic tangent saturate at larger input values. So, they do not discriminate well or train well in these regions. They are most adaptive in the middle where the curve is sloping since at either end, they are effectively flat, as can be seen at either end of the curves in

Fig. 9.7 ReLU activation function

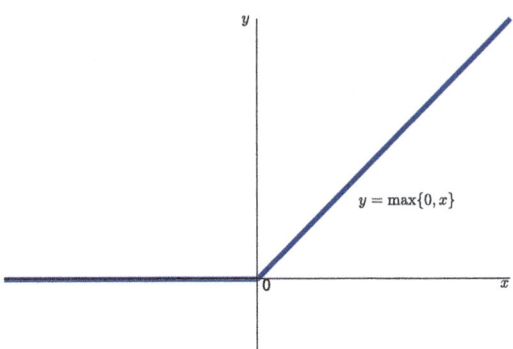

Fig. 9.6. So, these non-linear functions have negligible gradient once you get a little way away from the centre, and this means they have a limited ability to discriminate at these points. Together with the fact that in a multilayer network, these small errors are back-propagated a long way means the network does not train well in lots of circumstances. This problem is known as the vanishing gradient problem.

A different approach is needed, and recently the use of a relatively linear function, or piecewise linear function, has been tried and found to be exceptionally good at training and, as a bonus, is really quick to calculate. The main function used is illustrated in Fig. 9.7. It is in two linear parts; for $x \leq 0$, it is zero; for $x > 0$, it is the linear function $y = x$. It can be written as $y = \max\{0, x\}$. This function is non-linear since it acts differently with positive and negative values of x. Also, using biases, as described in Sect. 9.5.1, means that different units can use different parts of the activation function.

This relatively linear function, as shown in Fig. 9.7, is clearly not differentiable at $x = 0$, but, conveniently, it is continuous.

So, if the gradient at the point $x = 0$ is *defined* to be zero, then this function has a gradient at all points and has been found to work extremely well despite its native non-differentiability. This function is called a rectified linear activation function, and a unit with it as its activation function is called a rectified linear activation unit (ReLU). This sort of unit has become increasingly prevalent recently.

There are lots of other techniques that aid such deep neural networks to work effectively, especially in image processing, but these are outside the scope of this book.

Chapter 10
Probability

This chapter introduces the concept of probability, a way to deduce what is likely to happen when an *experiment* is performed. Probability is a value between zero and one. People also use other terms for probability: chance, percentage, likelihood, odds, or proportion. Usually, there are four ways to calculate the probability of an event, which are the following:

- The classical approach. This is a mathematical approach using counting rules. It is used on random processes with certain assumptions.
- The relative frequency approach. This is based on collecting data and finding the percentage of time that an event (E) occurred on that data.
- The subjective approach.
- The logical approach.

This book uses the first two approaches.

This chapter and the following two chapters develop enhancements to the basic algorithms developed so far, especially that of Chap. 8 on simple linear regression. Chapter 13 will complete this task by introducing the method of maximum likelihood.

The techniques in these three chapters also deal with statistical analysis and probabilistic measures of confidence associated with any scientific discipline that involves vast amounts of noisy real data.

10.1 Preliminary Knowledge: Combinatorial Analysis

10.1.1 Factorial Notation

The notation $n!$, read "n factorial", denotes the product of the positive integers from 1 to n (without repetition), inclusive:

$$n! = n \times (n-1) \times \cdots \times 2 \times 1. \tag{10.1}$$

Note that for completion, $1!$ and $0!$ are defined as

$$1! = 1,$$

and

$$0! = 1.$$

So,

$$n! = n \times (n-1)!.$$

10.1.2 Binomial Coefficients

The symbol $\binom{n}{r}$, read "en-see-are", is defined by

$$\binom{n}{r} = \frac{n!}{r!(n-r)!}, \tag{10.2}$$

where r and n are positive integers with $r \leq n$.
Since

$$\frac{n!}{(n-r)!} = \frac{n(n-1)\cdots(n-(r-1))(n-r)(n-(r+1))\cdots 3 \cdot 2 \cdot 1}{(n-r)(n-(r+1))\cdots 3 \cdot 2 \cdot 1}$$
$$= n(n-1)\cdots(n-(r-1)),$$

we have

$$\frac{n!}{r!(n-r)!} = \frac{n(n-1)\cdots(n-(r-1))}{r(r-1)\cdots 3 \cdot 2 \cdot 1}.$$

10.1 Preliminary Knowledge: Combinatorial Analysis

Table 10.1 Pascal's Triangle

$n=0$				1				
$n=1$			1		1			
$n=2$		1		2		1		
$n=3$	1		3		3		1	
$n=4$	1	4		6		4		1
$n=5$	1	5		10		10	5	1

The numbers $\binom{n}{r}$ are called the binomial coefficients since they appear as the coefficients in the expansion of $(a+b)^n$. That is,

$$(a+b)^n = \sum_{r=0}^{n} \binom{n}{r} a^{n-r} b^r.$$

Remark 10.1 Expanding $(a+b)^2$, we get $1a^2 + 2ab + 1b^2$, so the binomial coefficients are 1, 2, and 1. These are equal to $\binom{2}{0}$, $\binom{2}{1}$, and $\binom{2}{2}$, respectively.

Similarly, when expanding $(a+b)^3$, we get the binomial coefficients 1, 3, 3, and 1, being $\binom{3}{0}$, $\binom{3}{1}$, $\binom{3}{2}$, and $\binom{3}{3}$, respectively.

This pattern of binomial coefficients can be extended, and we get a structure known as Pascal's triangle for the coefficients. This gives a quick way to calculate the coefficients. All rows start with the number 1. We can add each consecutive pair of elements of each row and write their sum in the gap between them, but on the line below, to get the elements in the next row. See Table 10.1 for a diagram of Pascal's triangle.

♦

10.1.3 Permutation and Combination

10.1.3.1 Permutations of *n* Items Without Repetitions

There are n ways of picking the first item; then there are $n-1$ ways of picking the second item since we cannot have repetitions, and so on. Hence, there are $n!$ permutations of n objects.

Example 10.1 There are $3! = 3 \cdot 2 \cdot 1 = 6$ permutations of the three letters s, t and u, namely, stu, sut, tsu, tus, ust, and uts.

In permutations, the order of the items matters. So, in Example 10.1 stu is different from sut, and so on. Think of a password or key to unlock a phone, the order of the letters or digits is significant, and each permutation is a different password or key.

10.1.3.2 Permutations of k Items Out of n Without Repetition

The number of permutations of n objects taken k at a time where the order matters is denoted as $P(n, k)$ and is computed as

$$P(n, k) = \frac{n!}{(n - k)!} = n(n - 1) \cdots (n - (k - 1)).$$

Example 10.2 There are five numbers, 3, 4, 5, 6, and 7. The number of two digit numbers we can form by taking any two numbers from these five is

$$P(5, 2) = \frac{5!}{(5 - 2)!} = 5 \times (5 - 1) = 20,$$

which are 34, 35, 36, 37, 43, 45, 46, 47, 53, 54, 56, 57, 63, 64, 65, 67, 73, 74, 75, and 76.

Again, the order matters, so 34 is different from 43. Again, for a password, picking 8 out of 26 letters without repetition would give different passwords for each order. These are *permutations*. In this case, there are $\frac{26!}{(26-8)!}$ different passwords: a large number.

10.1.3.3 Combinations of k Items Out of n Without Repetitions

The number of combinations of n objects taken k at a time is denoted as $C(n, k)$ and is computed as

$$C(n, k) = \frac{n!}{k!(n - k)!}. \tag{10.3}$$

Here, the order does not count. Think of mixing colour lights—red and green give yellow, as do green and red. Combinations are just collections of items, and there are many more permutations than combinations since items in a different order would be a new permutation but not a new combination. Hence, when the order does not matter, or where the items are not picked in order, these are *combinations*. When doing an exercise, the first question to ask is whether changing the order gives a different answer. If the answer is yes, you want permutations; if no, then it is combinations.

Remark 10.2 Looking at Eqs. (10.3) and (10.2), we can see that using binomial coefficients is another notation for combinations. Thus,

$$C(n,k) = \binom{n}{k} = \frac{n!}{k!(n-k)!}.$$

♦

Example 10.3 Continue Example 10.2. Compute the number of combinations of two numbers if we take any two from those five numbers.

Solution The number of combinations is calculated as

$$C(5,2) = \frac{5!}{2!(5-2)!} = \frac{5 \times 4! \times 3!}{2!3!} = 10.$$

When counting the combinations, the order does not matter. For example, 43 and 34 are the same combinations.

Example 10.4 Compute the number of combinations of the letters s, t, u, and v taken three at a time.

Solution The number of combinations is calculated as

$$C(4,3) = \frac{4!}{3!(4-3)!} = \frac{4 \times 3!}{3!1!} = 4.$$

Four combinations are stu, stv, suv, and tuv. The following combinations are equal: stu, sut, tsu, tus, ust, and uts, since they are the combinations of three letters, s, t, u, and the order does not matter.

Exercises

10.1 (a) Given the letters in the word Wales, how many different five letter strings of letters, without any repetitions, can you make? (b) Given the letters in the word Scotland, how many different eight letter strings of letters, without any repetitions, can you make?

(continued)

10.2 Mary wants to take a photo with her four friends, and they all stand in one row. If Mary must stand in the middle, how many different photos can they take?

10.3 How many different passwords can you make by using *four* digits from the *ten* digits on your mobile phone screen, where you are not allowed to repeat any digits?

10.4 There are six numbers, 0, 3, 6, 7, 8, and 9. How many different five-digit numbers can be made where 0 cannot be the first digit?

10.5 How many different fruit salads can you make using four different fruits from the list: apple, orange, pear, banana, grapefruit, pineapple, and grapes?

10.6 There are 12 dots in a plane. No three dots are in the same line. How many triangles can be made?

10.2 Probability

10.2.1 Axiomatic Probability Theory

The probability of some event E occurring is the likelihood of that event happening. Initial illustrations usually consider flipping a coin (or coins) or rolling a die (or several dice). This is because there are an easily calculated set of possible outcomes. For a coin, there are just two possible faces, and for a die, we can only get one of six faces (ignoring landing on an edge or corner or disappearing under the sideboard—these are unstable or silly outcomes!). This makes it easy to calculate the likelihood of a particular number being thrown on a die or a particular side of a coin coming up. This can be codified into an axiomatic theory as follows.

Let (Ω, Σ, P) denote a probability space, where

- Ω is the set of all possible outcomes, known as the sample space.

Example 10.5 A fair six-sided die is rolled. The number of dots on each side is from one to six. We use $\Omega = \{1, 2, 3, 4, 5, 6\}$ to denote all possible numbers of dots on the top of the side after a roll.

10.2 Probability

- Σ is a collection of subsets of Ω, and each subset is called an event.

Example 10.6 An event (E) could show an odd number when the fair six-sided die lands. That is, $E = \{1, 3, 5\}$, and $E \subset \Omega$.

- P is a probability measure defined as a real-valued function of the elements of Σ satisfying the following axioms of probability:
  - Axiom 1: $0 \leq P(A) \leq 1$ for all $A \in \Sigma$.
  - Axiom 2: $P(\Omega) = 1$.
  - Axiom 3: If two events A and B are mutually exclusive, that is, no elements in common, then the probability of either A or B occurring is the probability of A occurring plus the probability of B occurring:

$$P(A \cup B) = P(A) + P(B).$$

Note: set union is defined in Sect. 2.1.3 of Chap. 2

If E is an event, $P(E)$ is the probability of the occurrence of the event, that is,

$$P(E) = \frac{\text{The number of elements in event E}}{\text{The size of the sample space}}. \qquad (10.4)$$

Note that the maximum probability of any event is one.

Example 10.7 Let us roll a fair six-sided die. The number of dots on each side is from one to six. Let A be the event showing an even number when it lands, B showing either three or five dots, and C showing a prime number. Calculate the following probabilities:
- $P(A \cup B)$.
- $P(A \cap C)$.

Set union and set intersection are defined in Sect. 2.1.3 of Chap. 2.

Solution Since

$$\Omega = \{1, 2, 3, 4, 5, 6\}, A = \{2, 4, 6\}, B = \{3, 5\}, \text{ and } C = \{2, 3, 5\},$$

(continued)

Example 10.7 (continued)
we have

$$A \cup B = \{2, 3, 4, 5, 6\},$$

and

$$A \cap C = \{2\}.$$

Therefore,

$$P(A \cup B) = \frac{\#(A \cup B)}{\#\Omega} = \frac{5}{6},$$

and

$$P(A \cap C) = \frac{\#(A \cap C)}{\#\Omega} = \frac{1}{6}.$$

Recall that # denotes the cardinality of a finite set (see Sect. 2.1.1 of Chap. 2).

Example 10.8 Suppose we roll two fair six-sided dice. What is the probability of getting two even numbers? Now, roll three fair six-sided dice. What is the probability of getting three even numbers?

Solution All possible outcomes are listed in Table 10.2, where elements in the event of getting two even numbers are shown in red. The sample space size is 36 since there are 36 possibilities, as shown in Table 10.2, and the number of elements in the event of getting two even numbers is 9. Therefore, P(getting two even numbers) $= \frac{9}{36} = 0.25$, or 25%.

It gets harder to count the events once we get to three dice. So let us look at the method again for two dice and then extend it to three. With two dice, there are six ways the first dice could fall and six for the second. So, there are $6 \times 6 = 36$ different possible dice rolls. Similarly, to get two even numbers, there are just three possible dice rolls (one of $\{2, 4, 6\}$) for the first dice and three for the second dice. So, there are $3 \times 3 = 9$ ways to get two even numbers. So, as before, we get $\frac{9}{36} = 0.25$ as the probability.

For three dice, we can see that there are $6 \times 6 \times 6 = 216$ possible dice rolls. To get three even numbers again, there are $3 \times 3 \times 3 = 27$ ways to do that. So, the probability is $\frac{27}{216} = \frac{1}{8} = 0.125$, or 12.5%.

10.2 Probability

Table 10.2 Results of rolling two fair six-sided dice: elements getting two even numbers are shown in red

	1	2	3	4	5	6
1	(1,1)	(1,2)	(1,3)	(1,4)	(1,5)	(1,6)
2	(2,1)	(2,2)	(2,3)	(2,4)	(2,5)	(2,6)
3	(3,1)	(3,2)	(3,3)	(3,4)	(3,5)	(3,6)
4	(4,1)	(4,2)	(4,3)	(4,4)	(4,5)	(4,6)
5	(5,1)	(5,2)	(5,3)	(5,4)	(5,5)	(5,6)
6	(6,1)	(6,2)	(6,3)	(6,4)	(6,5)	(6,6)

Example 10.9 Suppose we roll a fair six-sided die twice. What is the probability of getting two different numbers? And then, if you roll it again, what is the probability of getting three different numbers?

Solution We are using the principle that we need to find the number of elements in the event and divide by the total number of all the possibilities (see Eq. (10.4)). When rolling twice, you can get $6 \times 6 = 36$ different outcomes (as shown in Table 10.2).

To get two different numbers, the first number can be any digit, but the second can be only *one* of *five* different numbers. So, there are $6 \times 5 = 30$ different possibilities. In Table 10.2, this is all the outcomes apart from those on the main diagonal. Therefore, P (getting two different numbers) $= \frac{30}{36} = \frac{5}{6}$, or 83.3%.

Now, doing the same for three rolls, we have $6 \times 6 \times 6 = 216$ different rolls. But getting three different numbers requires *six* choices for the first roll, *five* for the second, and *four* for the third roll. This gives $6 \times 5 \times 4 = 120$ elements in this event. Therefore, P (getting three different numbers) $= \frac{120}{216} = \frac{5}{9}$, or 55.5%.

Note that getting three different numbers is, in fact, the number of permutations of picking *three* out of *six* items without repetition. That is, it is $P(n, k) = \frac{6!}{(6-3)!} = \frac{6!}{3!} = \frac{720}{6} = 120$. It is permutations rather than combinations because you are rolling the dice in order—so a 1 followed by a 2 is different from a 2 followed by a 1.

Exercises

10.7 A coin is flipped three times in succession. What is the probability of getting exactly two heads?

(continued)

10.8 Suppose we roll a fair six-sided die four times. What is the probability of getting four different numbers? (You might like to try rolling five times with five different numbers and rolling six times with six different numbers)

10.9 Telephone numbers include six digits from 0, 1, 2, $\cdots$, 8, and 9. What is the probability that all six digits are different if we randomly select a telephone number?

10.3 Discrete Random Variables

This section and the next define random variables. There are two types of random variables: discrete, dealt with in this section, and continuous, dealt with in the next section. A random variable is discrete if it can only take one of a countable set of values, for example, an integer value, such as the number of aces in a standard pack of cards or the number of people at a football match. A random variable is continuous if it can take an infinite number of different values, for example, a real number value. Most continuous random variables are measurements, for instance, a person's weight or height.

Any random variable is a map from the outcome space (Ω) to the real numbers. It is the result of some outcome of a random experiment.

For example, let us consider throwing two fair dice. The sample space Ω includes all those pairs (ω's) of numbers listed in Table 10.2. Suppose we are interested in the total value obtained. Then, we can write $X((1, 1)) = 2$, $X((2, 3)) = 5$, and so on, for all ω's, where $\omega \in \Omega$ and X is the random variable, mapping each ω to the sum of two dice values.

If X is a random variable and x, x_1, and x_2 are fixed real numbers, we may have the following events:

$$(X = x), \ (X \leq x), \ (X > x) \text{ or } (x_1 < X \leq x_2).$$

These events have probabilities that are denoted by

$$P(X = x), \ P(X \leq x), \ P(X > x) \text{ or } P(x_1 < X \leq x_2).$$

Definition 10.1 (Discrete Random Variables) A random variable X is called discrete if it only takes values in the integers or (possibly) some other countable set of real numbers.

10.3 Discrete Random Variables

Example 10.10 The idea of so-called intelligence tests is to ask each individual to try to solve a certain number of questions. Each question can be solved either correctly or incorrectly. Usually, the number of correct answers is used as an empirical measure of individual intelligence. Such a number or score is a discrete random variable.

The probability mass function, $f_X(x_k)$, is defined as the probability that X takes on a certain value x_k:

$$f_X(x_k) = P(X = x_k). \tag{10.5}$$

The cumulative distribution function, $F_X(x)$, is defined as the probability that the random variable, X, will take on a value that is lesser than or equal to a particular value, x. This is defined as follows:

$$F_X(x) = P(X \leq x) = \sum_{x_k \leq x} f_X(x_k). \tag{10.6}$$

$F_X(x)$ is a staircase function.

Example 10.11 Suppose we flip a fair coin three times. Let X be the number of heads in three tosses of the coin. Find the probability mass function and cumulative distribution function.

Solution The sample space is

$$\Omega = \{TTT, HTT, THT, TTH, HHT, HTH, THH, HHH\}.$$

x may be any fixed real numbers, though X may take values of 0, 1, 2, and 3 only. So, X is a discrete random variable. From Ω, we can see that the probability of getting 0 heads is $\frac{1}{8}$ since it occurs once in the eight possible outcomes. Similarly, by counting the elements in Ω, we can see that the probability of getting *one* head is $\frac{3}{8}$, of getting *two* heads is $\frac{3}{8}$, and of getting *three* heads is $\frac{1}{8}$. So, the probability mass function is a histogram as illustrated in Fig. 10.1.

(continued)

Example 10.11 (continued)

Considering the cumulative distribution function as given in Eq. (10.6), we can set up Table 10.3 to present the cumulative distribution, $F_X(x)$, and we can illustrate the cumulative distribution as shown in Fig. 10.2. Please note that the x-axis and y-axis scales are not consistent across Figs. 10.1 and 10.2. This was done to better illustrate specific features in each figure.

Table 10.3 is built up as follows. $F_X(-1) = 0$, since the number of elements where the number of heads observed is less than 0 is zero. $F_X(0) = \frac{1}{8}$, since there is only one way of getting zero heads out of the *eight* possible flips. $F_X(1) = \frac{4}{8}$ since there is one way of getting zero heads together with *three* ways of getting *one* head out of the *eight* possible flips. The table shows the rest of the values, and once we get to x being greater than 3, then the number of elements with more than *three* heads is zero, so $F_X(4)$ remains at 1.

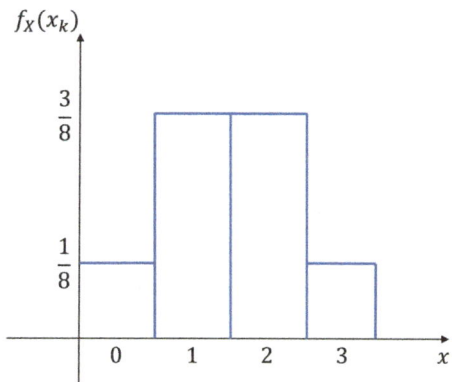

Fig. 10.1 Probability mass function of X, where X represents the number of heads in three tosses of a fair coin

Table 10.3 The cumulative distribution function of X, where X represents the number of heads in three tosses of a fair coin

x	Event ($X \leq x$)	$F_X(x)$
-1	$\emptyset$	0
0	{TTT}	$\frac{1}{8}$
1	{TTT, HTT, THT, TTH}	$\frac{4}{8}$
2	{TTT, HTT, THT, TTH, HHT, HTH, THH}	$\frac{7}{8}$
3	{TTT, HTT, THT, TTH, HHT, HTH, THH, HHH}	1
4	{TTT, HTT, THT, TTH, HHT, HTH, THH, HHH}	1

10.4 Continuous Random Variables

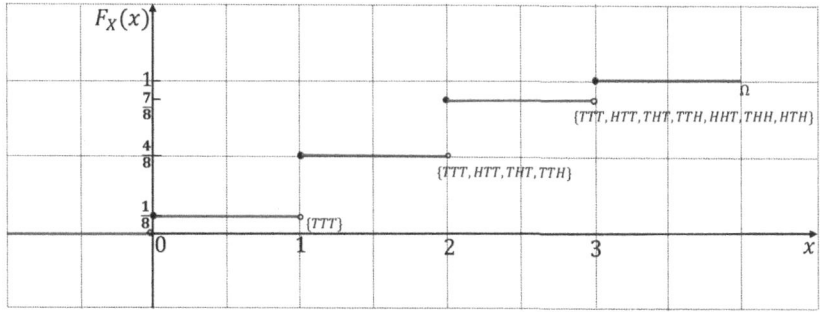

Fig. 10.2 The cumulative distribution function of X, where X represents the number of heads in three tosses of a fair coin

Exercises

10.10 If you throw two fair dice, we can define a discrete random number X as the total value obtained by the two dice. So, X takes values from $X((1, 1)) = 2$ up to $X((6, 6)) = 12$. With the help of Table 10.2, draw up a table of the probabilities of getting a total value from 2 to 12 after throwing two fair dice. Sketch the probability mass function and the cumulative distribution function.

10.11 Suppose we flip a fair coin four times. Let X be the number of heads in four tosses of the coin. Find the probability mass function and cumulative distribution function.

10.4 Continuous Random Variables

Definition 10.2 (Continuous Random Variables) A random variable X is called continuous if it takes values from a real-valued interval, either open or closed. That is, it can take one of an infinite number of values.

Example 10.12 The average incandescent bulb light span is approximately 1000 hours. The light span of incandescent bulbs can be considered as a continuous variable. This is because it has a lifespan that is not a whole number of hours, minutes, or seconds—the bulb could go at any point in time.

Similar to a discrete random variable, we can define a probability density function and a cumulative distribution function for a continuous random variable. However, the sum of the possible values of such functions is now calculated using an integral since it is a continuous (or piecewise continuous) curve.

So, if X is a continuous random variable, then there is a real-valued function, f_X, called the probability density function of X, which is a curve, and it satisfies the following:

- f_X is piecewise continuous. That is, the function is continuous except at finitely many points.
- $f_X(x) \geq 0$. That is, it does not go below the horizontal axis.
- $\int_{-\infty}^{\infty} f_X(x)dx = 1$. That is, the total area under the curve is 1.

Figure 10.3 shows a possible probability density function for the lifespan of a certain type of incandescent bulb produced from the same manufacturing plant. The area under the curve $\int_{-\infty}^{\infty} f_X(x)dx$ is the total probability of all lifespans and equals 1.

The cumulative distribution function is again defined as the probability that the random variable, X, will take on a value that is less than or equal to a particular value, x. So, the cumulative distribution function (cdf), $F_X(x)$, is a nondecreasing and continuous function and is defined as

$$F_X(x) = P(X \leq x) = \int_{-\infty}^{x} f_X(t)dt. \tag{10.7}$$

Referring again to Fig. 10.3, $F_X(a)$ is the cumulative probability of a lifespan less than or equal to a. This is represented as the shaded region in Fig. 10.3.

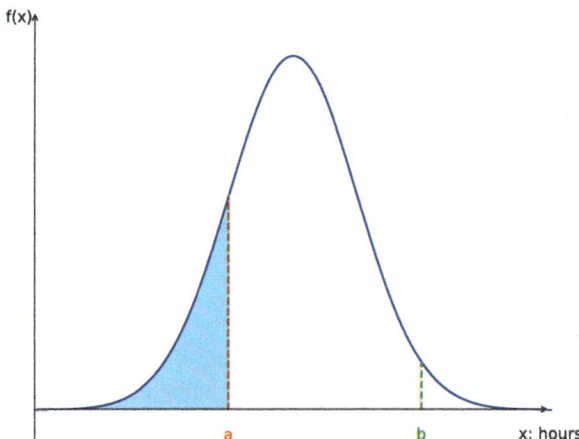

Fig. 10.3 Probability density function representing the lifespan of a certain type of incandescent bulb produced by a specific manufacturing plant

10.4 Continuous Random Variables

Based on Eq. (10.7), we have the following:

- $P(a < X \leq b) = F_X(b) - F_X(a)$.
  Recall that the second fundamental theorem of calculus (see Sect. 5.4.3 of Chap. 5) states that if f is continuous on $[a, b]$, and F is any antiderivative of f with respect to x, then $\int_a^b f_X(x)dx = F_X(b) - F_X(a)$. For example, this could be the area under the curve illustrated in Fig. 10.3 between lifespan values of a and b.
- $P(X > a) = 1 - F_X(a)$.

  Proof Since $P(X > a) = 1 - P(X \leq a)$, from Eq. (10.7), we have $P(X \leq a) = F_X(a)$. Therefore, $P(X > a) = 1 - F_X(a)$. □

In Fig. 10.3, $P(X > a)$ is the non-shaded region under the curve above the point a on the horizontal axis.

Example 10.13 Suppose the probability density function $f(x)$ of X is given by the following:

$$f(x) = \begin{cases} \frac{\cos(x)}{W}, & \text{if } |x| < \frac{\pi}{2}; \\ 0, & \text{otherwise.} \end{cases}$$

1. Find the value of W.
2. Find the cumulative distribution function $F_X(x)$.
3. - Find $P(X \leq 0)$.
   - Find $P(X \leq \frac{\pi}{4})$.
   - Find $P(X > \frac{\pi}{4})$.
   - Find $P(X > \frac{\pi}{2})$.

Solution

1. We use the fact that $\int_{-\infty}^{\infty} f_X(x)dx = 1$ to find W.

$$\int_{-\infty}^{\infty} f_X(x)dx = \int_{-\frac{\pi}{2}}^{\frac{\pi}{2}} \frac{\cos(x)}{W} dx = \frac{\sin(x)}{W} \Big|_{-\frac{\pi}{2}}^{\frac{\pi}{2}} = \frac{2}{W} = 1, \Rightarrow W = 2.$$

Figure 10.4 illustrates the probability density function curve.

2. Applying Eq. (10.7),

$$F_X(x) = \int_{-\infty}^{x} f_X(t)dt,$$

we have the following:

- If $x < -\frac{\pi}{2}$, $F_X(x) = \int_{-\infty}^{x} 0 dx = 0$.

(continued)

Example 10.13 (continued)

- If $-\frac{\pi}{2} \leq x < \frac{\pi}{2}$, $F_X(x) = \int_{-\frac{\pi}{2}}^{x} \frac{\cos(x)}{2} dx = \frac{1}{2} + \frac{1}{2}\sin(x)$.
- If $x \geq \frac{\pi}{2}$, $F_X(x) = 1$.

3. - $P(X \leq 0)$: Using the result for $F_X(x)$ given above with $x = 0$ gives $P(X \leq 0) = F_X(0) = \frac{1}{2} + 0 = \frac{1}{2}$.

    This makes sense since from the symmetry of the curve for the probability density function, as shown in Fig. 10.4, the cumulative probability up to $x = 0$ is exactly half the area under the curve.
   - $P(X \leq \frac{\pi}{4})$: $P(X \leq \frac{\pi}{4}) = F_X(\frac{\pi}{4}) = \frac{1}{2} + \frac{1}{2 \times \sqrt{2}} = 0.8536$.
   - $P(X > \frac{\pi}{4}) = 1 - P(X \leq \frac{\pi}{4}) = 1 - 0.8536 = 0.1464$.
   - $P(X > \frac{\pi}{2}) = 1 - P(X \leq \frac{\pi}{2}) = 1 - 1 = 0$, as expected.

Remark 10.3 If X is a continuous random variable, then the probability of X taking a specific value C is zero, that is, $P(X = C) = 0$.

Proof Suppose Δx is a tiny increase of C. We have

$$P(C < x \leq C + \Delta x) = F_X(C + \Delta x) - F_X(C) = \int_C^{C+\Delta x} f(x)dx,$$

and

$$0 \leq P(X = C) \leq P(C < X \leq C + \Delta x).$$

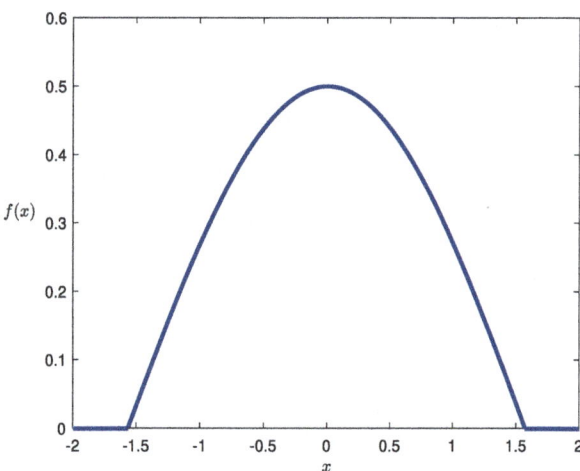

Fig. 10.4 The probability density function of X in Example 10.13

10.4 Continuous Random Variables

Since $F_X(x)$ is continuous from the right, when Δx approaches zero, we have

$$\lim_{\Delta x \to 0^+} \int_C^{C+\Delta x} f(x)dx = 0.$$

That gives us the following:

$$0 \le P(X = C) \le 0.$$

Therefore, we have $P(X = C) = 0$. □

An event $E = \{X = C\}$ may happen even though $P(X = C) = 0$. For example, the lifespan of incandescent bulbs is approximately from 800 hours to 1200 hours. The exact event $E = \{X = 900 \text{ hours}\}$ may occur, but $P(X = 900 \text{ hours}) = 0$. That is, the probability of getting an exact value from an infinity of possible answers must be zero or else the sum of all the probabilities of all the exact values would add up to infinity and not 1 as is required. So, the probability of getting exactly 900 hours is zero, though we might get 900 hours to the best of our ability to measure the time.

♦

Exercises

10.12 The probability density function of the random variable X is

$$f(x) = \begin{cases} a(2x - x^2), & 0 < x < 2, \\ 0, & \text{otherwise.} \end{cases}$$

Compute the following:

1. The value of a
2. The cumulative distribution function $F_X(x)$.
3. $P(X \le 1)$
4. $P(X \le \frac{1}{2})$
5. $P(X > \frac{1}{2})$

10.13 The probability density function of the random variable X is

$$f(x) = \begin{cases} 8x^7, & 0 < x < 1, \\ 0, & \text{otherwise.} \end{cases}$$

(continued)

Compute the following:

1. m so that $P(X > m) = P(X < m)$
2. n so that $P(X > n) = 0.05$

10.5 Mean and Variance of Probability Distributions

If we know the probability distribution of a random variable, we will know and be able to describe the properties of the random variable. However, obtaining an accurate probability distribution in real-world applications is hard. Moreover, we often only need to know some properties of a random variable but not all, such as the centre value, the value the random variable is most likely to take, and the correlation between two random variables. These properties can be parameters of probability distributions. This section introduces the two most important parameters: mean (or expected value) and variance.

10.5.1 Mean

Definition 10.3 (Mean) The mean (or expected value) of a random variable X, denoted by μ_X, or $E(X)$, is defined by

$$\mu_X = E(X) = \begin{cases} \sum_k x_k f_X(x_k), & X : \text{discrete}; \\ \int_{-\infty}^{\infty} x f_X(x) dx, & X : \text{continuous}. \end{cases} \quad (10.8)$$

The expected value should be regarded as the average (mean) value. If you compare this with the definition of the sample mean given in Sect. 4.2.1 of Chap. 4, you can see that this is a weighted mean, weighted by the probability, as will be illustrated in Example 10.14. Note that we may also denote the expected value as $\mu_X = E[X]$ in this book.

Example 10.14 Three products are selected randomly from nine products, of which two are defective. The sample space consists of the distinct, equally likely, samples of size 3. Let X be the random variable that counts the number of defective items in a sample. The possible values of X are 0, 1, and 2. What is the expected value of defective products in a sample of size 3?

(continued)

Example 10.14 (continued)
Solution

- Let $x_i = 0, 1, 2$ be the possible values of X.
- The number of ways of choosing x_i defectives from two defectives and choosing $3 - x_i$ non-defectives from seven non-defectives is $\binom{2}{x_i}$ and $\binom{7}{3-x_i}$, respectively. (Remember that $\binom{n}{r}$ is a notation for the number of combinations of n object taken r at a time.)
- The total number of possible outcomes (i.e., the number of combinations of three products out of) is $\binom{9}{3} = 84$.
- The probability of the value x_i of X is

$$p_i = \frac{\binom{2}{x_i}\binom{7}{3-x_i}}{\binom{9}{3}}, \quad (x_i = 0, 1, 2).$$

- Applying Eq. (10.8) for the discrete variable, we have

$$E(X) = 0 \times \frac{\binom{2}{0}\binom{7}{3}}{84} + 1 \times \frac{\binom{2}{1}\binom{7}{2}}{84} + 2 \times \frac{\binom{2}{2}\binom{7}{1}}{84}$$

$$= 0 \times \frac{1 \times 35}{84} + 1 \times \frac{2 \times 21}{84} + 2 \times \frac{1 \times 7}{84} = \frac{2}{3}.$$

Note that the three values above 84 in the last line above add up to the total of 84 combinations as expected. When each is divided by 84, these represent the probabilities of getting the three different outcomes. This means we have 0 defective products occurring in 35 combinations, 1 defective product in 42 combinations, and 2 defective products in 7 combinations. So, we have a list of numbers consisting of 35 0s, 42 1s, and 7 2s. Now, just think of these as a list of 84 numbers. The usual mean of such a list is to add up the numbers, that is, $35 \times 0 + 42 \times 1 + 7 \times 2 = 56$, and divide by the total numbers in the list, that is, 84. This division gives $\frac{2}{3}$.

This shows that this definition of mean corresponds with our previous definition given in Sect. 4.2.1 of Chap. 4.

Example 10.15 Let X be a random variable. Consider its distribution function on the interval [0, 1] has the probability density function:

$$f_X(x) = \begin{cases} 0, & \text{if } x < 0 \text{ or } x > 1; \\ 1, & \text{if } 0 \leq x \leq 1. \end{cases}$$

Compute $E(X)$.

Solution Applying Eq. (10.8) for the continuous variable, we have

$$E(X) = \int_{-\infty}^{\infty} x f_X(x) dx$$

$$= \int_{-\infty}^{0} x \times 0 dx + \int_{0}^{1} x \times 1 dx + \int_{1}^{\infty} x \times 0 dx$$

$$= 0 + \frac{1}{2}x^2 \Big|_{x=0}^{1} + 0$$

$$= \frac{1}{2}.$$

If you look at the probability density function, you can see it is a square of height one and width one, as shown in Fig. 10.5. Not surprisingly, the mean value is in the middle of the x-values at $x = \frac{1}{2}$.

Fig. 10.5 The probability density function shown in Example 10.15

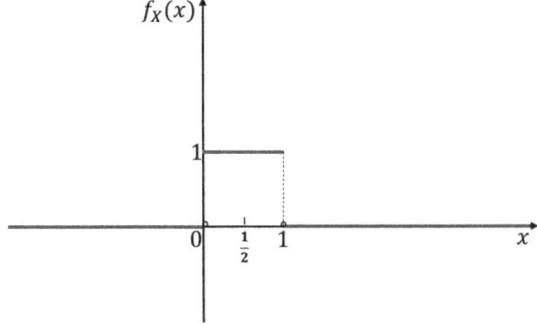

10.5 Mean and Variance of Probability Distributions

Table 10.4 Probability distributions of the random variables X_1 and X_2

X_1	1	2	X_2	1	2
P	$\frac{1}{3}$	$\frac{2}{3}$	P	$\frac{1}{2}$	$\frac{1}{2}$

10.5.1.1 Properties of Mean

To motivate the statement of these properties, we will do a simple example that illustrates them.

Example 10.16 Table 10.4 shows the probability distributions of two independent random variables X_1 and X_2,[a] both of which only take the values 1 and 2. Compute the following:

1. $E(X_1)$
2. $E(X_2)$
3. $E(aX_1)$, where a is a constant
4. $E(a + X_1)$, where a is a constant
5. $E(X_1 + X_2)$
6. $E(X_1 X_2)$

Solution

1. Applying Eq. (10.8) for the discrete variable, we have

$$E(X_1) = 1 \times \frac{1}{3} + 2 \times \frac{2}{3} = \frac{5}{3}.$$

2. Applying Eq. (10.8) for the discrete variable, we have

$$E(X_2) = 1 \times \frac{1}{2} + 2 \times \frac{1}{2} = \frac{3}{2}.$$

3. aX_1 takes the two values a and $2a$ with the probabilities as given in Table 10.4. So again applying Eq. (10.8) for the discrete variable, we have

$$E(aX_1) = 1a \times \frac{1}{3} + 2a \times \frac{2}{3} = a\frac{5}{3} = aE(X_1).$$

4. $a + X_1$ takes the two values $a + 1$ and $a + 2$ with the probabilities as given in Table 10.4. So again applying Eq. (10.8) for the discrete variable, we have

$$E(a + X_1) = (1 + a) \times \frac{1}{3} + (2 + a) \times \frac{2}{3} = \frac{5}{3} + a = E(X_1) + a.$$

(continued)

Example 10.16 (continued)
5. $X_1 + X_2$ takes just the values that are the sum of the ones in X_1 and X_2, namely, 2, 3, and 4. To get a 2, you must have a 1 in both X_1 and X_2, so that has probability $\frac{1}{3} \times \frac{1}{2} = \frac{1}{6}$. To get a 4, both have to be 2, so the probability is $\frac{2}{3} \times \frac{1}{2} = \frac{1}{3}$. To get a 3, we can have $X_1 = 1$ and $X_2 = 2$ with probability $\frac{1}{3} \times \frac{1}{2} = \frac{1}{6}$ or have $X_1 = 2$ and $X_2 = 1$ with probability $\frac{2}{3} \times \frac{1}{2} = \frac{1}{3}$. So, the total probability of a 3 is $\frac{1}{6} + \frac{1}{3} = \frac{1}{2}$.
We now have our probability distribution (2 with probability $\frac{1}{6}$, 3 with probability $\frac{1}{2}$, and 4 with probability $\frac{1}{3}$) and note that the probabilities add up to 1 as required.
Finally, we apply Eq. (10.8) for the discrete variable:

$$E(X_1 + X_2) = 2 \times \frac{1}{6} + 3 \times \frac{1}{2} + 4 \times \frac{1}{3} = \frac{19}{6}.$$

Note that this is the same as $E(X_1) + E(X_2)$.

6. This is similar to the last one. Now, $X_1 X_2$ can take values of 1, 2, and 4. We get a probability distribution by the same method as before to get a probability of 1 is $\frac{1}{6}$, a probability of 2 is $\frac{1}{2}$, and a probability of 4 is $\frac{1}{3}$. So, we apply Eq. (10.8) for the discrete variable:

$$E(X_1 X_2) = 1 \times \frac{1}{6} + 2 \times \frac{1}{2} + 4 \times \frac{1}{3} = \frac{5}{2}.$$

Note that this is the same as $E(X_1) E(X_2)$.

---

[a] Intuitively, two random variables X_1 and X_2 are independent if knowing the value of one of them does not change the probabilities for the other one. We will define independent random variables formally in Chap. 11.

We can now generalise these results and state the properties that apply to both discrete and continuous random variables:

(1) $E(a) = a$, where a is a constant.
(2) $E(aX) = aE(X)$, where a is a constant.
(3) $E(a + X) = E(X) + a$, where a is a constant.
(4) $E(\sum_{i=1}^{n} X_i) = \sum_{i=1}^{n} E(X_i)$, if $E(X_i)$, for $i = 1, \cdots, n$ exists.
(5) $E(XY) = E(X)E(Y)$, where X and Y are two independent random variables.

Now that we have these properties, we can use them to do examples without going through all the working used in Example 10.16.

10.5 Mean and Variance of Probability Distributions

Table 10.5 Probability distributions of the random variables X_3 and X_4

X_3	1	2	3	X_4	2	3	4
P	$\frac{1}{3}$	$\frac{1}{2}$	$\frac{1}{6}$	P	$\frac{1}{3}$	$\frac{1}{3}$	$\frac{1}{3}$

Remark 10.4 Note that $E(X^2) = E(XX)$ is not, in general, the same as $E(X)E(X)$ since X and X are not independent variables (indeed, they are the same). For example, using X_1 from Example 10.16, we see that the random variable $X_1^2 = X_1 X_1$ can only take the values 1 and 4, since X_1 is either 1 or 2, and it has the same probability distribution as X_1. That is, X_1^2 takes the value 1 with probability $\frac{1}{3}$ and the value 4 with probability $\frac{2}{3}$.

Hence, we have that $E(X_1)$ was $1 \times \frac{1}{3} + 2 \times \frac{2}{3} = \frac{5}{3}$, as we saw in Example 10.16. And therefore $E(X_1^2)$ is $1 \times \frac{1}{3} + 4 \times \frac{2}{3} = 3$.

So, in general, for a discrete random variable, we have

$$E(X^2) = \sum_k x_k^2 f_X(x_k).$$

By analogy, for a continuous random variable, we have

$$E(X^2) = \int_{-\infty}^{\infty} x^2 f_X(x) dx.$$

We will find that these last two results are important in Sect. 10.5.2. In fact, if X is a continuous (or discrete) random variable with a probability density function (or probability mass function) $f(x)$, the expected value of any function $g(X)$, denoted as $E(g(X))$, can be computed. Readers can refer to the details provided in [10].

♦

Example 10.17 Tables 10.4 and 10.5 show the probability distributions of the independent random variables X_1, X_2, X_3, and X_4. Compute the following:

1. $E(X_1 + X_2 + X_3 + X_4)$.
2. $E(2X_3 + 4)$.
3. $E(X_3 X_4)$.
4. $E(X_1 X_3)$.
5. $E(X_3^2)$.

(continued)

Example 10.17 (continued)
Solution

1. We already have $E(X_1) = \frac{5}{3}$ and $E(X_2) = \frac{3}{2}$. Applying Eq. (10.8) for the discrete variables X_3 and X_4, we have

$$E(X_3) = 1 \times \frac{1}{3} + 2 \times \frac{1}{2} + 3 \times \frac{1}{6} = \frac{11}{6}.$$

$$E(X_4) = 2 \times \frac{1}{3} + 3 \times \frac{1}{3} + 4 \times \frac{1}{3} = 3.$$

Finally, using the fourth property of means, we have

$$E(X_1 + X_2 + X_3 + X_4) = \frac{5}{3} + \frac{3}{2} + \frac{11}{6} + 3 = 8.$$

2. Using the second and third properties of means, we have

$$E(2X_3 + 4) = 2 \times \frac{11}{6} + 4 = \frac{23}{3}.$$

3. Using the fifth property of means, we have

$$E(X_3 X_4) = \frac{11}{6} \times 3 = \frac{11}{2}.$$

4. Using the fifth property of means, we have

$$E(X_1 X_3) = \frac{5}{3} \times \frac{11}{6} = \frac{55}{18}.$$

5. X_3^2 has a probability distribution of 1, with probability $\frac{1}{3}$, 4 with probability $\frac{1}{2}$, and 9 with probability $\frac{1}{6}$. So,

$$E(X_3^2) = 1 \times \frac{1}{3} + 4 \times \frac{1}{2} + 9 \times \frac{1}{6} = \frac{23}{6}.$$

10.5 Mean and Variance of Probability Distributions

Table 10.6 The probability distribution of the random variable Z

Z	-1	0	$\frac{1}{2}$	1	2
p	$\frac{1}{6}$	$\frac{5}{12}$	$\frac{1}{12}$	$\frac{1}{6}$	$\frac{1}{6}$

Exercises

10.14 Table 10.6 shows the probability distribution of the random variable Z. Compute the following:

(1) $E(Z)$.
(2) $E(-Z + 2)$.
(3) $E(Z^2)$.

10.15 Using Tables 10.4, 10.5, and 10.6 and assuming the variables are independent, compute the following:

(1) $E(Z + X_1 + X_3)$.
(2) $E(ZX_4)$.
(3) $E(3Z - 5)$.
(4) $E(X_1 Z)$.
(5) $E(X_2 + X_4 + Z)$.

10.16 Take any two numbers from 1, 2, 3, and 4. What is the mean value of the absolute difference between the two numbers?

10.17 The probability density function of the random variable X is

$$f(x) = \begin{cases} 8x^7, & 0 < x < 1, \\ 0, & \text{otherwise}. \end{cases}$$

Compute the following:

(1) $E(X)$.
(2) $E(X^2)$.

10.5.2 Variance

Definition 10.4 (Variance) The variance of a random variable X, denoted by σ_X^2, or $Var(X)$, is defined by

$$\sigma_X^2 = Var(X) = \begin{cases} \sum_k (x_k - \mu_X)^2 f_X(x_k), & X : \text{discrete}; \\ \int_{-\infty}^{\infty} (x - \mu_X)^2 f_X(x) dx, & X : \text{continuous}. \end{cases} \quad (10.9)$$

This definition is basically the same as that given in Sect. 4.2.1 of Chap. 4, but weighted by the probabilities, as was the case for the mean given in the previous section.

Comparing the definition of variance with the definition of mean, $E(X)$, given in Eq. (10.8), we can see that an alternate definition is

$$\sigma_X^2 = Var(X) = E((X - E(X))^2). \tag{10.10}$$

Using the properties of mean, we have the following:

$$\begin{aligned} Var(X) &= E((X - E(X))^2) \\ &= E(X^2 - 2XE(X) + (E(X))^2) \\ &= E(X^2) - 2E(X)E(X) + (E(X))^2 \\ &= E(X^2) - (E(X))^2. \end{aligned} \tag{10.11}$$

The variance measures the average difference of the actual values from the average. For example, the average light span of 500 incandescent bulbs is 1000 hours. All these bulbs' light spans may be between 950 and 1050 hours. It is also possible that half of them have a light span of about 1400 hours, and the other half of them have about 600 hours only. To assess the quality of these 500 incandescent bulbs, we need to measure not only the mean value of the light span but also its variance. If the variance value is small, the quality is stable.

Example 10.18 If X_1 is again the discrete random variable with a probability distribution given by Table 10.4, find $Var(X_1)$.

Solution $E(X_1) = \frac{5}{3}$ from before. Also, from before, X_1^2 just takes the values 1 and 4 with probability $\frac{1}{3}$ and $\frac{2}{3}$ respectively. So, $E(X_1^2) = 1 \times \frac{1}{3} + 4 \times \frac{2}{3} = 3$. Applying Eq. (10.11), we have

$$Var(X_1) = 3 - \left(\frac{5}{3}\right)^2 = \frac{2}{9}.$$

Alternatively, we could go back to the definition, namely, Eq. (10.9) and use that. In this case,
$k = 2, x_1 = 1, x_2 = 2, \mu_{X_1} = \frac{5}{3}, f_{X_1}(x_1) = \frac{1}{3}$, and $f_{X_1}(x_2) = \frac{2}{3}$, giving

$$\sum_k (x_k - \mu_{X_1})^2 f_{X_1}(x_k) = \left(1 - \frac{5}{3}\right)^2 \times \frac{1}{3} + \left(2 - \frac{5}{3}\right)^2 \times \frac{2}{3} = \frac{2}{9}.$$

(continued)

10.5 Mean and Variance of Probability Distributions

Example 10.18 (continued)
again. Obviously, the first method is the quickest. The long method was just used to illustrate that the two methods are equivalent.

Example 10.19 What is $Var(X)$ if X is the outcome of a fair six-sided die with numbers from one to six?

Solution Since

$$E(X) = 1 \times \frac{1}{6} + 2 \times \frac{1}{6} + 3 \times \frac{1}{6} + 4 \times \frac{1}{6} + 5 \times \frac{1}{6} + 6 \times \frac{1}{6} = \frac{7}{2},$$

and

$$E(X^2) = 1^2 \times \frac{1}{6} + 2^2 \times \frac{1}{6} + 3^2 \times \frac{1}{6} + 4^2 \times \frac{1}{6} + 5^2 \times \frac{1}{6} + 6^2 \times \frac{1}{6} = \frac{91}{6},$$

applying Eq. (10.11), we have

$$Var(X) = \frac{91}{6} - \left(\frac{7}{2}\right)^2 = \frac{35}{12}.$$

Example 10.20 Let X be a random variable. It has the probability density function

$$f_X(x) = \begin{cases} \frac{1}{b-a}, & a \leq x \leq b; \\ 0, & \text{otherwise.} \end{cases}$$

Compute the following:

1. $E(X)$
2. $Var(X)$

Solution $E(X) = \int_{-\infty}^{+\infty} x \frac{1}{b-a} dx = \left.\frac{x^2}{2(b-a)}\right|_a^b = \frac{b+a}{2}.$
Also $E(X^2) = \int_{-\infty}^{+\infty} x^2 \frac{1}{b-a} dx = \frac{1}{3}(b^2 + ab + a^2).$

(continued)

Example 10.20 (continued)
Applying Eq. (10.11), we have

$$Var(X) = E(X^2) - (E(X))^2$$
$$= \frac{1}{3}(b^2 + ab + a^2) - \left(\frac{a+b}{2}\right)^2$$
$$= \frac{(b-a)^2}{12}.$$

Again, the long way round is to go back to the definition of variance for a continuous random variable (Eq. 10.9) using the value of $\mu_X = E(X) = \frac{b+a}{2}$ we have just found:

$$Var(X) = \int_{-\infty}^{\infty} (x - \mu_X)^2 f_X(x) dx$$
$$= \int_{-\infty}^{\infty} \left(x - \frac{b+a}{2}\right)^2 \frac{1}{b-a} dx$$
$$= \frac{1}{b-a} \frac{1}{3} \left(x - \frac{b+a}{2}\right)^3 \Big|_a^b$$
$$= \frac{1}{3(b-a)} \left(\left(\frac{b-a}{2}\right)^3 - \left(\frac{a-b}{2}\right)^3\right)$$
$$= \frac{(b-a)^2}{12},$$

as before. But the quicker method is the best. The long method was just used to illustrate that the two methods are equivalent.

10.5.2.1 Properties of Variance

Since $Var(X) = E(X^2) - (E(X))^2$ (Eq. 10.11), the following properties of variance are just consequences of the properties of the mean (see Sect. 10.5.1.1).

(1) $Var(X) = 0$, if X takes a constant value.
(2) $Var(aX + b) = a^2 Var(X)$, where a and b are constants.
(3) $Var(X + Y) = Var(X) + Var(Y)$, if X and Y are two independent variables.

Exercise

10.18 Suppose the probability density function of X is given by

$$f_X(x) = \begin{cases} 2, & 0 < x < \frac{1}{2}, \\ 0, & \text{others.} \end{cases}$$

Compute the following:

(1) $E(X^2)$ and $E(X^4)$
(2) $Var(2X^2)$
(3) $Var(2X^2 + 5)$

10.6 Special Univariate Distributions

In this section, we will look at some important and famous single-variable probability distributions, first discrete ones and then continuous ones.

10.6.1 Discrete Random Variables

10.6.1.1 Discrete Uniform Distribution

This distribution gives the same probability for each value of the random variable.

So if the random variable X takes integer values from a to b, inclusive, then the probability mass function of the discrete uniform distribution is defined as follows:

$$f_X(x) = \begin{cases} \frac{1}{n} & a \leq x \leq b; \\ 0, & \text{otherwise;} \end{cases} \quad (10.12)$$

where $n = b - a + 1$. For example, if $a = 3$ and $b = 9$, then there are $n = 7$ values, each of $\frac{1}{7}$.

Figure 10.6 illustrates an example of a discrete uniform distribution in the range of $[a, b]$.

When $n = b - a + 1$ is satisfied, the mean and variance of a discrete uniform distribution are given as follows:

$$\mu_X = E(X) = \frac{a+b}{2},$$

Fig. 10.6 An illustration of a discrete uniform distribution

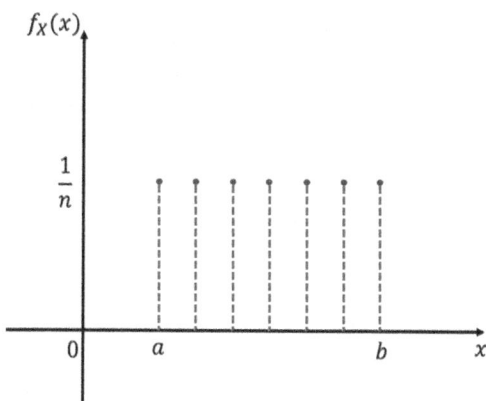

and

$$\sigma_X^2 = Var(X) = \frac{n^2 - 1}{12}.$$

For our example where $a = 3$ and $b = 9$, we get $\mu_X = \frac{12}{2} = 6$, the middle number. Also, $\sigma_X^2 = \frac{49-1}{12} = 4$. You can check these results by using the original definitions in $\mu_X = \sum_k x_k f_X(x_k)$ from Eq. (10.8), and $\sigma_X^2 = \sum_k (x_k - \mu_X)^2 f_X(x_k)$ from Eq. (10.9).

However, if $n = b - a + 1$ is not valid for a discrete uniform distribution, the original definitions for computing the mean and variance should be used. For example, four numbers, 3, 5, 7, 9, between $a = 3$ and $b = 9$. The mean value is still equal to $\frac{3+5+7+9}{4} = 6$, while $\frac{(3-6)^2}{4} + \frac{(5-6)^2}{4} + \frac{(7-6)^2}{4} + \frac{(9-6)^2}{4} = 5$.

> **Exercise**
>
> **10.19** There are ten balls labelled from zero to nine, respectively, in a bag. Randomly take a ball from the pack, write down the number on it, and then put it back in the bag. After many runs, what is the approximate mean value of those numbers?

10.6.1.2 Bernoulli Distribution

If X is a random variable taking two values, $x = 1$ and $x = 0$ only, with a probability of p and $1 - p$, respectively, the probability mass function f of this distribution is

10.6 Special Univariate Distributions

defined as follows:

$$f_X(x; p) = \begin{cases} p & \text{if } x = 1 \\ q = 1 - p & \text{if } x = 0, \end{cases} \quad (10.13)$$

then X has a Bernoulli distribution.

The key here is that the random variable can only take one of *two* values. Some event happens or does not happen. These are called Bernoulli trials and a set of Bernoulli trials gives the distribution.

Equation (10.13) can also be expressed as follows:

$$f_X(x; p) = p^x(1-p)^{1-x}, \text{ where } x = 0 \text{ or } 1. \quad (10.14)$$

The mean and variance of a Bernoulli distribution are given as follows:

$$\mu_X = E(X) = p,$$

and

$$\sigma_X^2 = Var(X) = p(1-p).$$

Again, you can check these results by using the original definitions from Eqs. (10.8) and (10.9) for the discrete random variable.

Example 10.21 Figure 10.7 shows a simulation result of generating 5000 random numbers from a Bernoulli distribution with a success probability of $P(X = 1) = 0.2$. As can be seen from the figure, about 1000 numbers have a value of 1, and about 4000 numbers have a value of 0. The only two outcomes are a 1 with probability 0.2 and a 0 with probability $1 - 0.2 = 0.8$.

10.6.1.3 Binomial Distribution

Let us consider the example of tossing coins again. Suppose we have n independent coins with a probability p of heads-up. For fair coins, then $p = \frac{1}{2}$, but generally, the probabilities could be p for heads-up and, therefore, $1 - p$ for tails. We flip them all simultaneously and check the number of heads-up coins. Alternatively, we can flip one coin n times and check the number of heads-ups in total. These two scenarios are equivalent because we assume these coins are independent. Both can be considered as n independent and identical Bernoulli trials or distributions (two outcomes: heads with probability p and tails with probability $1 - p$). The total

Fig. 10.7 The simulation result of a Bernoulli distribution

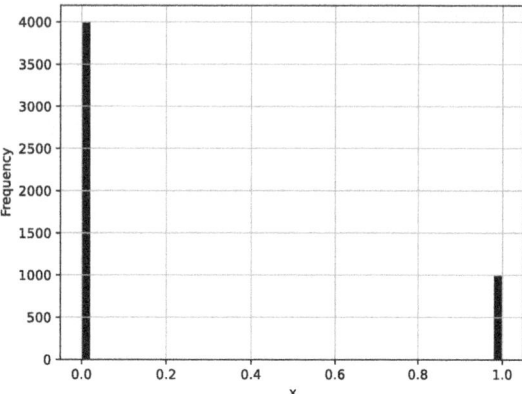

number of heads-ups can be modelled from a binomial distribution. For instance, if we tossed 100 fair coins, we might expect 50 to be heads-up. But we might want to know the probability of getting exactly 50 heads-ups (this is actually ≈ 0.080). Or perhaps exactly 49 heads-ups (≈ 0.078), 48 heads-ups (≈ 0.074), and so on. This sort of question is given by considering a binomial distribution, where we want to know the probability of getting x events in n trials.

A random variable X is called a binomial random variable with parameters (n, p) if its probability mass function is given as follows:

$$p_X(x) = P(X = x) = \frac{n!}{x!(n-x)!} p^x (1-p)^{n-x}. \tag{10.15}$$

Here, n is the number of independent trials, p is the probability of success on each trial, and x is the number of successes in those trials.

Note that the factors $\frac{n!}{x!(n-x)!}$ are $\binom{n}{x}$, which are the binomial coefficients, which is why this is called a binomial distribution. A random variable having a binomial distribution can be denoted as $X \sim B(n, p)$.

Using this formula, if we want to get the probability of getting exactly 50 heads up out of 100 fair trials, we set $n = 100$, $x = 50$, and $p = \frac{1}{2}$. So,

$$p_X(50) = P(X = 50) = \frac{100!}{50!50!} (\frac{1}{2})^{50} (\frac{1}{2})^{50} \approx 0.080,$$

as we stated previously.

The mean and variance of a binomial distribution can be obtained as follows:

$$\mu_X = E(X) = E(X_1 + \cdots + X_n) = p + \cdots + p = np,$$

and

$$\sigma_X^2 = Var(X) = Var(X_1 + \cdots + X_n) = p(1-p) + \cdots + p(1-p) = np(1-p),$$

where $X_1, \ldots, X_n$ are the outcome of each of the n independent Bernoulli trials.

> **Example 10.22** We want to know the probability of having two heads-ups when flipping a fair coin three times using the Bernoulli distribution and the binomial distribution separately. Suppose $p(\text{heads-up}) = \frac{1}{2}$. Let X be the random variable of seeing heads-up.
>
> 1. Solution—using Bernoulli distribution three times
>   We know that the sample space is
>
> $$\Omega = \{TTT, HTT, THT, TTH, HHT, HTH, THH, HHH\}$$
>
>   and there are three elements in the event that have two heads-ups, namely, HHT, HTH, and THH:
>
> $$P(X(HHT)) = \frac{1}{2} \times \frac{1}{2} \times \frac{1}{2},$$
>
> $$P(X(HTH)) = \frac{1}{2} \times \frac{1}{2} \times \frac{1}{2},$$
>
>   and
>
> $$P(X(THH)) = \frac{1}{2} \times \frac{1}{2} \times \frac{1}{2}.$$
>
>   Therefore, we have
>
> $$P(X = 2 \text{ heads-ups}) = \frac{1}{8} + \frac{1}{8} + \frac{1}{8} = \frac{3}{8}.$$
>
>   Note that there is one element in the sample space that has three heads-ups, three with two heads-ups, three with one heads-up, and one with zero heads-up. The numbers 1, 3, 3, and 1 are where the binomial coefficients come from in the next solution.
> 2. Solution—using binomial distribution
>   Substituting $n = 3$ and $x = 2$ to Eq. (10.15), we have
>
> $$p_X(2) = P(X = 2) = \frac{3!}{2!(3-2)!}\left(\frac{1}{2}\right)^2\left(1-\frac{1}{2}\right)^{3-2} = \frac{3}{8},$$
>
> (continued)

Example 10.22 (continued)

where the binomial coefficient returns the number of combinations of two heads-ups among three tosses and $(\frac{1}{2})^2(1-\frac{1}{2})^{3-2} = \frac{1}{8}$ gives the probability of one desired combination.

Example 10.23 If we flip 20 biased coins simultaneously, what is the probability that we see 12 heads-ups? Note that the probability of heads-up of each coin is 0.6.

Solution Applying Eq. (10.15), where $n = 20$, $x = 12$ and $p = 0.6$, we have

$$P(X = 12) = \frac{20!}{12!(20-12)!} 0.6^{12}(1-0.6)^{20-12} \approx 0.1797.$$

Exercise

10.20 A module's passing rate is 85%. Find the probability that:

(1) Exactly seven students out of ten pass the module.
(2) Exactly eight students out of ten pass the module.
(3) Exactly nine students out of ten pass the module.
(4) Exactly ten students out of ten pass the module.

10.6.1.4 Poisson Distribution

The Poisson distribution is characterised by the number of events that happen within some *interval*. A random variable X is called a Poisson random variable with parameter λ if its probability mass function is given as follows:

$$p_X(x) = P(X = x \text{ events in an interval}) = e^{-\lambda} \frac{\lambda^x}{x!}, \quad x = 0, 1, \cdots, \quad (10.16)$$

where $\lambda > 0$ is the average number of events per interval and e is Euler's number $2.71828\ldots$.

The mean and variance of a Poisson distribution are given by

$$\mu_X = E(X) = \lambda,$$

10.6 Special Univariate Distributions

and

$$\sigma_X^2 = Var(X) = \lambda.$$

The Poisson distribution is actually a limiting case of the binomial distribution as $n \to \infty$. Usually, with a large n and a small p, the Poisson distribution is a useful and very good approximation to the binomial distribution. So, it is used when n is large or unknown and p is small or unknown. If the mean value $\lambda = np$ is known, often in the form of a known average number of events in some interval, then we can use the Poisson distribution. λ is the only parameter in the Poisson distribution, whereas the binomial distribution has two parameters (n and p).

Example 10.24 Let us use a Poisson distribution to model the number of patients a doctor can see in one hour. Suppose a doctor was able to see three patients an hour on average. Find the probability that this doctor can see five patients in the next hour.

Solution Applying Eq. (10.16), where $\lambda = 3$ and $x = 5$, we have

$$P(X = 5) = e^{-3}\frac{3^5}{5!} = 0.1008.$$

Exercises

10.21 A doctor can see six patients an hour on average. Use a Poisson distribution to find:

(1) The probability that the doctor can see five patients in the next hour.
(2) The probability that the doctor can see six patients in the next hour.
(3) The probability that the doctor can see seven patients in the next hour.
(4) The probability that the doctor can see eight patients in the next hour.

10.22 The number of calls a helpdesk receives per minute follows a Poisson distribution with $\lambda = 4$.

(1) What is the probability that the number of calls is exactly eight in one minute?
(2) What is the probability that the number of calls is more than five per minute?

10.6.2 Continuous Random Variables

10.6.2.1 Continuous Uniform Distribution

The discrete uniform distribution can be easily extended to a continuous case. A random variable X is called a continuous uniform variable in the range of $[a, b]$ if its probability density function is given by

$$f_X(x) = \begin{cases} \frac{1}{b-a} & a \leq x \leq b; \\ 0 & \text{otherwise.} \end{cases} \qquad (10.17)$$

The value $\frac{1}{b-a}$ makes sure that the total area under the curve is 1 so that the total probability is 1.

To find the cumulative distribution, we need to find the area under the curve up to some random point x. So, we compute the following:

$$F_X(x) = \int_a^x \frac{1}{b-a} dz = \frac{z}{b-a}\Big|_a^x = \frac{x-a}{b-a}.$$

Therefore, the cumulative distribution function $F_X(x)$ is given by

$$F_X(x) = \begin{cases} 0, & x < a; \\ \frac{x-a}{b-a}, & a \leq x \leq b; \\ 1, & x > b. \end{cases} \qquad (10.18)$$

Figure 10.8 shows the sketch of the probability density function and the cumulative distribution function for the continuous uniform distributions. The mean and

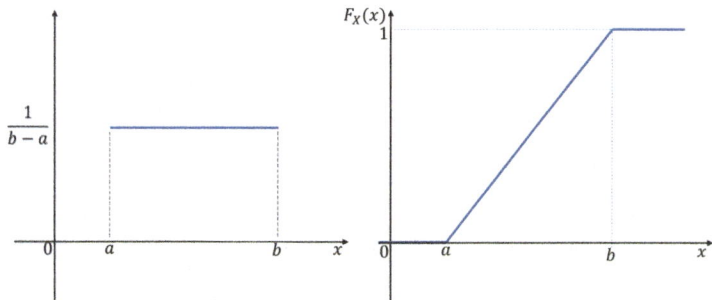

Fig. 10.8 A sketch of the probability density function and the cumulative distribution function for the continuous uniform distribution

10.6 Special Univariate Distributions

variance of a continuous uniform distribution are given as follows (see worked Example 10.20):

$$\mu_X = E(X) = \frac{a+b}{2},$$

and

$$\sigma^2 = Var(X) = \frac{(b-a)^2}{12}.$$

Example 10.25 Buses arrive at five-minute intervals from 5 pm to 6 pm. A student arrives at the bus stop at a random time X, which follows a uniform distribution, between 5 pm and 5:20 pm. What is the probability that the student waits less than two minutes for a bus?

Solution X is a random variable having a continuous uniform distribution, where $a = 0$ and $b = 20$(minutes).

Set E as the event the student waits less than two minutes to get on a bus. We have

$$E = \{3 < X < 5\} + \{8 < X < 10\} + \{13 < X < 15\} + \{18 < X < 20\}.$$

So,

$$P(E) = \int_3^5 \frac{dz}{20-0} + \int_8^{10} \frac{dz}{20-0} + \int_{13}^{15} \frac{dz}{20-0} + \int_{18}^{20} \frac{dz}{20-0}$$
$$= \frac{1}{20-0}(2+2+2+2)$$
$$= 0.4.$$

Exercise

10.23 Buses arrive at nine-minute intervals from 5 pm to 6 pm. A student arrives at the bus stop at a random time X, which follows a uniform distribution, between 5:20 pm and 5:45 pm. What is the probability that the student waits less than three minutes for a bus?

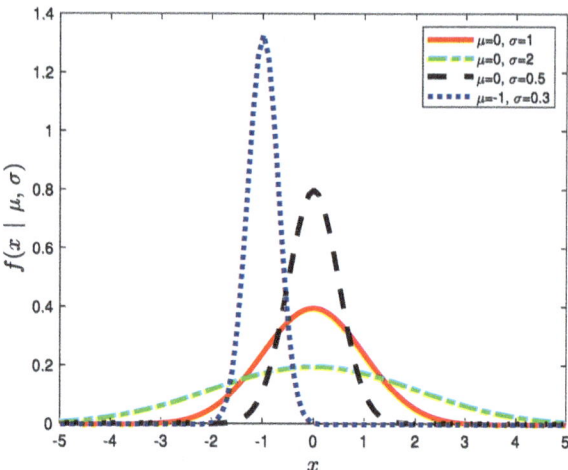

Fig. 10.9 Examples of Gaussian distributions

10.6.2.2 Gaussian or Normal Distribution

A random variable X is distributed normally with mean μ and variance σ^2 if its probability density function is given by

$$f(x|\mu,\ \sigma^2) = \frac{1}{\sqrt{2\pi\sigma^2}} e^{-\frac{(x-\mu)^2}{2\sigma^2}}. \tag{10.19}$$

The normal distribution is also called Gaussian distribution, denoted as $X \sim \mathcal{N}(\mu, \sigma^2)$. Figure 10.9 shows four normal distributions, each with a different mean value and standard deviation. As we can see, normal distributions have a bell shape and are symmetrical in their mean values. The standard deviation controls the shape of each distribution: the smaller the standard deviation, the narrower the bell curve and the higher the probability density value at the mean point; on the other hand, the larger the standard deviation, the wider the curve and the lower the probability density value at the mean point.

It is convenient to define a "standard" normal distribution with zero mean and a standard deviation of 1. As we will see later, this enables us to have just a single table of values to look up information, rather than one for each mean and standard deviation. All normal distributions can be standardised as shown below.

Definition 10.5 (Standard Normal Distribution) The random variable Z with zero mean and unit standard deviation, that is, $Z \sim \mathcal{N}(0, 1)$, is called the standard normal distribution. That is,

$$f(z|0,\ 1) = \frac{1}{\sqrt{2\pi}} e^{-\frac{z^2}{2}}. \tag{10.20}$$

10.6 Special Univariate Distributions

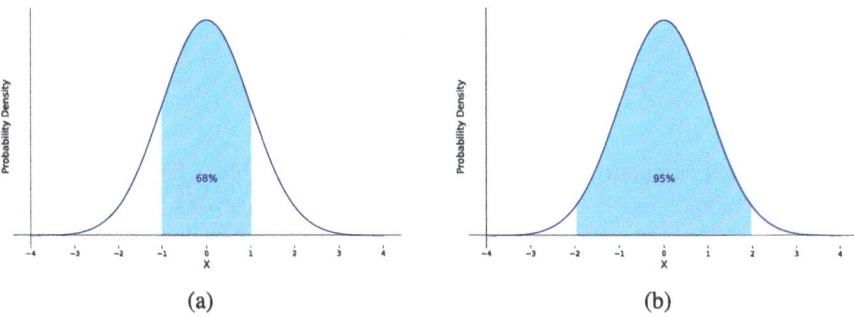

Fig. 10.10 The standard normal distribution: (**a**) 68% of values are within 1 standard deviation of the mean; (**b**) 95% of values are within about 2 (or more accurately 1.96) standard deviations of the mean

A general normal distribution $X \sim \mathcal{N}(\mu, \sigma^2)$ can be standardised to a standard normal distribution by computing the z-score for x as follows:

$$z = \frac{x - \mu}{\sigma}. \qquad (10.21)$$

A standard normal distribution has the following properties:

- About two-thirds (68%) of the total observations lie within one standard deviation on either side of the mean: one-third on one side and one-third on the other (see Panel (a) of Fig. 10.10).
- 95% of the total observations are located within about two (or more accurately 1.96) standard deviations on either side of the mean (see Panel (b) of Fig. 10.10).

These properties are valid for all normal distributions.

Exercises

10.24 About what percentage of the observations in a normal distribution will have values greater than one standard deviation above the mean?

10.25 The distribution of scores collected from a module in the past five years, including 1000 students, is approximately normal. The mean score is 54, and the standard deviation is 4. If the passing score of the module is 50, approximately how many students failed the module?

Again, to find the cumulative distribution, we need to find the area under the curve up to some random point z. The cumulative distribution function of Z

satisfying Eq. (10.20) is therefore

$$F_Z(z) = P(Z \le z) = \frac{1}{\sqrt{2\pi}} \int_{-\infty}^{z} e^{\frac{-\xi^2}{2}} d\xi.$$

This is the total cumulative probability of the random variable Z being less than z. (It is quite common to use ξ as the integration variable in this work.)

By convention, the cumulative distribution function of Z is denoted by Φ. So, we have

$$\Phi(z) = \frac{1}{\sqrt{2\pi}} \int_{-\infty}^{z} e^{\frac{-\xi^2}{2}} d\xi. \tag{10.22}$$

The cumulative distribution function of a random variable X, where $X \sim \mathcal{N}(\mu, \sigma^2)$ and where X can be converted to Z via $Z = \frac{X-\mu}{\sigma}$, can be obtained as follows:

$$\begin{aligned} F_X(x) &= P(X \le x) \\ &= P(\sigma Z + \mu \le x) \\ &= P\left(Z \le \frac{x-\mu}{\sigma}\right). \end{aligned}$$

That is,

$$F_X(x) = \Phi\left(\frac{x-\mu}{\sigma}\right) = \Phi(z). \tag{10.23}$$

$\Phi\left(\frac{x-\mu}{\sigma}\right)$ has the standard normal distribution. The term $\frac{x-\mu}{\sigma}$ is commonly referred to as the z-score. Figure 10.11 shows the standard normal probability density distribution and its corresponding cumulative distribution. The cumulative distribution gives us the area under that probability density function for the interval of negative infinity to a specific z-score. Obviously, if $\mu = 0$ and $\sigma = 1$, then we already have a standard normal distribution and the conversion is unnecessary.

The following lists some properties of Φ:

- $\lim_{x \to -\infty} \Phi(x) = 0$ and $\lim_{x \to \infty} \Phi(x) = 1$.
- $\Phi(0) = \frac{1}{2}$, since half the probability is to the left of the mean.
- $\Phi(-z) = 1 - \Phi(z)$, since it is symmetric about the mean.

In addition, since $P(a < X \le b) = F_X(b) - F_X(a)$ (see Sect. 10.4 of this chapter), we have

$$P(x_1 < X \le x_2) = \Phi\left(\frac{x_2-\mu}{\sigma}\right) - \Phi\left(\frac{x_1-\mu}{\sigma}\right). \tag{10.24}$$

10.6 Special Univariate Distributions

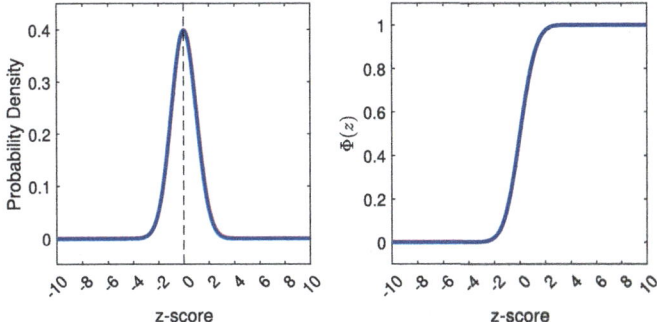

Fig. 10.11 The standard normal distribution. The left panel shows the probability density distribution; the right panel presents its cumulative distribution

Note that the integration in Eq. (10.22) cannot be evaluated analytically in a closed form. It requires numerical analysis approximation methods. However, since the normal distribution is one of the most widely used distributions, people have constructed a mathematical table for the standard normal distribution.[1]

Reading a Normal Distribution Table A normal distribution table is usually composed of three parts:

- The heading of rows, which is the left column, contains the integer part and the first decimal place of the z-score.
- The heading of columns contains the second decimal place of the z-score.
- The values within the table are the probabilities, which are the area under the normal curve from the starting point to z. The starting point could be zero for cumulative from the mean, negative infinity for full cumulative, or positive infinity for complementary cumulative. (Different books have different tables with different starting points.)

Example 10.26 Find the cumulative probability of $z = 0.56$. That is, find $\Phi(0.56)$.

Solution We find the row starting with 0.5 and then across the column to 0.06 in a standard normal distribution table. It gives us a probability of 0.2123 for a cumulative from the mean table (i.e., 0.2123 above halfway) or 0.7123 from a full cumulative table.

---

[1] The example table we have used is at https://en.wikipedia.org/wiki/Standard_normal_table.

Example 10.27 Suppose $X \sim \mathcal{N}(3, 4^2)$. Compute $P(X < 10.84)$.

Solution Using Eq. (10.23),

$$F_X(10.84) = \Phi\left(\frac{10.84 - 3}{4}\right) = \Phi(1.96).$$

This gives $z = 1.96$. Looking this up in a full cumulative table, we get 0.9750. In fact, the value 1.96 on either side of the mean represents the distance from the mean that gives 95% of the total (see Fig. 10.10). Of the other 5% half is on the left, so the cumulative probability up to 1.96 is 95% + 2.5% = 97.5% or 0.9750, which is the value we obtained.

Exercises

10.26 Suppose $X \sim \mathcal{N}(2, 3^2)$. Compute, using an appropriate table, the following:

(1) $P(X < 8)$, $P(X < 0.5)$, and hence $P(0.5 < X < 8)$
(2) $P(-1 < X < 5)$
(3) The value of C so that $P(X > C) = P(X \leq C)$

10.27 The lifespan (in years) of refrigerators designed by manufacturer A follows a normal distribution with a mean value μ of 16 years. If $P(12 < X \leq 20) = 0.8$, what is the largest standard deviation value σ?

The normal distribution is extremely widely used in both real-world applications and theoretical work in research. For example, the heights of adult men or adult women may be approximately normally distributed. Residual errors from a regression model may have a normal distribution, and this normality is an assumption when formulating the linear regression method within the probability framework (see Chap. 13). There are many more examples of its use.

Chapter 11
Further Probability

The previous chapter introduced probability, probability distributions for both discrete and continuous random variables, and properties of probability distributions like mean and variance. We then illustrated some common and important probability distributions, like the binomial distribution, the Poisson distribution, and the normal distribution. All of the introductory probability concerned material related to single random variables.

In this chapter, we continue with some more advanced topics centering around multiple random variables and conditional probability, starting with an important theorem in probability called the central limit theorem.

11.1 The Law of Large Numbers and the Central Limit Theorem

In practice, people have observed that when the number of experiments approaches a large number, the probability of the occurrence of a specific event (E) will become stable. For instance, the number of head-ups obtained when flipping a fair coin will get nearer to half of the total as more flips are performed. In addition, people have also found that many observed random variables are the sum of other independent random variables, some of which may not be measurable. Research shows that when the number of independent random variables approaches infinity, the distribution of the random variable of the sum of these independent random variables approximates the normal distribution.

This topic is usually associated with taking samples from a population and how the mean of the sample relates to the mean of the population. This section introduces the law of large numbers and the central limit theorem without giving any proof.

11.1.1 The Law of Large Numbers

Intuitively, the law of large numbers is obvious! It basically says that if you wish to know the value of a variable, like the average height of an adult, then you do not just look at one person; you need to look at lots of people, and the more you look, the better.

Formally, we have the following. Let $X_1, \ldots, X_n$ be a sequence of independent, identically distributed random variables with $E(X_i) = \mu$, $i = 1, \ldots, n$, and define the sample mean as follows:

$$\bar{X}_n = \frac{1}{n}\sum_{i=1}^{n}(X_1 + \cdots + X_n).$$

Then, for any $\epsilon > 0$,

- The weak law of large numbers states that

$$\lim_{n\to\infty} P(|\bar{X}_n - \mu| < \epsilon) \to 1. \tag{11.1}$$

It says that as $n \to \infty$, the probability of the difference between the sample mean and the expected mean being less than any small value ϵ approaches 1.

- The strong law of large numbers states that

$$P(\lim_{n\to\infty} \bar{X}_n = \mu) = 1. \tag{11.2}$$

It says that as $n \to \infty$, we know almost for sure ($P = 1$) that the sequence $\bar{X}_n$ converges to the expected mean.

Remark 11.1 The weak law tells us how a sequence of probabilities converges. In other words, for any small value $\epsilon > 0$, the probability of the difference between the sample mean $\bar{X}_n$ and the population mean μ being less than ϵ tends to 1. The strong law states that $\bar{X}_n$ approach μ as $n \to \infty$ with probability 1.

The differences between the weak law of large numbers and the strong law of large numbers are subtle and not important for this book.

In practice, we can apply the principle of the law of large numbers to find the approximated expected value of a distribution by repeating a procedure over and over and then computing the average result, which will be close to the expected value.

For instance, if X_i is the attendance at a football match, then X_i is a discrete random number taking integer values from zero to the stadium size. If n is the number of matches in a month, then we have a sample size of, say, four (one match each week), and then the average attendance may be reasonably close to the real average attendance. But if we take n as the number of matches in a year, we will have a larger sample size, and the mean attendance will be closer to the real mean. ♦

Fig. 11.1 The result of a simulation to estimate the probability of obtaining 12 heads in 20 coin flips, where the event occurred 1826 times out of 10,000 trials

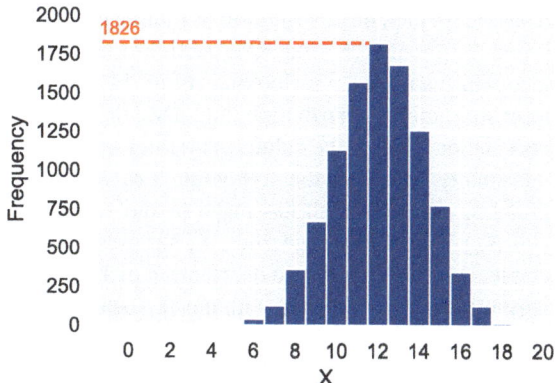

Example 11.1 Continue Example 10.23 from the previous chapter, Chap. 10. We can do a simulation via a computer program to estimate the probability of seeing 12 heads if we flip 20 coins simultaneously. Let E be the event where we observe the number of heads-up, $X = 12$. We have used Python programming to implement the simulation in this example. In one trial, 20 independent experiments can be run, each having a Bernoulli distribution with the probability of coming-up heads of each coin being 0.6. The number of heads-up can be obtained in these 20 experiments; this is noted. We repeat this procedure with a large number of trials. Each time the trial of 20 experiments is performed and the number of heads-up is noted. Figure 11.1 shows the simulated result with 10,000 trials. That is, we have done the 20 experiments 10,000 times, each time noting the number of heads obtained. Specifically, the number of our events, E, that is, 12 heads observed, in 10,000 repetitions, is 1826. Therefore, the probability that we see 12 heads if we flip 20 coins simultaneously is approximated as 0.1826 based on the law of large numbers. It is larger than 0.1797, the theoretical value computed in the example shown in Example 10.23. To make the estimation more accurate, we may increase the number of trials in the simulation.

11.1.2 Central Limit Theorem

So far, we have that if we take a sample of size n from a population with an unknown mean, then the mean of the sample is a reasonable estimate of the mean of the population, and this estimate gets better as n gets larger. The central limit theorem takes the analysis further and says that if we repeatedly find the mean of an n-sized sample, we expect to get as many below the population (real) mean as there are above. In fact, the distribution of sample means approaches a normal distribution around the population mean, irrespective of the sort of distribution we had in the first place. This can be seen by looking at Fig. 11.1. As n gets larger, the distribution gets closer to a normal distribution, and also, the standard deviation of the distribution gets smaller. That is, the values get tighter around the population mean.

Formally, we have the following. Let $X_1, \ldots, X_n$ be n independent random variables, each of which has mean μ and standard deviation σ. Let $Y_n = (X_1 + \cdots + X_n)/n$ be the average; thus, Y_n has mean μ and standard deviation $\sigma/\sqrt{n}$. If n is large, then the cumulative distribution of Y is very nearly equal to the cumulative distribution of the Gaussian with mean μ and standard deviation $\sigma/\sqrt{n}$. That is,

$$\lim_{n \to \infty} \frac{Y_n - \mu}{\sigma/\sqrt{n}} \sim \mathcal{N}(0, 1), \qquad (11.3)$$

or,

$$\lim_{n \to \infty} P\left(\frac{Y_n - \mu}{\sigma/\sqrt{n}} \leq z\right) = \Phi(z). \qquad (11.4)$$

Example 11.2 In a doctor's surgery over a long period, it is noted that the average length of a patient's appointment is 10 minutes with a standard deviation of 5 minutes. So, patients are scheduled every 10 minutes for the $2\frac{1}{2}$ hours of each morning's surgery time. Unfortunately, one morning there is an emergency appointment, so there are 16 patients to see rather than the normal 15. What is the chance that the doctor can finish on time?

Solution Each patient is assumed to be independent and drawn from the distribution with a mean time of 10 minutes and a standard deviation of 5 minutes. This means the variance is $5^2 = 25$. Let each of these 16 appointment lengths be denoted X_i for $1 \leq i \leq 16$. For the total time taken, we have

$$Y_{16} = X_1 + \cdots + X_{16}.$$

The mean for Y_{16} is $10 \times 16 = 160$, using property (4) of means (see Sect. 10.5.1.1 of Chap. 10), and the variance of Y_{16} is 16×25 using property (3) of variance (see Sect. 10.5.2.1 of Chap. 10), and so the standard deviation for Y_{16} is $4 \times 5 = 20$.

From the central limit theorem, this means that the distribution of Y_{16} is approximately a Gaussian with a mean of 160 and a standard deviation of 20.

(continued)

11.1 The Law of Large Numbers and the Central Limit Theorem

Example 11.2 (continued)
We want $P(y \leq 150)$ for the doctor to finish on time. So that we can use the tables of values for a normal distribution, we have to standardise the normal distribution using z-scores as in Eq. (10.21). This gives

$$P(y \leq 150) = P\left(\frac{y-160}{20} \leq \frac{150-160}{20}\right) = P(z \leq -\frac{1}{2}),$$

which using the central limit theorem, Eq. (10.23), and the table (refer to the information provided in the footnote of Sect. 10.6.2.2) gives us

$$P(z \leq -\frac{1}{2}) = \Phi(-\frac{1}{2}) = 0.3085.$$

A special case of the central limit theorem is the *De Moivre-Laplace* theorem. Let η_n be a binomial random variable with parameters (n, p). Then, *De Moivre-Laplace* theorem states:

$$P(a \leq \eta_n \leq b) \approx \Phi\left(\frac{b-np}{\sqrt{np(1-p)}}\right) - \Phi\left(\frac{a-np}{\sqrt{np(1-p)}}\right), \quad (11.5)$$

where np and $np(1-p)$ are the mean and variance of a binomial distribution (see Sect. 10.6.1.3 of Chap. 10). These values are needed to make the normal distribution into a standard normal distribution so we can use a table of values. See also Eq. (10.24) in Chap. 10, which is where the above equation comes from.

Example 11.3 There are 100 computers running independently in a PC lab. The probability of the actual working time of each PC is 80% of the total lab opening time each day. Compute the probability that there are between 70 and 86 computers working at any lab opening time.

Solution Suppose each computer has two statuses: working or not working. Since computers work independently, we consider 100 computers as 100 Bernoulli trials. Repeated Bernoulli trials give us a binomial distribution— see Sect. 10.6.1.3 of Chap. 10. Suppose the number of working computers is $\eta_n \sim B(100, 0.8)$. Since the mean of this binomial distribution is $np = 80$ and the standard deviation of this binomial distribution is $\sqrt{np(1-p)} = 4$,

(continued)

Example 11.3 (continued)
we can standardise the normal distribution using z-scores as in Eq. (10.21) and have

$$P(70 \leq \eta_n \leq 86) = P\left(\frac{70-80}{4} \leq \frac{\eta_n - 80}{4} \leq \frac{86-80}{4}\right)$$

$$= P\left(-2.5 \leq \frac{\eta_n - 80}{4} \leq 1.5\right).$$

Applying Eq. (11.5), we have

$$P\left(-2.5 \leq \frac{\eta_n - 80}{4} \leq 1.5\right) = \Phi(1.5) - \Phi(-2.5) = 0.9270.$$

Exercises

11.1 For the doctor's surgery introduced in Example 11.2:

(1) Confirm that if the expected number of 15 patients turn up for a morning appointment, then the probability of the doctor finishing on time is 0.5.
(2) One of the doctors deals with only older patients, and it is found that these require an appointment with an average length of 15 minutes and a standard deviation of 10 minutes. Again, if an emergency patient turns up one morning so that there are 11 instead of the 10 expected patients to see, find the chance that the doctor can finish on time.
(3) Now assume that the older patients still take an average of 15 minutes, but now the standard deviation is 5 minutes. Again, 11 instead of 10 patients arrive, finding the chance that the doctor can finish on time.

11.2 Compute the probability of coming-up heads is greater than 60 times when flipping a fair coin 100 times.

Remark 11.2 The law of large numbers (LLN) and the central limit theorem (CLT) are used for different purposes. LLN addresses the convergence of sample means to the population mean, while CLT concerns the convergence of sample means to a normal distribution.

♦

11.2 Multiple Random Variables

In many real-world applications, it is not enough to compute the mean and variance of one random variable. Studying two or more random variables defined on the same sample space is important. For example, consider tossing one fair coin five times and repeat this experiment many times. We use X to denote the random variable of observing the number of coming-up heads in the first two tosses, and Y denotes the number of heads-up in the last three. Then, we can compute the joint probability of (X, Y). That is, we are interested in the probability distribution of both random variables, X and Y.

11.2.1 *Joint Probability Distributions: Discrete Random Variables*

Let us start with two examples to bear in mind while reading this.

First: We could roll a fair six-sided die and define $X = 1$ if an even number is thrown and $X = 0$ otherwise. We could also define $Y = 1$ if a square number is thrown (i.e., a one or a four) and $Y = 0$ otherwise. Here, we can look at joint probabilities such as the chance of throwing an even number and a square number, that is, both $X = 1$ and $Y = 1$.

Second: Consider an ordinary pack of cards and turn over a card to look at it. Define $X = 1$ if it is a red card and $X = 0$ if it is a black card. Also, define $Y = 1$ if it is a "high" card (defined to be an Ace, a King, a Queen, or a Jack) so that $Y = 0$ for all other cards (2, 3, 4, 5, 6, 7, 8, 9, 10). Here, we can consider the probability of it being black and a high card ($X = 0$ and $Y = 1$), for instance.

11.2.1.1 Joint Probability Mass Functions and Cumulative Distribution Functions

Consider two discrete random variables, X and Y. The joint probability mass function, $p_{XY}(x, y)$, can be given by

$$p_{XY}(x_i, y_j) = P(X = x_i, Y = y_j), \qquad (11.6)$$

where (x_i, y_j) denotes pairs of values X and Y can take. In the case of the die roll, both X and Y can only take the values 0 and 1. So, $p_{XY}(x_i, y_j)$ is a table of probabilities with the four combinations, which are

$P(X = 0, Y = 0)$, $P(X = 1, Y = 0)$, $P(X = 0, Y = 1)$, and $P(X = 1, Y = 1)$.

Properties of $p_{XY}(x_i, y_j)$

1. Each joint probability is between 0 and 1:

$$0 \leq p_{XY}(x_i, y_j) \leq 1.$$

2. The sum of all the joint probabilities adds up to 1:

$$\sum_{x_i}\sum_{y_j} p_{XY}(x_i, y_j) = 1,$$

where the summation is taken over all possible pairs of (x_i, y_j).

The joint cumulative distribution function of X and Y, denoted by $F_{XY}(x, y)$, is again the sum of all the probabilities, in this case for x_i up to x and for y_j up to y. The function is defined by

$$F_{XY}(x, y) = P(X \leq x, Y \leq y) = \sum_{x_i \leq x}\sum_{y_j \leq y} p_{XY}(x_i, y_j). \tag{11.7}$$

11.2.1.2 Marginal Probability Distributions

Consider two discrete random variables X and Y. If

$$P(X = x_i) = p_X(x_i) = \sum_{y_j} P(X = x_i, Y = y_j), \tag{11.8}$$

that is, the summation is taken over all possible values (x_i, y_j) with x_i fixed. In this case, Eq. (11.8) is called the marginal probability mass function of X. So, you could find the sum of the probabilities for Y when X is fixed, for instance, at $X = 0$. Similarly, the marginal probability mass function of Y is given by

$$P(Y = y_j) = p_Y(y_j) = \sum_{x_i} P(X = x_i, Y = y_j). \tag{11.9}$$

Remark 11.3 Note that X and Y are independent random variables, if

$$p_{XY}(x_i, y_j) = p_X(x_i)p_Y(y_j). \tag{11.10}$$

♦

We can visualise the joint and marginal probability distributions through a table.

11.2 Multiple Random Variables

Example 11.4 Flipping two fair coins. Let

$$X = \begin{cases} 0, & \text{The first coin tails-up,} \\ 1, & \text{The first coin heads-up,} \end{cases}$$

and

$$Y = \begin{cases} 0, & \text{The second coin tails-up,} \\ 1, & \text{The second coin heads-up.} \end{cases}$$

Compute the probability distribution of (X, Y).

Solution Possible values of (X, Y) are $(0, 0)$, $(0, 1)$, $(1, 0)$, and $(1, 1)$. Assume two flips are independent.

Applying Eqs. (11.6) and (11.10), we can obtain p_{11}, p_{12}, p_{21}, and p_{22}. For example,

$$p_{11} = P(X = 0, Y = 0) = P(X = 0)P(Y = 0) = \frac{1}{2} \times \frac{1}{2} = \frac{1}{4}.$$

Therefore, the joint probability mass distribution of (X, Y) is shown in Table 11.1.

Table 11.2 shows the joint cumulative distribution function of (X, Y). The value in each cell is obtained by summing over values of the joint probability mass function (i.e., to apply Eq. (11.7)). For example, consider $F_{XY}(X = 0, Y = 1)$:

$$F_{XY}(X = 0, Y = 1) = P(X = 0, Y = 0) + P(X = 0, Y = 1) = \frac{1}{4} + \frac{1}{4} = \frac{1}{2}.$$

The marginal probability distribution of (X, Y) is computed by applying Eq. (11.8) or Eq. (11.9). For example, $P(X = 0) = p_X(X = 0) = (P(X = 0, Y = 0) + P(X = 0, Y = 1)) = \frac{1}{4} + \frac{1}{4} = \frac{1}{2}$. Table 11.3 shows the marginal probability distribution of (X, Y).

Table 11.1 The joint probability mass function of X and Y as given in Example 11.4

	X=0	X=1
Y=0	$\frac{1}{4}$	$\frac{1}{4}$
Y=1	$\frac{1}{4}$	$\frac{1}{4}$

Table 11.2 The joint cumulative distribution function of X and Y as given in Example 11.4

	X=0	X=1
Y=0	$\frac{1}{4}$	$\frac{1}{2}$
Y=1	$\frac{1}{2}$	1

Table 11.3 The marginal probability of X and Y (as given in Example 11.4) are shown in the last row and the last column of the table

	X=0	X=1	$P(Y = y_i)$
Y=0	$\frac{1}{4}$	$\frac{1}{4}$	$\frac{1}{2}$
Y=1	$\frac{1}{4}$	$\frac{1}{4}$	$\frac{1}{2}$
$P(X = x_i)$	$\frac{1}{2}$	$\frac{1}{2}$	–

Example 11.5 Now, let us do the dice-throwing example where

$$X = \begin{cases} 0, & \text{An odd number is thrown,} \\ 1, & \text{An even number is thrown,} \end{cases}$$

and

$$Y = \begin{cases} 0, & \text{A non-square number is thrown,} \\ 1, & \text{A square number is thrown.} \end{cases}$$

Compute the joint probability distribution and the marginal probability distribution for (X, Y).

Solution Possible values of (X, Y) are $(0, 0)$, $(0, 1)$, $(1, 0)$, and $(1, 1)$. Assume two throws are independent.

Applying Eqs. (11.6) and (11.10), we can obtain p_{11}, p_{12}, p_{21}, and p_{22}. For example,

$$p_{22} = P(X = 1, Y = 1) = P(X = 1)P(Y = 1) = \frac{1}{2} \times \frac{1}{3} = \frac{1}{6}.$$

Therefore, the joint probability mass distribution of (X, Y) is shown in Table 11.4.

The marginal probability distribution of (X, Y) is computed by applying Eq. (11.8). For example, $P(X = 0) = p_X(X = 0) = P(X = 0, Y = 0) + P(X = 0, Y = 1) = \frac{1}{3} + \frac{1}{6} = \frac{1}{2}$. Table 11.4 also shows the marginal probability distribution of (X, Y).

11.2 Multiple Random Variables

Table 11.4 The joint probability mass function of X and Y (as given in Example 11.5) with the marginal probability of X and Y shown in the last row and the last column of the table

	X=0	X=1	$P(Y=y_i)$
Y=0	$\frac{1}{3}$	$\frac{1}{3}$	$\frac{2}{3}$
Y=1	$\frac{1}{6}$	$\frac{1}{6}$	$\frac{1}{3}$
$P(X=x_i)$	$\frac{1}{2}$	$\frac{1}{2}$	–

Table 11.5 An example of a joint probability mass function of X (games) and Y (weather) as given in Example 11.6

X\Y	Sunny	Cloudy	Rainy
Badminton	0	0	$\frac{1}{9}$
Swimming	$\frac{3}{9}$	$\frac{1}{9}$	$\frac{1}{9}$
Football	$\frac{2}{9}$	$\frac{1}{9}$	0

Table 11.6 An example of a joint probability mass function of X (games) and Y (weather) as given in Example 11.6

X\Y	Sunny	Cloudy	Rainy	$P(X=x_i)$
Badminton	0	0	$\frac{1}{9}$	$\frac{1}{9}$
Swimming	$\frac{3}{9}$	$\frac{1}{9}$	$\frac{1}{9}$	$\frac{5}{9}$
Football	$\frac{2}{9}$	$\frac{1}{9}$	0	$\frac{3}{9}$
$P(Y=y_i)$	$\frac{5}{9}$	$\frac{2}{9}$	$\frac{2}{9}$	–

Example 11.6 Suppose we have two random variables. X denotes the game Jack wants to play; Y denotes the weather condition. Table 11.5 shows the joint probability mass distribution of (X, Y), where the total probability of this table is equal to 1. The probability of Jack playing football when it is sunny is $P(X = football, Y = sunny) = \frac{2}{9}$ as shown in the table.

We need to apply Eq. (11.7) to obtain the cumulative distribution. However, X and Y are two categorical variables, and we cannot simply say $Y =$ Sunny $< Y =$ Rainy. Therefore, we can only set up the cumulative distribution table if given more information.

Table 11.6 shows the marginal probability distribution. If we sum over a row in Table 11.5, we are looking at all pairs (x_i, y_j), where Y can take on all three values with the value of X fixed. For example, the probability of Jack swimming is $P(X = Swimming) = \frac{3}{9} + \frac{1}{9} + \frac{1}{9} = \frac{5}{9}$. Similarly, if we sum over a column, we consider all possible values of X, and then we obtain the marginal probability mass value of Y.

Exercises

11.3 Given the pack of cards example described earlier, that is,

$$X = \begin{cases} 0, & \text{The card is black,} \\ 1, & \text{The card is red,} \end{cases}$$

and

$$Y = \begin{cases} 0, & \text{The card is a "low" card,} \\ 1, & \text{The card is a "high" card,} \end{cases}$$

find the joint probability distribution of (X, Y) and the marginal probability distribution of (X, Y).

11.4 There are five balls, three red and two green, in a bag. Take two balls one by one from the bag without putting them back. Define X and Y as follows:

$$X = \begin{cases} 0, & \text{The first one is green,} \\ 1, & \text{The first one is red,} \end{cases}$$

and

$$Y = \begin{cases} 0, & \text{The second one is green,} \\ 1, & \text{The second one is red.} \end{cases}$$

Compute the joint probability distribution of (X, Y) and the marginal probability of (X, Y).

11.5 There are ten balls, five red, three green, and two yellow, in a bag. Take a ball from the bag and put it back, and then take a second ball. Define X and Y as follows:

$$X = \begin{cases} 0, & \text{The first one is yellow,} \\ 1, & \text{The first one is green,} \\ 2, & \text{The first one is red,} \end{cases}$$

(continued)

11.2 Multiple Random Variables

and

$$Y = \begin{cases} 0, & \text{The second one is yellow,} \\ 1, & \text{The second one is green,} \\ 2, & \text{The second one is red.} \end{cases}$$

Compute the joint probability distribution of (X, Y) and the marginal probability of (X, Y).

11.2.2 Joint Probability Distributions: Continuous Random Variables

This is similar to the previous section, except, since the random variables are continuous, we need to integrate rather than add up probabilities. Also, since we have two random variables, the probability density function is a surface "above" the two variables plotted horizontally. To find probabilities, we need to use double integration to find the volume "under" the surface as in Sect. 6.3 of Chap. 6.

11.2.2.1 Joint Probability Mass Functions and Cumulative Distribution Functions

Consider two continuous random variables X and Y. The joint probability density function, $f_{XY}(x, y)$, can be given by

$$f_{XY}(x, y) = \frac{\partial^2 F_{XY}(x, y)}{\partial x \partial y}, \quad (11.11)$$

where $F_{XY}(x, y)$ is the cumulative distribution function given by

$$F_{XY}(x, y) = \int_{-\infty}^{x} \int_{-\infty}^{y} f_{XY}(\eta, \xi) d\eta d\xi. \quad (11.12)$$

Properties of $f_{XY}(x, y)$

1. $f_{XY}(x, y) \geq 0$, probability is always positive.
2. $\int_{-\infty}^{\infty} \int_{-\infty}^{\infty} f_{XY}(x, y) dx dy = 1$, the total probability is always 1.
3. $f_{XY}(x, y)$ is continuous for all values of (x, y), or except for a finite set.
4. $P(X, Y \in D) = \iint_D f(x, y) dx dy$.

Example 11.7 Let us start with a really simple example: The joint probability density function of (X, Y) is given by

$$f_{XY}(x, y) = \begin{cases} 1, & 0 < x < 1,\ 0 < y < 1 \\ 0, & \text{otherwise.} \end{cases}$$

This is a uniform distribution. The two random variables, x and y, go from 0 to 1, and the probability density function is just a flat surface "above" at a constant "height" of 1. (See the equivalent distribution with one variable given in Sect. 10.6.2.1 and illustrated on the left in Fig. 10.8 in Chap. 10.). Here, it would be illustrated by a cube of size 1. Show that
$\int_{-\infty}^{\infty} \int_{-\infty}^{\infty} f_{XY}(x, y) dx dy = 1$.
Find:

1. The cumulative distribution function $F_{XY}(x, y)$
2. $P(0 \leq X < \frac{1}{2},\ 0 \leq Y < \frac{1}{2})$
3. $P(X + Y < 1)$

Solution $\int_{-\infty}^{\infty} \int_{-\infty}^{\infty} f_{XY}(x, y) dx dy = \int_0^1 \int_0^1 1 dx dy = \int_0^1 x \Big|_0^1 dy = \int_0^1 1 dy = y \Big|_0^1 = 1$.

1. Applying Eq. (11.12), when $0 < x < 1$ and $0 < y < 1$, we have
$\int_0^x \int_0^y 1 d\eta d\xi = \int_0^y \eta \Big|_0^x d\xi = \int_0^y x d\xi = x\eta \Big|_0^y = xy$.
Hence, we have

$$F_{XY}(x, y) = \begin{cases} xy, & 0 < x < 1,\ 0 < y < 1 \\ 0, & \text{otherwise.} \end{cases}$$

2. $P(0 \leq X < \frac{1}{2},\ 0 \leq Y < \frac{1}{2}) =$
$\int_0^{\frac{1}{2}} \int_0^{\frac{1}{2}} 1 dx dy = \int_0^{\frac{1}{2}} x \Big|_0^{\frac{1}{2}} dy = \int_0^{\frac{1}{2}} \frac{1}{2} dy = \frac{1}{2} y \Big|_0^{\frac{1}{2}} = \frac{1}{4}$.

This answer makes sense if you think about it. With $0 < x < \frac{1}{2}$ and $0 < y < \frac{1}{2}$, looking "down" on the cube, we are taking just a quarter of the horizontal area. So, since the probability density function is uniform, this gives a quarter of the volume, that is, a quarter of the total probability.

3. $P(X + Y < 1)$. This is more complicated since the limits of the integration involve the line $x + y = 1$. Looking back at the examples in Sect. 6.3 of

(continued)

Example 11.7 (continued)

Chap. 6, we can use the y limits of 0 to $1-x$ and the x limits of 0 to 1. This gives
$P(X + Y < 1) =$

$\int_0^1 \left(\int_0^{1-x} 1 dy \right) dx = \int_0^1 y \Big|_0^{1-x} dx = \int_0^1 1 - x dx = x - \frac{1}{2}x^2 \Big|_0^1 = \frac{1}{2}.$

This again makes sense, since the line $x + y = 1$ cuts the horizontal area in half, and so, since the probability density function is uniform, the volume and probability are also a half.

Example 11.8 The joint probability density function of (X, Y) is given by

$$f_{XY}(x, y) = \begin{cases} Ce^{-(2x+3y)}, & x > 0, \ y > 0 \\ 0, & \text{otherwise} \end{cases}$$

where C is a constant. Find the following:

1. The value of C.
2. The cumulative distribution function $F_{XY}(x, y)$.
3. $P(0 \leq X < 1, 0 \leq Y < 2)$.

Solution

1. Since $\int_{-\infty}^{\infty} \int_{-\infty}^{\infty} f_{XY}(x, y) dx dy = 1$, we have

$$\int_0^{\infty} \int_0^{\infty} Ce^{-(2x+3y)} dx dy = C \int_0^{\infty} e^{-2x} dx \int_0^{\infty} e^{-3y} dy$$

$$= C \times (-\frac{1}{2})$$

$$\times (-\frac{1}{3}) \int_0^{\infty} e^{-2x} d(-2x) \int_0^{\infty} e^{-3y} d(-3y)$$

$$= \frac{C}{6} e^{-2x} \Big|_0^{\infty} e^{-3y} \Big|_0^{\infty}$$

$$= \frac{C}{6} = 1.$$

Therefore, $C = 6$.

(continued)

Example 11.8 (continued)

2. Applying Eq. (11.12), when $x > 0$ and $y > 0$, we have

$$\int_0^x \int_0^y 6e^{-(2\eta+3\xi)}d\eta d\xi = e^{-2\eta}\Big|_0^x \, e^{-3\xi}\Big|_0^y = (e^{-2x}-1)(e^{-3y}-1).$$

Hence, we have

$$F_{XY}(x,y) = \begin{cases} (e^{-2x}-1)(e^{-3y}-1), & x>0,\ y>0 \\ 0, & \text{otherwise}. \end{cases}$$

3. $P(0 \leq X < 1,\ 0 \leq Y < 2) = 6\int_0^1 e^{-2x}dx \int_0^2 e^{-3y}dy = e^{-2x}\Big|_0^1 e^{-3y}\Big|_0^2 = (e^{-2}-1)(e^{-6}-1) \approx 0.8625.$

Exercise

11.6 The joint probability density function of (X, Y) is given by

$$f_{XY}(x,y) = \begin{cases} A(x+y), & 0<x<4,\ 0<y<4 \\ 0, & \text{otherwise} \end{cases}$$

where A is a constant. Find the following:

(1) The value of A.
(2) The cumulative distribution function $F_{XY}(x, y)$.
(3) $P(0 \leq X < 2,\ 0 \leq Y < 2)$.
(4) $P(X + Y < 4)$.

11.2.2.2 Marginal Probability Distributions

Consider two continuous random variables X and Y. The marginal probability density function of X or Y is given by

$$f_X(x) = \int_{-\infty}^{\infty} f_{XY}(x,y)dy, \tag{11.13}$$

or

$$f_Y(y) = \int_{-\infty}^{\infty} f_{XY}(x, y)dx, \qquad (11.14)$$

respectively. This is like just adding up the values in a row or a column in the discrete case.

Example 11.9 The joint probability distribution function of random variables X and Y is given by

$$f_{XY}(x, y) = \begin{cases} 2e^{-(2x+y)}, & x > 0, y > 0 \\ 0, & \text{otherwise.} \end{cases}$$

Compute the marginal probability distribution of X.

Solution Applying Eq. (11.13), the marginal probability distribution of X is

$$f_X(x) = 2e^{-2x} \times \int_0^{\infty} e^{-y} dy$$

$$= -2e^{-2x} e^{-y} \Big|_0^{\infty}$$

$$= 2e^{-2x}.$$

Exercises

11.7 The joint probability distribution function of random variables X and Y is given by

$$f_{XY}(x, y) = \begin{cases} \frac{1}{x} e^{\frac{-y}{x}} e^{-x}, & x > 0, y > 0 \\ 0, & \text{otherwise.} \end{cases}$$

Compute the marginal probability distribution of X.

11.8 The joint probability density function of (X, Y) is given by

$$f_{XY}(x, y) = \begin{cases} A\cos(x - y), & 0 < x < \frac{\pi}{2}, 0 < y < \frac{\pi}{2} \\ 0, & \text{otherwise} \end{cases}$$

(continued)

where A is a constant. Find the following:

(1) The value of A.
(2) The cumulative distribution function $F_{XY}(x, y)$.
(3) $P(0 \leq X < \frac{\pi}{4}, 0 \leq Y < \frac{\pi}{4})$.
(4) The marginal probability distribution of X.
(5) The marginal probability distribution of Y.

Remark 11.4 Note that X and Y are independent random variables, if

$$F_{XY}(x, y) = F_X(x) F_Y(y),$$

or

$$f_{XY}(x, y) = f_X(x) f_Y(y).$$

♦

11.2.2.3 Covariance

Consider two jointly distributed continuous random variables X and Y. We can define the covariance via the expected value as follows:

$$cov(X, Y) = E\Big((X - E(X))(Y - E(Y))\Big). \tag{11.15}$$

Remark 11.5 We can see that this definition fits the shape of the original sample covariance definition, which was in Sect. 4.2.1 of Chap. 4, namely,

$$cov(x_h, x_k) = \frac{\sum_{i=1}^{n}(x_{i,h} - \bar{x}_h)(x_{i,k} - \bar{x}_k)}{n - 1}.$$

As can be seen, it basically multiplies each value of one variable, with its mean subtracted, and with each value of the other variable, with its mean subtracted. It then sums all of these together and divides the result by a constant. The final summing is basically finding a mean again. Hence, the given definition is of the correct form.

♦

11.2 Multiple Random Variables

Equation (11.15) can be further expanded using properties of the expected value given in Sect. 10.5.1.1 of Chap. 10 as follows:

$$cov(X, Y) = E\Big((X - E(X))(Y - E(Y))\Big)$$
$$= E\Big(XY - E(X)Y - XE(Y) + E(X)E(Y)\Big) \quad (11.16)$$
$$= E(XY) - E(X)E(Y) - E(X)E(Y) + E(X)E(Y)$$
$$= E(XY) - E(X)E(Y).$$

Hence, if $cov(X, Y) = 0$, that is, if X and Y are uncorrelated, then we have $E(XY) = E(X)E(Y)$.

We have the two following properties of uncorrelated variables:

- If X and Y are independent random variables, they are also uncorrelated. This can be proved as follows:

$$E(XY) = \int\int xy f_{XY}(x, y) dx dy$$
$$= \int\int xy f_X(x) f_Y(y) dx dy$$
$$= \int x f_X(x) \left(\int y f_Y(y) dy\right) dx \quad (11.17)$$
$$= \left(\int x f_X(x) dx\right)\left(\int y f_Y(y) dy\right)$$
$$= E(X)E(Y).$$

- However, if X and Y are uncorrelated, they can still be dependent as the following example illustrates.

Example 11.10 This relationship is usually illustrated by using variables that have clearly got zero mean, so that $cov(X, Y)$ is also zero. A common example is the following:

Let X be uniformly distributed in the interval $[-1, 1]$. It clearly has a mean value at the centre, that is, $E(X) = 0$. Now, define Y so that $y = x^2$ in the interval $[-1, 1]$ and zero elsewhere. This also obviously has a mean value at the centre, since it is evenly distributed either side of zero. So, E(Y) = 0. Now, $E(XY) = E(X^3)$, which again has its mean value at the centre. So $E(XY) = 0$. We now have $cov(X, Y) = E(XY) - E(X)E(Y) = 0$. So, X and Y are uncorrelated.

(continued)

Example 11.10 (continued)
However, from the way that Y was defined, it is clearly dependent on X. So, X and Y are uncorrelated and also dependent.

We can now do an example of finding the covariance between two continuous random variables. We will start with the easy example described in Example 11.7.

Example 11.11 Start with the example started in Example 11.7, namely, the joint probability distribution function of random variables X and Y given by

$$f_{XY}(x, y) = \begin{cases} 1, & 0 < x < 1, \ 0 < y < 1 \\ 0, & \text{otherwise.} \end{cases}$$

Find the covariance of X and Y.

Solution To find the covariance, we will use Eq. (11.16). This requires finding $E(X)$ and $E(Y)$, which means we need to know $f_X(x)$ and $f_Y(y)$ as is seen from the definition of mean given in Eq. (10.8) in Sect. 10.5 of Chap. 10. But $f_X(x)$ and $f_Y(y)$ are just the marginal probability distributions of X and Y. To find $f_X(x)$, we use Eq. (11.13):

$$f_X(x) = \int_0^\infty 1 \, dy = y \Big|_0^1 = 1.$$

To find $f_Y(y)$, we use Eq. (11.14):

$$f_Y(y) = \int_0^\infty 1 \, dx = x \Big|_0^1 = 1.$$

We can now find $E(X)$ and $E(Y)$ using Eq. (10.8):

$$E(X) = \int_{-\infty}^{\infty} x f_X(x) dx = \int_0^1 x \, dx = \frac{x^2}{2} \Big|_0^1 = \frac{1}{2}$$

By a similar calculation, we find that

$$E(Y) = \int_{-\infty}^{\infty} y f_Y(y) dy = \int_0^1 y \, dy = \frac{y^2}{2} \Big|_0^1 = \frac{1}{2}.$$

Then, we find $E(XY)$: $E(XY) = \int_{-\infty}^{\infty} \int_{-\infty}^{\infty} xy f_{XY}(x,y) dx dy =$
$\int_0^1 \int_0^1 xy \, dx \, dy = \int_0^1 y \frac{x^2}{2} \Big|_0^1 dy = \int_0^1 \frac{1}{2} y \, dy = \frac{y^2}{4} \Big|_0^1 = \frac{1}{4}$.
Finally, $cov(X, Y) = E(XY) - E(X)E(Y) = \frac{1}{4} - \frac{1}{2} \times \frac{1}{2} = 0$.

Example 11.12 We will continue with the example started in Example 11.9, namely, the joint probability distribution function of random variables X and Y given by

$$f_{XY}(x, y) = \begin{cases} 2e^{-(2x+y)}, & x > 0, \ y > 0 \\ 0, & \text{otherwise.} \end{cases}$$

Find the covariance of X and Y.

Solution To find the covariance, we will use Eq. (11.16). Again, this requires finding $E(X)$ and $E(Y)$, which means we need to know $f_X(x)$ and $f_Y(y)$ as is seen from the definition of mean given in Eq. (10.8) in Sect. 10.5 of Chap. 10. $f_X(x)$ and $f_Y(y)$ are the marginal probability distributions of X and Y. In fact, Example 11.9 already found $f_X(x)$. It was $f_X(x) = 2e^{-2x}$.
To find $f_Y(y)$, we use Eq. (11.14):

$$f_Y(y) = 2e^{-y} \times \int_0^\infty e^{-2x} dx = -e^{-y} e^{-2x} \Big|_0^\infty = e^{-y}.$$

We can now find $E(X)$ and $E(Y)$ using Eq. (10.8):

$$E(X) = \int_{-\infty}^\infty x f_X(x) dx = 2 \int_0^\infty x e^{-2x} dx$$

This integral is calculated using integration by parts (Sect. 5.5.2 of Chap. 5) with $u = x$ and $\frac{dv}{dx} = e^{-2x}$, giving

$$2x \frac{e^{-2x}}{-2} \Big|_0^\infty - 2 \int_0^\infty \frac{e^{-2x}}{-2} \times 1 dx = 0 + \frac{e^{-2x}}{-2} \Big|_0^\infty = \frac{1}{2}.$$

By a similar calculation, we find that

$$E(Y) = \int_{-\infty}^\infty y f_Y(y) dy = \int_0^\infty y e^{-y} dy = 1.$$

Finally, we find $E(XY)$:

$$E(XY) = \int_{-\infty}^\infty \int_{-\infty}^\infty xy f_{XY}(x, y) dx dy$$

$$= \int_0^\infty \int_0^\infty xy 2 e^{-(2x+y)} dx dy$$

$$= 2 \int_0^\infty y e^{-y} \left(\int_0^\infty x e^{-2x} dx \right) dy$$

(continued)

Example 11.12 (continued)

$$= \frac{1}{2}\int_0^\infty ye^{-y}dy$$

$$= \frac{1}{2}.$$

Both integrals use integration by parts again (and in fact are the same as the integrals calculated to find $E(X)$ and $E(Y)$).
Finally,

$$cov(X, Y) = E(XY) - E(X)E(Y) = \frac{1}{2} - \frac{1}{2} \times 1 = 0.$$

Exercises

11.9 The joint probability distribution function of random variables X and Y is given by

$$f_{XY}(x, y) = \begin{cases} 6e^{-(2x+3y)}, & x > 0, \ y > 0 \\ 0, & \text{otherwise.} \end{cases}$$

Find $cov(X, Y)$.

11.10 The joint probability distribution function of random variables X and Y is given by

$$f_{XY}(x, y) = \begin{cases} \dfrac{(x+y)}{64}, & 0 < x < 4, \ 0 < y < 4 \\ 0, & \text{otherwise} \end{cases}$$

Find $cov(X, Y)$.

11.2.3 Multinomial Distribution

A binomial distribution (see Sect. 10.6.1.3 of Chap. 10) is used to study the probability distribution of multiple independent trials, each with two possible outcomes. The multinomial distribution gives the probability of an event over

11.2 Multiple Random Variables

multiple trials when we have more than two possible outcomes for each trial. So, this is a generalisation of a binomial distribution.

Consider n repeated and independent trials. Suppose each trial has k possible outcomes. Let X_i be the discrete random variable taking values of x_i, which is the number of occurrences of outcome i, and $i = 1, 2, \cdots, k$. Then, we have $x_1, x_2, \cdots, x_k \in [0, 1, \cdots, n]$, such that $x_1 + x_2 + \cdots + x_k = n$. Let $p_1, p_2, \cdots, p_k \in [0, 1]$ and $\sum_{i=1}^{k} p_i = 1$.

For example, in a bag containing three types of fruit (apples, oranges, and pears), if you pick one out of the bag, there are just three outcomes (an apple, an orange, or a pear). If a trial consists of taking one fruit out without looking, noting its type, and replacing it, then the probabilities remain the same for each trial. Suppose you do ten trials. Then, x_1 is the number of times an apple is picked, x_2 is the number of times an orange is picked, and x_3 is the number of times a pear is picked. Obviously, each x_i is a number between 1 and 10 (you could pick an apple each time, in which case x_1 is ten and x_2 and x_3 are zeros). Whatever the types of fruit picked, the total must add up to ten, that is, $x_1 + x_2 + x_3 = 10$. The probabilities of picking each fruit, p_1, p_2, p_3, depend on the numbers of each in the bag (this does not count any differences in feel between the three fruits!).

The probability mass function for the multinomial distribution is given by

$$p_{X_1 X_2 \cdots X_k}(x_1, x_2, \cdots, x_k) = \frac{n!}{x_1! x_2! \cdots x_k!} p_1^{x_1} p_2^{x_2} \cdots p_k^{x_k}. \tag{11.18}$$

This is just a generalisation of the equation for the binomial mass function given in Eq. (10.15) of Chap. 10. In that equation, there were just two outcomes with probability p and $1 - p$, where the first outcome happened x times, so the other happened $n - x$ times. Now, we have k different occurrences happening $x_1, \cdots, x_k$ times, respectively, to replace the x and $n - x$.

The mean and variance of X_i are given by

$$E(X_i) = np_i, \tag{11.19}$$

and

$$Var(X_i) = np_i(1 - p_i), \tag{11.20}$$

respectively.

Example 11.13 Suppose that a fair die is rolled eight times. Find the probability that one, two, three, and four dots appear once each and five dots and six dots twice each.

Solution Applying Eq. (11.18), we have

$$P(X_1 = 1, X_2 = 1, X_3 = 1, X_4 = 1, X_5 = 2, X_6 = 2)$$
$$= \frac{8!}{1!1!1!1!2!2!}\left(\frac{1}{6}\right)^1\left(\frac{1}{6}\right)^1\left(\frac{1}{6}\right)^1\left(\frac{1}{6}\right)^1\left(\frac{1}{6}\right)^2\left(\frac{1}{6}\right)^2$$
$$\approx 0.006.$$

Exercise

11.11 There are 100 marbles of the same size but four different colours in a bag. The ratio of red, black, white, and yellow is 2:3:4:1. Jack takes six marbles from the bag without looking, replacing the marble each time. Compute the probability that he takes two red, one black, and three yellow.

11.2.4 Multivariate Normal Distribution

Let $X_1, \ldots, X_d$ be independent normally distributed with means $\mu_1, \ldots, \mu_d$ and variance $\sigma_1^2, \ldots, \sigma_d^2$. Then, the joint density of $X_1, \ldots, X_d$ is

$$f_\mathbf{X}(x_1, \ldots, x_d) = \frac{1}{(2\pi)^{\frac{d}{2}}} \prod_{i=1}^{d}\left(\frac{1}{\sigma_i}\right) \exp\left(-\frac{1}{2}\sum_i^d \frac{(x_i - \mu_i)^2}{\sigma_i^2}\right), \qquad (11.21)$$

where $\mathbf{X}$ is a d-dimensional random vector $[X_1, \ldots, X_d]$.

Let us look at this for $d = 2$:

$$f_\mathbf{X}(x_1, x_2) = \frac{1}{(2\pi)} \frac{1}{\sigma_1 \sigma_2} \exp\left(-\frac{1}{2}\left(\frac{(x_1-\mu_1)^2}{\sigma_1^2}\right) - \frac{1}{2}\left(\frac{(x_2-\mu_2)^2}{\sigma_2^2}\right)\right)$$
$$= \frac{1}{(2\pi)} \frac{1}{\sigma_1 \sigma_2} e^{-\frac{(x_1-\mu_1)^2}{2\sigma_1^2}} \times e^{-\frac{(x_2-\mu_2)^2}{2\sigma_2^2}}$$
$$= \left(\frac{1}{\sqrt{2\pi}} \frac{1}{\sigma_1} e^{-\frac{(x_1-\mu_1)^2}{2\sigma_1^2}}\right) \times \left(\frac{1}{\sqrt{2\pi}} \frac{1}{\sigma_2} e^{-\frac{(x_2-\mu_2)^2}{2\sigma_2^2}}\right).$$

11.2 Multiple Random Variables

If you look back at Eq. (10.19) in Chap. 10, you can see that this is just two normal distribution equations in two different dimensions multiplied together.

Since Eq. (11.21) is an equation in vectors, it can also be expressed as follows:

$$f_{\mathbf{X}}(\mathbf{x}) = \frac{1}{(2\pi)^{\frac{d}{2}} |\mathbf{\Sigma}_0|^{\frac{1}{2}}} \exp[-\frac{1}{2}(\mathbf{x} - \boldsymbol{\mu})^T \mathbf{\Sigma}_0^{-1}(\mathbf{x} - \boldsymbol{\mu})], \qquad (11.22)$$

where $\boldsymbol{\mu}$ is the mean vector and $\mathbf{\Sigma}_0$ is a diagonal matrix:

$$\mathbf{\Sigma}_0 = \begin{bmatrix} \sigma_1^2 & 0 & \ldots & 0 \\ 0 & \sigma_2^2 & \ldots & 0 \\ \vdots & \vdots & \ddots & \vdots \\ 0 & 0 & \ldots & \sigma_d^2 \end{bmatrix}.$$

Again you can check this by expanding out the $d = 2$ case, where

$$\mathbf{\Sigma}_0 = \begin{bmatrix} \sigma_1^2 & 0 \\ 0 & \sigma_2^2 \end{bmatrix},$$

and

$$\mathbf{\Sigma}_0^{-1} = \frac{1}{\sigma_1^2 \sigma_2^2} \times \begin{bmatrix} \sigma_2^2 & 0 \\ 0 & \sigma_1^2 \end{bmatrix}.$$

In fact, $\mathbf{\Sigma}_0$ in Eq. (11.22) is a d-dimensional covariance matrix with no cross-correlation between any of $X_1, \ldots, X_d$. The multivariate normal distribution can be denoted as $\mathbf{X} \sim \mathcal{N}(\boldsymbol{\mu}, \mathbf{\Sigma}_0)$.

A more general formula for the joint density of a random vector $\mathbf{X} = [X_1, \ldots, X_d]$ of size d, which is distributed according to a multivariate normal distribution and does have a cross-correlation between the $X_1, \ldots, X_d$, is given as follows:

$$f_{\mathbf{X}}(\mathbf{x}) = \frac{1}{(2\pi)^{\frac{d}{2}} |\mathbf{\Sigma}|^{\frac{1}{2}}} \exp[-\frac{1}{2}(\mathbf{x} - \boldsymbol{\mu})^T \mathbf{\Sigma}^{-1}(\mathbf{x} - \boldsymbol{\mu})], \qquad (11.23)$$

where $\boldsymbol{\mu}$ is the mean vector and $\mathbf{\Sigma}$ is a general covariance matrix. It can be denoted as $\mathbf{X} \sim \mathcal{N}(\boldsymbol{\mu}, \mathbf{\Sigma})$.

Figure 11.2 shows two contour plots of bivariate normal distributions with 100 samples, each displayed as the plus signs. The left panel shows a bivariate distribution as

$$\mathcal{N}\left(\begin{bmatrix} 2 \\ 4 \end{bmatrix}, \begin{bmatrix} 1 & 0 \\ 0 & 1 \end{bmatrix}\right),$$

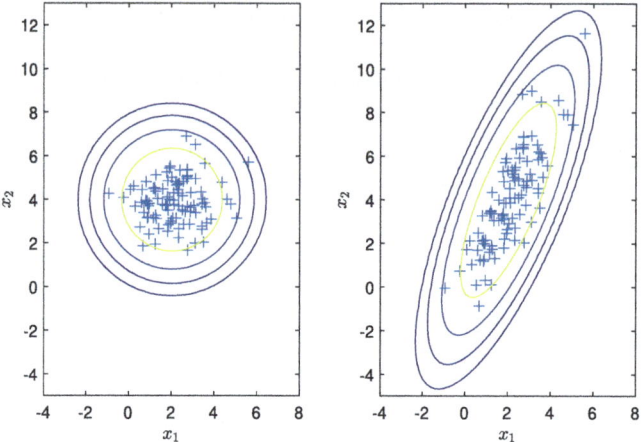

Fig. 11.2 Examples of bivariate normal distribution. In the left panel, the two normal random variables are uncorrelated; in the right panel, the two normal random variables are positively correlated

where the covariance between X_1 and X_2 is 0, that is, X_1 and X_2 are uncorrelated. The right panel shows a bivariate distribution as

$$\mathcal{N}\left(\begin{bmatrix} 2 \\ 4 \end{bmatrix}, \begin{bmatrix} 1 & 1.5 \\ 1.5 & 4 \end{bmatrix}\right),$$

where the covariance between X_1 and X_2 is 1.5. As can be seen in the plot, it shows that as values of X_1 increase, values of X_2 also increase. That is, the two variables are positively correlated.

Figure 11.3 shows another example of a two-dimensional multivariate normal distribution:

$$\mathcal{N}\left(\begin{bmatrix} 0 \\ 0 \end{bmatrix}, \begin{bmatrix} 0.49 & 0.4 \\ 0.4 & 1 \end{bmatrix}\right).$$

The left panel presents the probability density distribution. The right displays its corresponding cumulative distribution.

Variables with a multivariate normal distribution with a mean vector μ and a covariance matrix Σ have the following properties:

- Every single variable has a univariate normal distribution.
- Any subset of the variables also has a multivariate normal distribution.
- Any linear combination of the variables has a univariate normal distribution.
- Any conditional distribution for a subset of the variables dependent on known values for another subset of variables is a multivariate distribution.

We will discuss conditional probability in the following section.

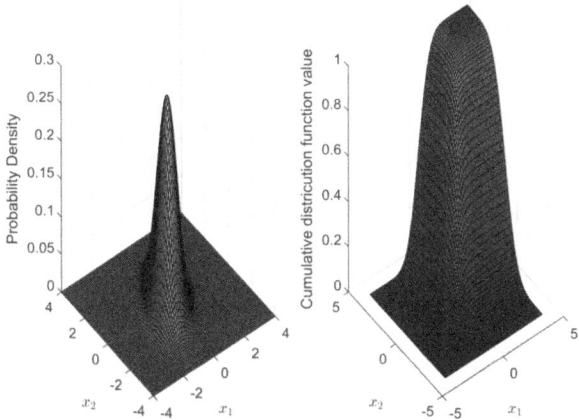

Fig. 11.3 An example of a two-dimensional multivariate normal distribution. The left panel presents the probability density distribution. The right displays its corresponding cumulative distribution

11.3 Conditional Probability and Corresponding Rules

11.3.1 Conditional Probability

So far, we have discussed the probability of an event (E) occurring without any conditions. Sometimes, a specific event may happen under certain conditions. Let us consider the following example first.

> **Example 11.14** 100 people showed up for a new test for bowel cancer. 30 of them have bowel cancer. 40 people had a positive test result, of which 25 had bowel cancer. Calculate the following:
>
> - The probability of people having bowel cancer and testing positive.
> - The likelihood of people having bowel cancer or being tested positive.
>
> **Solution** Let us define event A, people have bowel cancer, and event B, the test result of people was positive. Figure 11.4 shows the Venn diagram of this example. In this figure, A has 35 people, B has 40 people, and 25 people are in both A and B, denoted as $A \cap B$ in the diagram. So, this means there are ten people in A but not in $A \cap B$, and 15 people in B but not in $A \cap B$.
>
> So, from the diagram, among the total 100 people, 25 people have bowel cancer and tested positive, that is, $A \cap B$. Thus, the probability of people who
>
> (continued)

Example 11.14 (continued)
actually have bowel cancer and who tested positive is

$$P(A \cap B) = \frac{25}{100} = 25\%.$$

To calculate the probability of people having bowel cancer or being tested positive, we need to know the number of people in set A and B, that is, $P(A \cup B)$. But if we add A and B together, we get the overlapping area of $A \cap B$ twice, so we need to subtract the overlapping area. We, therefore, have $35 + 40 - 25 = 50$. (Alternatively, we add the number in A but not in $A \cap B$, that is, $A \setminus (A \cap B) = 10$, plus the number in B but not in $A \cap B$, that is, $B \setminus (A \cap B) = 15$, and the number in $A \cap B = 25$. This gives 50, as before). Thus, the probability of people either having bowel cancer or being tested positive is

$$P(A \cup B) = \frac{50}{100} = 50\%.$$

Alternatively, we have, as a formula (see Sect. 11.3.3),

$$P(A \cup B) = P(A) + P(B) - P(A \cap B) = \frac{35}{100} + \frac{40}{100} - \frac{25}{100} = 50\%.$$

11.3.1.1 Conditional Probability for Two Discrete Random Variables

Conditional probability is the probability of some event happening given that some other event has already occurred. So, the probability is conditional on the other event having occurred. The probability that $X = x_i$ given that $Y = y_j$ is usually written as $p_{X|Y}(x_i|y_j)$ or as $p(X = x_i|Y = y_j)$.

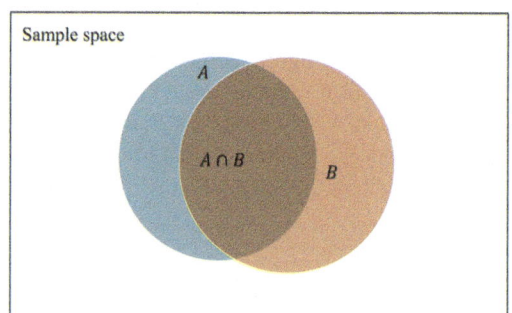

Fig. 11.4 The Venn diagram of Example 11.14. A represents having cancer, and B represents testing positive

11.3 Conditional Probability and Corresponding Rules

Definition 11.1 Suppose (X, Y) are two discrete random variables with joint probability mass function $p_{XY}(x_i, y_i)$. The conditional probability mass function of X given that $Y = y_j$ is defined as

$$p_{X|Y}(x_i|y_j) = \frac{p_{XY}(x_i \cap y_j)}{p_Y(y_j)}, \quad \text{where } p_Y(y_j) > 0. \tag{11.24}$$

Note that $0 \le p_{X|Y}(x_i|y_j) \le 1$ and $\sum_{x_i} p_{X|Y}(x_i|y_j) = 1$.
This definition makes sense since, whereas normally we are dividing by the whole population to get a probability, now we are restricted to just those events that occur when $Y = y_j$. So, we divide by just the relevant ones, that is, all the $Y = y_j$'s.

Example 11.15 Continue Example 11.14. What percent of those who tested positive have cancer?

Solution Here, we have the condition that we are only interested in those who tested positive. That is, we want the number who have cancer **given that** they have already tested positive. Let $X = A$ denote people having bowel cancer, and $Y = B$, the test result of people being positive. So, in Fig. 11.4, we are interested in those people with cancer inside circle B, that is, $p(A \cap B)$ as a proportion of all of B.

Applying Eq. (11.24), we have

$$p(A|B) = \frac{p(A \cap B)}{p(B)} = \frac{25\%}{40\%} = 62.5\%.$$

In Example 11.14, we were given all the numbers, but often we have to work them out first, as in the next example.

Example 11.16 A bag contains six red balls and four green balls. A ball is taken out but not put back, and then a second ball is taken out. Calculate the probability of picking a red ball as the second ball, given that the first ball was green.

Solution Let A be the probability that the second ball is red and B be the probability that the first ball is green. We are interested in $p(A|B)$. So, we need the probability of the first ball being green, $P(B)$, which is $\frac{4}{10} = \frac{2}{5}$.

(continued)

Example 11.16 (continued)
We now need $P(A \cap B)$. This is the probability that it is both A and B. So, we need the probability that the first ball is green and then the second is red, which is $\frac{4}{10} \times \frac{6}{9} = \frac{4}{15}$. Hence, using Eq. (11.24), we have

$$p(A|B) = \frac{p(A \cap B)}{p(B)} = \frac{\frac{4}{15}}{\frac{2}{5}} = \frac{2}{3}.$$

Note that this is not equal to the probability of getting a red ball second. Doing this includes the case of getting a red first and then a red second, which is $\frac{6}{10} \times \frac{5}{9} = \frac{1}{3}$ as well as the case of getting a green followed by a red, which is $\frac{4}{15}$ from above. So, the probability of getting a red second is $\frac{1}{3} + \frac{4}{15} = \frac{3}{5}$.

Remark 11.6 Note that

$$p(\bar{A}|B) = 1 - p(A|B), \qquad (11.25)$$

where $\bar{A}$ is the complement of A.

This can be seen by considering the following illustration. If you are a man, then the probability you have a beard could be $\frac{1}{5}$, so that the probability that you do not have a beard is $\frac{4}{5}$. If you define A as the probability of having a beard and B as the probability of being a man, then $p(A|B)$ is the probability that you have a beard given that you are a man, and this is $\frac{1}{5}$. So, $p(\bar{A}|B)$ is the probability of not having a beard given that you are a man, which is $1 - p(A|B) = 1 - \frac{1}{5} = \frac{4}{5}$.

♦

Exercises

11.12 The percentage of all adults in America who are women and will have an episode of depression by the age of 65 is 16.667%. The percentage of all adults in America who are men and will have an episode of depression by the age of 65 is 10%. What is the probability that a given woman will have an episode of depression by the age of 65 in America? What is the probability that a given man will have an episode of depression by the age of 65 in America? (You can assume that the probabilities of an adult being male or female are both 50%.)

11.13 In a certain university, there are 1000 students, of which 540 are female. Of the female students, 300 take humanities subjects, and the rest take

(continued)

science subjects. Of the male students, 180 take humanities subjects, and the rest take science subjects. For a given female student, what is the probability that they do science? For a given science student, what is the probability of them being female? For a given male student, what is the probability that they do science?

11.14 Suppose we flip a fair coin three times. What is the probability of coming up precisely two heads, given the first flip is a heads-up?

11.15 Ann has two children. You learn that she has a daughter, Sarah. What is the probability that Sarah's sibling is a brother? (You can assume that the probabilities of a child being male or female are both 50%.)

11.3.1.2 Conditional Probability for Two Continuous Random Variables

This again follows the discrete variable case.

Definition 11.2 Suppose (X, Y) are two continuous random variables with joint probability density function $f_{XY}(x, y)$. The conditional probability density function of X given that $Y = y$ is defined as

$$f_{X|Y}(x|y) = \frac{f_{XY}(x, y)}{f_Y(y)}, \quad \text{where } f_Y(y) > 0. \tag{11.26}$$

Note that $f_{X|Y}(x|y) \geq 0$ and $\int_{-\infty}^{\infty} f_{X|Y}(x|y)dx = 1$.

Example 11.17 Suppose X and Y are two continuous random variables, and their joint probability density function is given as follows:

$$f_{XY}(x, y) = \begin{cases} 2xy, & 0 < x < 1, \ 0 < y < \sqrt{2} \\ 0, & \text{otherwise.} \end{cases}$$

Find the conditional probability density function $f_{X|Y}(x|y)$.

Solution To apply Eq. (11.26), we need to compute $f_Y(y)$ first:

$$f_Y(y) = \int_0^1 2xy\,dx = y.$$

(continued)

Example 11.17 (continued)
Substituting $f_Y(y) = y$ and $f_{XY}(x, y) = 2xy$ into Eq. (11.26), we have

$$f_{X|Y}(x|y) = \frac{2xy}{y} = 2x$$

for $0 < x < 1$ and $0 < y < \sqrt{2}$.

Exercises

11.16 Suppose X and Y are two continuous random variables and their joint probability density function is given as follows:

$$f_{XY}(x, y) = \begin{cases} \frac{1}{2}, & 0 < y \leq x < 2 \\ 0, & \text{otherwise.} \end{cases}$$

Find the conditional probability density function $f_{Y|X}(y|x)$.

11.17 Suppose X and Y are two continuous random variables and their joint probability density function is given as follows:

$$f_{XY}(x, y) = \begin{cases} A \sin x \cos y, & 0 < x, y < \frac{\pi}{2} \\ 0, & \text{otherwise.} \end{cases}$$

(1) Find the value of A.
(2) Find the conditional probability density function $f_{X|Y}(x|y)$.
(3) Find the conditional probability density function $f_{Y|X}(y|x)$.

11.3.2 Conditional Means and Conditional Variances

11.3.2.1 Conditional Means and Conditional Variances for Two Discrete Random Variables

If X and Y are two discrete random variables with joint probability mass function $p_{XY}(x_i, y_i)$, then the conditional mean of Y given $X = x_i$ is defined by

$$E(Y|x_i) = \sum_{y_i} y_i \, p_{Y|X}(y_i|x_i), \tag{11.27}$$

11.3 Conditional Probability and Corresponding Rules

and the conditional variance of Y given $X = x_i$ is defined by

$$Var(Y|x_i) = \sum_{y_i} \left(y_i - E(Y|x_i)\right)^2 p_{Y|X}(y_i|x_i). \quad (11.28)$$

Note that these follow the same construction as those given in Eq. (10.8) for mean and Eq. (10.9) for variance in Chap. 10, except that in these, we are summing over y_i rather than x_k since the x_i's are given and so effectively constant.

11.3.2.2 Conditional Means and Conditional Variances for Two Continuous Random Variables

If X and Y are two continuous random variables with joint probability density function $f_{XY}(x, y)$, then the conditional mean of Y given $X = x$ is defined by

$$E(Y|x) = \int_{-\infty}^{\infty} y f_{Y|X}(y|x) dy, \quad (11.29)$$

and the conditional variance of Y given $X = x$ is defined by

$$Var(Y|x) = \int_{-\infty}^{\infty} \left(y - E(Y|x)\right)^2 f_{Y|X}(y|x) dy. \quad (11.30)$$

The variance can also be written as

$$Var(Y|x) = E(Y^2|x) - [E(Y|x)]^2. \quad (11.31)$$

Again see Eq. (10.8) for mean and Eq. (10.9) for variance in Chap. 10.

Example 11.18 Suppose X and Y are two continuous variables. The joint density function is given by

$$f_{XY}(x, y) = \begin{cases} 24x^2 y, & 0 < x < 1, \ 0 < y < \frac{1}{2} \\ 0, & \text{otherwise.} \end{cases}$$

Find $E(Y|x)$.

(continued)

Example 11.18 (continued)
Solution Since $f_X(x)$ is independent of y, Eq. (11.29) can be further written as follows:

$$E(Y|x) = \int_{-\infty}^{\infty} y \frac{f_{XY}(x,y)dy}{f_X(x)} = \frac{\int_{-\infty}^{\infty} y f_{XY}(x,y)dy}{f_X(x)}.$$

First, find $f_X(x)$:

$$f_X(x) = \int_{-\infty}^{\infty} f_{XY}(x,y)dy = \int_0^{\frac{1}{2}} 24x^2 y\, dy = 12x^2 y^2 \Big|_0^{\frac{1}{2}} = 3x^2 \text{ for } 0 < x < 1.$$

Therefore, we have

$$E(Y|x) = \frac{1}{3x^2} \int_0^{\frac{1}{2}} 24x^2 y^2 dy = \frac{8}{3} y^3 \Big|_0^{\frac{1}{2}} = \frac{1}{3}.$$

Exercise

11.18 Suppose X and Y are two continuous variables. The joint density function is given by

$$f_{XY}(x,y) = \begin{cases} 12x^3 y^2, & 0 < x < 1, \ 0 < y < 1 \\ 0, & \text{otherwise.} \end{cases}$$

Find $Var(Y|x)$ (Hint: apply Eq. (11.31)).

11.3.3 Mutual Exclusivity

Suppose A and B are two events. Then, they are mutually exclusive if they cannot co-occur, that is, $A \cap B = \emptyset$, and $P(A \cap B) = 0$. Then, we have

$$P(A \cup B) = P(A) + P(B) - P(A \cap B) = P(A) + P(B).$$

For example, the event of a student who passes or fails the same module is mutually exclusive.

11.3.4 The Multiplication Rule

The multiplication rule is defined by

$$P(A \cap B) = P(A)P(B|A). \tag{11.32}$$

We can obtain this rule by rearranging the definition of conditional probability. For two events A and B, if they can occur at the same time, that is, the overlapping area in a Venn diagram, then the probability of this is given by the product of the probability of event A occurring and the probability of B occurring given that A happens. The general multiplication rule is a handy way to find the probability that two events, A and B, occur if we can easily calculate the conditional probability $P(B|A)$ and the probability of A.

11.3.5 Independence

Event B is considered independent of event A if $P(B|A) = P(B)$. It says that learning that event A happened provides us with no additional information about event B. Using Eq. (11.32), this relationship is more usually written equivalently as

$$P(A \cap B) = P(A)P(B). \tag{11.33}$$

This is a nice result because we can find the probability of two events occurring without dealing with conditional probability calculations. The definition of independence of two events can be extended to describe three or more events.

> **Example 11.19** Flipping a fair coin three times. Let T_1, T_2, and T_3 be the events that the first, second, and third flips have tails-up. The probability that we see three flips coming up tails is computed as $P(T_1 \cap T_2 \cap T_3) = \frac{1}{2} \times \frac{1}{2} \times \frac{1}{2} = \frac{1}{8}$, assuming three flips are independent.

Similarly, if X and Y are two continuous independent random variables, then $f_{X|Y}(x|y) = f_X(x)$.

Exercises

11.19 Consider throwing two fair six-sided dice. Let A be the event that the first die is odd, B the event that the second die is odd, and C the event that the product of numbers on these two throws is odd. Determine whether these events are pairwise independent, that is, if (i) A and B are independent, (ii) A and C are independent, and (iii) B and C are independent.

11.20 A fair four-sided die has numbers 1, 2, 3, or 4 on its faces, respectively. Let A be the event that the face with an even number landed, that is, $A = \{2, 4\}$. List all possible solutions for event B such that A and B are independent. (Hint: start by listing all the real subsets of $\{1, 2, 3, 4\}$, including the full set $\{1, 2, 3, 4\}$, as possible solutions for event B and check each in turn.)

11.3.6 The Law of Total Probability

Suppose $B_1, B_2, \cdots, B_n$ is a partition of the sample space S, such that any two partitions are disjoint, that is, $B_i \cap B_j = \emptyset$, and the union of all the $B_i's$ is the entire sample space, that is, $\cup_i B_i = S$. Then, for any event A, we have

$$P(A) = \sum_i^n P(A \cap B_i) = \sum_i^n P(A|B_i)P(B_i). \tag{11.34}$$

This rule is used to find the probability of an event A when we do not know enough about A's probability to calculate it directly. Instead, we take related events B_i and use them to calculate the probability for A.

Example 11.20 40% of people watched the film *The Lord of the Rings*. Among them, 35% have read the book before seeing the movie. Out of 60% of people who do not watch the movie, only 3% have read the book. What is the probability a random person has not read the book?

Solution Consider event A being a random person who has not read the book, B_1 the person who watched the movie, and B_2 not watched the movie. Since B_1 and B_2 together are mutually disjoint, and their union is all of the sample space (all the people), then we can use the law of total probability. The known probabilities are shown in Fig. 11.5.

(continued)

11.3 Conditional Probability and Corresponding Rules

Example 11.20 (continued)
Applying the law of total probability (Eq. (11.34)), we have

$$(1 - 0.35) \times 0.4 + (1 - 0.03) \times 0.6 = 0.842.$$

Example 11.21 A phone-making company has five pipelines of production. Each pipeline has a different rate of working and each has errors associated with it: the first produces 10% of the phones with an error rate of 1%, the second produces 15% of the phones with an error rate of 1.2%, the third produces 20% of the phones with an error rate of 1.4%, the fourth produces 25% of the phones with an error rate of 1.6%, and the final one produces 30% of the phones with an error rate of 2%. What is the probability that a random phone is faulty?

Solution Let $B_1, \cdots, B_5$ be the working rate in the five pipelines and A the probability of a phone being faulty. Since $B_1, \cdots, B_5$ together are mutually disjoint and their union is all of the sample space (all of the production), then we can use the law of total probability.

Consider the first pipeline. We need $P(A|B_1)$, and this says, "Given that we have the first pipeline, what is the probability of error?" and that is given as $1\% = 0.01$. Also, we need $P(B_1)$, which we also know as $10\% = 0.1$. So, the first term in the sum to find $P(A)$ is $0.01 \times 0.1 = 0.001$. We can work out the other four terms similarly.

Hence, applying the law of total probability (Eq. (11.34)), we have $P(A)$ as

$$0.01 \times 0.1 + 0.012 \times 0.15 + 0.014 \times 0.20 + 0.016 \times 0.25 + 0.02 \times 0.30 = 0.0156,$$

which gives a probability for $P(A)$ of 1.56%.

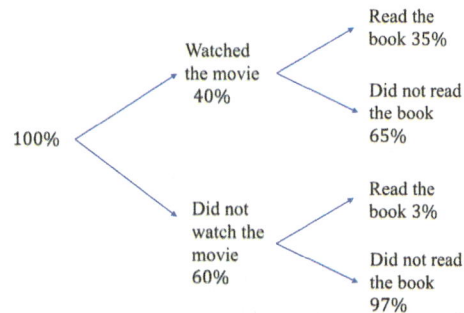

Fig. 11.5 Tree diagram illustrating the partitions of the sample space and their associated probabilities in Example 11.20

Exercises

11.21 We have four bags containing red and black balls. The first contains six red and four black balls; the second contains seven red and eight black balls; the third contains six red and six black balls; and the fourth contains 16 red and four black balls. A bag is selected randomly, and then a ball within it is selected at random. What is the probability that the ball is red? Compare this to the chances of picking a red ball if all the balls were just in one bag.

11.22 There are five balls in a bag: two white and three red. Take the first ball out of the bag without replacing it, and then take out the second one. What is the probability that the second ball is white?

11.4 Bayes' Theorem

Recall that the multiplication rule says the probability that events A and B both occur is the probability that A occurs multiplied by the probability that B happened, given that A already occurred. That is,

$$P(A, B) = P(A \cap B) = P(A)P(B|A).$$

Alternatively, we have

$$P(B, A) = P(B \cap A) = P(B)P(A|B).$$

Since

$$P(A, B) = P(B, A),$$

we have

$$P(A)P(B|A) = P(B)P(A|B).$$

Bayes' theorem is given by

$$P(A|B) = \frac{P(A)P(B|A)}{P(B)}, \qquad (11.35)$$

where $P(B) \neq 0$.

$P(B)$ in Eq. (11.35) may need to be computed by applying the law of total probability, for example, $P(B) = P(A)P(B|A) + P(\bar{A})P(B|\bar{A})$ and $\bar{A}$ is the

11.4 Bayes' Theorem

complement of A. This gives an alternate form for Bayes' theorem as

$$P(A|B) = \frac{P(A)P(B|A)}{P(A)P(B|A) + P(\bar{A})P(B|\bar{A})}. \quad (11.36)$$

In even more generality, if we denote the data we have as D and the hypothesis about the given data as H_i, Bayes' theorem can be further written in a more general way as follows:

$$P(H_i|D) = \frac{P(H_i)P(D|H_i)}{P(D)}, \quad (11.37)$$

where:

- $P(D) = \sum_i P(D \cap H_i) = \sum_i P(H_i)P(D|H_i)$
- $P(H_i)$ defines the prior probability, that is, the probability of the hypothesis before the new data is observed
- $P(D|H_i)$ defines the likelihood, that is, the probability of the data under the given hypothesis
- $P(H_i|D)$ is called the posterior, that is, the probability of the hypothesis after the data is observed

Example 11.22 Mrs. Wright runs a small company providing domestic cleaning services with two employees—Sarah and Ann. Each cleaner has a very similar workload. From past performances, 85% of customers rank Sarah's work with five stars (the highest rating) and only 50% for Ann's work. A new review with five stars comes to Mrs. Wright without mentioning the cleaner's name. What is the probability that it is a review of Ann's work?

Solution Let us use D to denote a review with five stars, H_1 to denote the cleaning done by Ann and H_2 by Sarah. Since cleaners have a similar workload, we have $P(H_1) = P(H_2) = 0.5$. To compute $P(H_1|D)$, we apply Eq. (11.37) and the first bullet point below the equation to obtain the following:

$$P(H_1|D) = \frac{0.5 \times 0.5}{0.5 \times 0.5 + 0.85 \times 0.5} = \frac{0.25}{0.675} \approx 37\%.$$

Example 11.23 In the UK, men have a one in eight chance of having prostate cancer at some point in their life. There is a simple first test that can be done to determine if a man needs further testing; this is the PSA test. However, this

(continued)

Example 11.23 (continued)

test is not very accurate. In fact, $\frac{3}{4}$ of the results are false positives; that is, it gives a positive result when the man does not have the condition. Also, there is a $\frac{1}{7}$ chance of a false negative, which gives a negative result when the man has the condition. What are the chances that a man has the condition given that he gets a positive test result?

Solution Let A be the probability that you have the condition and B be the probability that you test positive. We require $P(A|B)$, namely, the probability you have the condition given that you test positive.

We know that $P(A) = \frac{1}{8}$; $P(B|\bar{A}) = \frac{3}{4}$, namely, that the probability of testing positive given that you do not have the condition; and $P(\bar{B}|A) = \frac{1}{7}$, namely, the probability that the test is negative given that you have the condition.

To find $P(A|B)$, we will use Bayes' theorem in the form given in Eq. (11.36) since we do not know $P(B)$ directly, namely,

$$P(A|B) = \frac{P(A)P(B|A)}{P(A)P(B|A) + P(\bar{A})P(B|\bar{A})}.$$

Now $P(\bar{A}) = \frac{7}{8}$, and using Eq. (11.25), we have $P(B|A) = \frac{6}{7}$. Plugging these into the equation, we get

$$P(A|B) = \frac{\frac{1}{8} \times \frac{6}{7}}{(\frac{1}{8} \times \frac{6}{7}) + (\frac{7}{8} \times \frac{3}{4})} \approx 0.14.$$

We can show this result using a diagram; see Fig. 11.6. Assuming we have 1120 men (the number is chosen so that it divides nicely), then in the first split, we have $1120 \times \frac{1}{8} = 140$ men that have the condition and the rest, $1120 \times \frac{7}{8} = 980$, do not. In the top half, it splits again with $140 \times \frac{1}{7} = 20$ who have the condition and test negative, and the rest, $140 - 20 = 120$, testing positive. Similarly, in the bottom half, $980 \times \frac{3}{4} = 735$ test positive but do not have the condition, and the rest, $980 \times \frac{1}{4} = 245$, test negative.

We want the number who test positive and have the condition compared to the total that tests positive. Namely, $\frac{120}{120+735} \approx 0.14$.

11.4 Bayes' Theorem

Fig. 11.6 Illustration of the data presented in Example 11.23

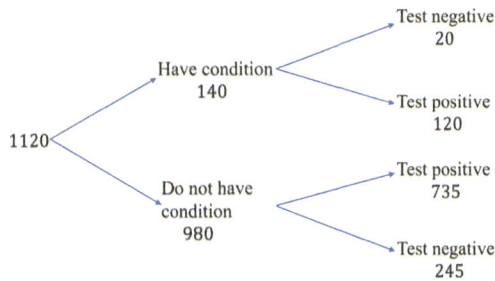

Exercises

11.23 In Example 11.23, change the numbers to the following. UK men have a 1 in 12 chance of having prostate cancer at some point in their life. An improved PSA test has a false positive rate of $\frac{1}{3}$ and a false negative rate of $\frac{1}{20}$. Again, calculate the chances that a man has the condition given that he gets a positive test result.

11.24 A type of product sold in a local shop is produced by factories A, B, and C. Among them, 5/10, 3/10, and 2/10 are from A, B, and C, respectively. The defective rate of these products is 1/10, 1/15, and 1/20, respectively. One of these products is drawn randomly from the shop. If it is non-defective, what is the probability that it is produced by factory A?

11.25 Let A be the event that a positive test result shows up and B the event that the person has breast cancer. Based on previous clinical records, we have $P(A|B) = 0.95$ and $P(\bar{A}|\bar{B}) = 0.98$. A census is taking place, and we know the probability of suffering from breast cancer among this population is 0.003. Compute $P(B|A)$.

11.26 You have four bags of multicoloured balls. Bag X has 50 balls, Bag Y has 60 balls, Bag Z has 60 balls, and Bag W has 30 balls. The probability of picking a red ball from bag X is $\frac{1}{10}$, from bag Y is $\frac{1}{20}$, from bag Z is $\frac{1}{30}$, and from bag W is $\frac{1}{5}$. A bag is picked with the probability given by the relative number of balls in it and a ball is picked at random from this bag. If the ball is not red, what is the probability of it coming from Bag Z?

Chapter 12
Elements of Statistics

As we have seen in Chap. 10, probability deduces what is likely to happen when an experiment is performed. The entire pool of subjects in an investigation is called a population. It may be challenging to involve the whole population in the experiment. Instead, random samples may be chosen so that every member of a population has an equal chance of being selected as any other member. People use statistics obtained from these samples to describe the entire population.

We introduce statistics in this chapter. First, we present descriptive statistics: methods used to describe or summarise observations. Then we briefly bring in elementary sampling theory. Especially, we introduce two more sampling distributions: Student's t-distribution and the Chi-square distribution. Finally, we focus on inferential statistics, that is, to infer by what mechanism the observation, the outcome of an experiment, has been generated.

Because probability and statistics are related, there is quite a lot of overlap between this chapter and both the two previous chapters. Some topics, such as average and standard deviation, have been introduced even earlier in the book. In this chapter, we collect all the relevant results together even if they have been introduced before, and we make back references where appropriate.

12.1 Descriptive Statistics

Statistics is the most basic and important concept and tool to estimate characterisations of the probability distribution of a population. There are lots of statistics, like mean, variance, and interquartile range, all of which will be introduced here. Some of these have been introduced before, but here we bring them all together. Formally, suppose $x_1, x_2, \ldots, x_n$ is a sample of a random variable X. Then the function $g(x_1, x_2, \ldots, x_n)$ is called a statistic of the sample if there are no unknown parameters in g.

Example 12.1 Suppose X is a continuous random variable and $X \sim \mathcal{N}(\mu, \sigma^2)$. Remember this means that X is a normal or Gaussian distribution as defined in Sect. 10.6.2.2 of Chap. 10. If μ is known but not σ^2, then $\sum_{i=1}^{n}(x_i - \mu)^2$ is a statistic, but $\frac{\sum_{i}^{n} x_i}{\sigma}$ is not.

So a statistic is a numerical quantity calculated from a set of observations,[1] and statistics is the collection and analysis of such data.

12.1.1 Measures of Centre

The sample mean, median, and mode are three measures of central tendency.

12.1.1.1 The Arithmetic Mean

Sum up all the values and divide them by the number of data points.

Example 12.2 The set of five numbers 3, 5, 7, 8, and 10 has a mean of
$$\frac{3 + 5 + 7 + 8 + 10}{5} = 6.6.$$

The general definition of expected value or mean can be viewed in Sect. 10.5.1 of Chap. 10.

12.1.1.2 Median

That is the number found at the dataset's middle when the dataset is sorted into order. When the number of data is even, we use the average of the two middle numbers as the median.

---

[1] Strictly speaking, a statistic is a function of random variables, while a numerical quantity is its realisation based on a specific sample.

Example 12.3

- The set of numbers 3, 5, 7, 8, and 10 has a median of 7.
- The set of numbers 3, 5, 6, 7, 8, and 10 has a median of $\frac{6+7}{2} = 6.5$.

12.1.1.3 Mode

That is the value that occurs most often in the dataset.

Example 12.4

- The set of numbers 2, 3, 3, 5, 7, 8, 8, 8, 8, and 11 has a mode of 8.
- The set of numbers 3, 5, 6, 7, 8, and 10 has no mode.
- The set of numbers 1, 3, 3, 3, 5, 7, 8, 8, 8, and 11 has two modes, 3 and 8, and is called bimodal.

The mode is most likely to be used when data concerns categories.

Example 12.5 Suppose we have a survey result of the sports students like the most in Year 7 of a local school in the St Albans area, shown in Table 12.1.

Which form of average should we use for this type of data?

In this example, *Football* would be considered average, the modal average. There would be no point in finding a mean number of students for each listed sport or looking for a median value.

Remark 12.1 When we take many samples from the same population, their means are likely to differ less than their medians or modes. That is, the mean is relatively stable. The median is preferable if outliers (extremely high or low values) are observed.

Table 12.1 A survey result of the sport a student likes the most in Year 7 of a local school in the St Albans area

Sports that students like most	Number of students
Swimming	60
Tennis	30
Football	70
Basketball	40

Fig. 12.1 An illustration of the mean, mode, and median for a symmetrical distribution

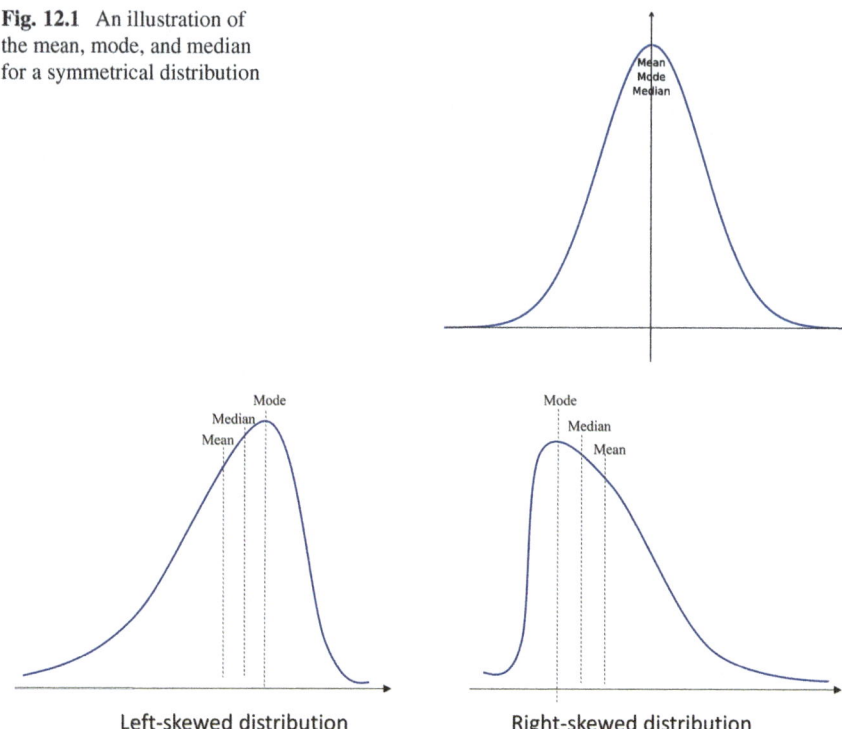

Fig. 12.2 An illustration of the relationships between the mean, mode, and median for asymmetrical distributions

The mean, median, and mode are the same in a perfectly symmetrical distribution (see Fig. 12.1). On the other hand, the effect of extreme values can distort the mean and pull it far from the centre of the distribution. The left panel of Fig. 12.2 shows a left-skewed (or negatively skewed) distribution. More values occur around the distribution's left tail, and its mean value is less than its median value, which is less than the mode value. In contrast, the right panel presents a right-skewed (or positively skewed) distribution, where more values occur around the distribution's right tail. Its mean value is greater than its median, which is greater than the mode value.

♦

Exercise

12.1 Find the mean, median, and mode of the following sets:

1. 1, 5, 8, 7, 6, 6, 6, and 11.
2. 0, 0, 3, 5, 5, and 10.
3. 1.5, 7.4, 3.7, 5.5, 5.5, 8.3, 2.1, and 6.8.
4. 5.8, 6.2, 5.7, 6.1, 6.0, 5.6, 6.0, 5.9, and 5.8.
5. 3, 9, 10, 7, 4, 12, 10, 11, and 6.

12.1.2 Measures of Variation

Variation measures how spread out the data we collect is. It is a helpful way to identify if our data has many outliers.

12.1.2.1 Standard Deviation and Variance

Usually, people use standard deviation and variance to measure the variation. We have introduced them in Sect. 4.2.1 of Chap. 4. For the convenience of readers, we show the corresponding equations again as follows.

Suppose $\mathbf{X}$ is a data matrix including n data observations with d dimensions (variables, features, or attributes). Each element of $\mathbf{X}$ is denoted as $x_{i,j}$, where $i = 1, \ldots, n$ and $j = 1, \ldots, d$. The sample standard deviation for each dimension x_j is defined as

$$s(x_j) = \sqrt{\frac{\sum_{i=1}^{n}(x_{i,j} - \bar{x}_j)^2}{n-1}},$$

where $\bar{x}_j$ is the sample mean of the jth dimension:

$$\bar{x}_j = \frac{1}{n}\sum_{i=1}^{n} x_{i,j}.$$

The squared sample standard deviation is called sample variance:

$$var(x_j) = (s(x_j))^2.$$

Readers can view the general definition of variance in Sect. 10.5.2 of Chap. 10.

Exercise

12.2 Find the standard deviation of the following:

(1) 1, 5, 8, 7, 6, 6, 6, and 11.
(2) 0, 0, 3, 5, 5, and 10.
(3) 1.5, 7.4, 3.7, 5.5, 5.5, 8.3, 2.1, and 6.8.
(Hint: You have already found the means for these three in the previous exercise.)
(4) Find the standard deviation of each dimension for the following data:

$$\mathbf{D} = \begin{bmatrix} 5.8 & 3 & 10.7 \\ 6.2 & 9 & 10.6 \\ 5.7 & 10 & 10.8 \\ 6.1 & 7 & 10.9 \\ 6.0 & 4 & 10.8 \\ 5.6 & 12 & 10.9 \\ 6.0 & 10 & 10.1 \\ 5.9 & 11 & 10.2 \\ 5.8 & 6 & 10.4 \end{bmatrix}$$

Hint: You have already found the means for the first two columns in the previous exercise.

Remark 12.2 The standard deviation, divided by $n - 1$, calculates the sample standard deviation. Readers may see a different version from other resources as follows:

$$\sigma_j = \sqrt{\frac{\sum_{i=1}^{n}(x_{i,j} - \mu_j)^2}{n}},$$

which computes the population standard deviation. There are two differences compared with the sample standard deviation equation. First, the mean value μ_j is the population mean, not the sample mean $\bar{x}_j$. Second, the denominator under the square root is n, not $n - 1$. Statisticians found that the sample mean after taking a large number of samples is smaller than the population mean. To compensate for this difference, a smaller denominator $n - 1$ is used when computing the sample standard deviation so that $s(x_j)$ is as close to σ_j as possible.

◆

12.1.2.2 Covariance and Pearson Correlation Coefficient

In addition, we introduced the sample covariance measure and Pearson correlation coefficient in Sect. 4.2.1 of Chap. 4, which is a quantitative measure that describes the strength of association between two variables. Again for the convenience of readers, we show the corresponding equations. The sample covariance is given by

$$cov(x_h, x_k) = \frac{\sum_{i=1}^{n}(x_{i,h} - \bar{x}_h)(x_{i,k} - \bar{x}_k)}{n-1},$$

and Pearson correlation coefficient r is given by

$$r = \frac{cov(x_h, x_k)}{\sqrt{var(x_h)var(x_k)}}.$$

Exercise

12.3 Find the covariance and Pearson correlation coefficient between:

(1) Column 1 and column 2 of matrix **D** from Exercise 12.2 (4).
(2) Column 1 and column 3 of matrix **D** from Exercise 12.2 (4).
(3) Column 2 and column 3 of matrix **D** from Exercise 12.2 (4).
(4) Now for **D** from Exercise 12.2 (4) form a new matrix where all the columns are put in order separately, from the lowest value at the top to the highest value at the bottom of each column. Repeat the previous calculations of covariance and Pearson correlation coefficient for columns 1 and 2, 1 and 3, and 2 and 3 for the new matrix.

12.1.2.3 Coefficient of Variation

Another useful way to measure the variation is to use the coefficient of variation, a ratio of the sample standard deviation to the mean, as shown as follows:

$$\text{coefficient of variation} = \frac{s}{\bar{x}}. \quad (12.1)$$

We cannot compare standard deviations with different measurement units. However, since the coefficient of variation is independent of measurement units, we can use it to compare variations in different datasets with varying units of measurement. A drawback of the coefficient of variation is that it fails to be useful when $\bar{x}$, the sample mean value, is close to zero.

Example 12.6 A fridge manufacturer has two fridge models, M and N. The mean lifetimes of the two models are $\bar{x}_M = 20$ years and $\bar{x}_N = 15$ years, and standard deviations are $s_M = 3.5$ years and $s_N = 3$, respectively. Which model has the greater relative dispersion?

Solution Applying Eq. (12.1) to models M and N, respectively, we have:

$$\text{coefficient of variation} = \frac{3.5}{20} = 17.5\% \text{ for model M}$$

and

$$\text{coefficient of variation} = \frac{3}{15} = 20\% \text{ for model N}.$$

Therefore, model N has a greater relative dispersion.

Exercise

12.4 Two class students attended a maths competition. The mean score of class A is $\bar{x}_A = 50$ and its standard deviation is $s_A = 10$, and the mean score of class B is $\bar{x}_B = 60$ and its standard deviation is $s_B = 12$. Which class has the greater relative dispersion?

12.1.3 The Range and the Interquartile

Definition 12.1 (Range) The range equals the highest value minus the lowest value in a given real-valued dataset.

Definition 12.2 (Interquartile Range) The interquartile range (IQR) equals the upper quartile, denoted as Q_3, minus the lower quartile, denoted as Q_1. The upper quartile (Q_3) is a number such that the integral of the probability density function from $-\infty$ to this number (Q_3) equals 0.75, and the lower quartile (Q_1) is another number such that the integral of the probability density function from $-\infty$ to this number (Q_1) equals 0.25. Figure 12.3 illustrates the positions of Q_1, Q_2 (the median), and Q_3 in a dataset, where values are sorted in ascending order.

12.1 Descriptive Statistics

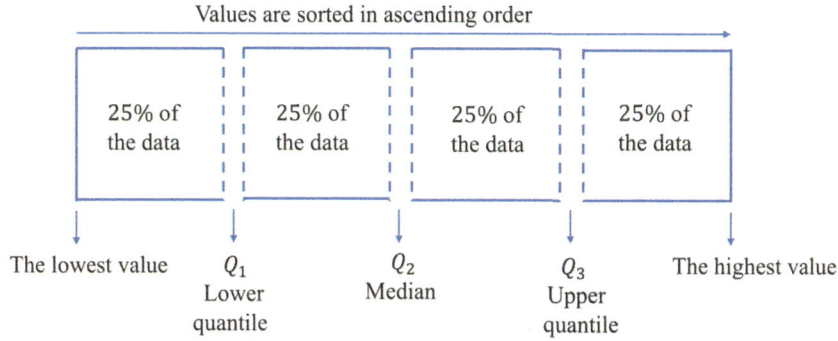

Fig. 12.3 The positions of quantiles within the dataset

Fig. 12.4 Boxplot representing the distribution of a dataset

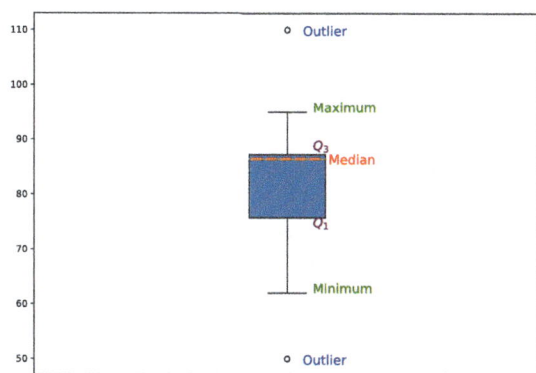

12.1.3.1 Boxplot

A boxplot shows the quartiles of a dataset as a box, and its "whiskers" (lines extending upwards and downwards from the box) show the extent of the rest of the distribution, except for points that are determined to be "outliers" using a method that is a function of the interquartile range.

Figure 12.4 illustrates a boxplot: Q_1 and Q_3 control the width of the box, which equals the interquartile range (IQR); the solid line within the box, highlighted with an overlaid dashed line, represents the median of the data; the bar on the top indicates the maximum value; the bar at the bottom shows the minimum value; two circle signs display the outliers. Outliers are data items outside 1.5 times the IQR above the upper quartile and below the lower quartile, that is,

$$\text{outlier values} \leq Q_1 - 1.5 \times IQR \quad \text{or} \quad \text{outlier values} \geq Q_3 + 1.5 \times IQR. \quad (12.2)$$

Please note that there are different methods that can be used to compute quartiles. We consider the median-based method only in this book.

Exercise

12.5 Find the following manually:

(1) Mode
(2) Median
(3) IQR
(4) Outliers

for both of the following two datasets:

- 90, 86, 87, 88, 50, 66, 95, 87, 72, 78, 77, 87, 86, 62, 87, 110
- 55, 72, 52, 45, 58, 55, 30, 52, 38, 55, 42, 65, 53, 55, 80, 48

Note that Fig. 12.4 is generated from the first dataset.

12.2 Elementary Sampling Theory

12.2.1 Random Sampling with and Without Replacement

Random sampling is the process of collecting a representative sample from a population. It means each member of the population has an equal chance of being selected and is independent of other members selected in the sample.

Sampling where each member of the population is allowed to appear only once is called random sampling without replacement. In contrast, if each member is allowed to appear more than once, it is called random sampling with replacement.

Let us consider samples of the same size collected from the same population. We can calculate the mean for each sample. The probability distribution of mean values obtained from these samples is called a sampling distribution of means. Similarly, we can get a sampling distribution of standard deviations or sampling distributions of other statistics.

12.2.2 Sampling Distributions of Means

This topic is related to the material we introduced in Chap. 11 on the central limit theorem. At that point, we talked about population means and the fact that the distribution of sample means approximated a normal distribution. So here we are talking about the sampling distribution of sample means, which again approximates a normal distribution.

Let us denote the population mean and standard deviation by μ and σ, respectively, and the sample mean and standard deviation by $\bar{x}$ and s, respectively.

12.2 Elementary Sampling Theory

The key to this work is that: The sampling distribution of means is approximately a normal distribution with mean $\mu_{\bar{x}}$ and standard deviation $\sigma_{\bar{x}}$ for a large value of the sample size n, $n \geq 30$, irrespective of the population distribution as long as the population size is at least twice the sample size and the mean and the standard deviation of the population distribution are finite.

Based on the central limit theorem (see Sect. 11.1.2 of Chap. 11), the mean and standard deviation of a sampling distribution of means can be calculated as follows:

- If all samples of size n are drawn from an infinite population or if sampling is with replacement, then we have

$$\mu_{\bar{x}} = \mu, \tag{12.3}$$

and

$$\sigma_{\bar{x}} = \frac{\sigma}{\sqrt{n}}. \tag{12.4}$$

So the sampling distribution of means, $\bar{x}$, has Gaussian distribution:

$$\mathcal{N}(\mu, (\frac{\sigma}{\sqrt{n}})^2).$$

- If all samples of size n are drawn from a finite population of size n_p without replacement ($n_p > n$), then we have

$$\mu_{\bar{x}} = \mu,$$

which is the same as Eq. (12.3). The formula for the standard deviation of all sample means must be modified by including a finite population correction. That is,

$$\sigma_{\bar{x}} = \frac{\sigma}{\sqrt{n}} \sqrt{\frac{n_p - n}{n_p - 1}}. \tag{12.5}$$

So the sampling distribution of means, $\bar{x}$, has Gaussian distribution:

$$\mathcal{N}\left(\mu, \left(\frac{\sigma}{\sqrt{n}} \sqrt{\frac{n_p - n}{n_p - 1}}\right)^2\right).$$

Example 12.7 Randomly take samples from $\mathcal{N} \sim (52, 6.3^2)$. If the size of a sample is 36, compute the probability of $P(50.4 \leq \bar{x} \leq 53.2)$.

Solution Consider all samples are drawn from the infinite population. Applying Eqs. (12.3) and (12.4) we have $\mu_{\bar{x}} = \mu = 52$, $\sigma_{\bar{x}} = \frac{6.3}{\sqrt{36}}$. So that $\bar{x}$ has a Gaussian distribution $\mathcal{N}(52, (\frac{6.3}{\sqrt{36}})^2)$. Now use Eq. (10.24) to get

$$P(50.4 \leq \bar{x} \leq 53.2) = \Phi(\frac{53.2 - 52}{6.3/\sqrt{36}}) - \Phi(\frac{50.4 - 52}{6.3/\sqrt{36}})$$
$$\approx \Phi(1.14) - \Phi(-1.52) = 0.8729 - 0.0643 = 0.8086.$$

Here we have used the values from a standard normal distribution table.[a] to look up $\Phi(1.14)$ and $\Phi(-1.52)$.

---

[a] The example table we have used is at https://en.wikipedia.org/wiki/Standard_normal_table.

Example 12.8 Suppose that the heights of 2000 female students at a university are normally distributed with a mean of 165 centimetres (cm) and a standard deviation of 8 cm. If 100 samples of 15 students each are collected, what is the expected mean and standard deviation of the resulting sampling distribution of means if the sampling was done without replacement?

Solution Applying Eqs. (12.3) and (12.5), we have

$$\mu_{\bar{x}} = 165 \text{ cm},$$

and

$$\sigma_{\bar{x}} = \frac{8}{\sqrt{15}} \sqrt{\frac{2000 - 15}{2000 - 1}} \approx 2.06 \text{ cm}.$$

Exercises

12.6 Randomly take a sample from $\mathcal{N} \sim (60, 10^2)$. If the size of the sample is 100, compute the probability of $P(|\mu - \bar{x}| < 0.3)$.

(continued)

12.7 Marks in a certain exam at A level are distributed according to a normal distribution with a mean of 57 and a standard deviation of 12. A random sample of students of size 64 is taken, compute the probability that the sample mean is within 1 of the population mean.

12.8 The mean weight and standard deviation of a set of four hundred marble balls are 4.58 grams and 0.3 grams, respectively. A random sample of 50 marble balls is taken out one at a time, collected together, and weighed. Compute the probability that this random sample of 50 has a total weight of more than 230 grams.

12.2.3 Sampling Distributions of Proportions

We are now going to look at the sampling distribution of sample proportions. Here we are looking at the proportion of heads-ups of a toss of a fair coin, the proportion of defective items from a product line, or the proportion of people in a university with a particular brand of phone. Here the event is a Bernoulli event: The coin is heads-up or not, an item is defective or non-defective, and a person at a university has that phone brand or not. So the distribution for multiple Bernoulli events is binomial, and the formulae for mean and standard deviation are derived for that distribution (see Sect. 10.6.1.3 of Chap. 10).

The key to this section is that, using the central limit theorem (see Sect. 11.1.2 of Chap. 11), the sampling distribution of sample proportions will follow a normal distribution.

Suppose that a population is infinite and that the probability of an event occurring is p and not occurring is $1 - p$. The number (x) of the event occurring in a sample with a size of n can be modelled using a binomial distribution, where the mean is $\mu_X = np$ and the variance is $\sigma_X^2 = np(1-p)$ (from Sect. 10.6.1.3 of Chap. 10). So the number x of defective items in a sample could be 6, for example. The proportion of the event occurring x times in the sample is $\frac{x}{n}$. So if the sample size is 100, then the proportion of defective items in the sample is $\frac{x}{n} = \frac{6}{100}$.

Consider all possible samples of size n drawn from the same population and denote the mean and standard deviation of the sampling distribution of proportions by μ_p and σ_p, respectively; then we have

$$\mu_p = p, \qquad (12.6)$$

and

$$\sigma_p = \sqrt{\frac{p(1-p)}{n}}. \qquad (12.7)$$

which can be obtained by dividing the mean, np, and standard deviation, $\sqrt{np(1-p)}$, of the binomial distribution by the sample size n.

Again, for a large value of n, $n \geq 30$, the sampling distribution of sample proportions follows closely to a normal distribution.

Note that Eqs. (12.6) and (12.7) are also valid for sampling with replacement in a finite population. For finite populations where sampling is without replacement, we have

$$\mu_p = p,$$

and

$$\sigma_p = \sqrt{\frac{p(1-p)}{n}} \sqrt{\frac{n_p - n}{n_p - 1}}, \quad (12.8)$$

where n_p is the size of the population and $n_p > n$.

Example 12.9 Find the probability that there will be $\frac{8}{15}$ or more heads-ups in 150 tosses of a fair coin.

Solution Consider the 150 tosses of the coin to be a sample from an infinite population of all possible tosses of the coin.

Applying Eqs. (12.6) and (12.7), we have

$\mu_p = 0.5$ and $\sigma_p = \sqrt{\frac{0.5(1-0.5)}{150}} \approx 0.0408$.

Using the normal approximation to the binomial, we convert $\frac{8}{15} \approx 0.5333$ to the standard z-score $= \frac{0.5333 - \mu_p}{\sigma_p} = \frac{0.5333 - 0.5}{0.0408} \approx 0.82$ (see Eq. (10.21) in Chap. 10) so that we can use the standard normal distribution table.

We require $P(z > 0.82)$, which is the probability given by the area under the normal curve to the right of z-score $= 0.82$. The value in the table (refer to the information provided in the footnote of Sect. 10.6.2.2) is 0.7939, so the probability to the right is $1 - 0.7939 = 0.2061$.

Exercises

12.9 It has been found that 2% of the products produced by manufacturer A are defective. What is the probability that a sample of 400 products from the manufacturer 3% or less will be defective?

(continued)

12.10 It is known that 25% of the student population have a particular brand of phone. What is the probability that in a sample of 80 students:

(1) At least 24% of them have that brand of phone?
(2) No more than 30% of them have that brand of phone?

12.2.4 Standard Errors

The standard deviation of a statistic's sampling distribution is often called its standard error. For example, Eq. (12.4) is $\bar{x}$'s (with replacement) standard error, and Eq. (12.7) is the proportion's (with replacement) standard error.

12.2.5 Degrees of Freedom

The value of degrees of freedom is the maximum number of independent values used to calculate the estimate that are free to vary. Suppose we want to use a sample of size n to compute one estimate. The quantity $n - 1$ is the number of degrees of freedom of an estimate, since the last one is fixed by the constraint of producing the answer.

Example 12.10 Consider selecting five students ($n = 5$) to attend a maths competition. The average score of these five students in the most recent maths test must be 60. Theoretically, four students can be chosen randomly, with the fifth student having to have a specific maths score so that the final average is 60. Therefore, the degrees of freedom is $5 - 1 = 4$.

The formula for calculating the number of degrees of freedom is different if we have a different number of samples or if the number of estimates is more than one. For example, if we have two samples whose sizes are n_1 and n_2, we want to estimate the *mean*. Then the degrees of freedom is $n_1 + n_2 - 2$. Another example can be viewed in Sect. 12.3.3.2 of this chapter, where the degrees of freedom of a two-way classification table is (the number of rows $- 1$) $\times$ (the number of columns $- 1$).

12.2.6 Two Specific Sampling Distributions

We have introduced some essential distributions in Chap. 10. We continue by introducing two more important and useful distributions in the following subsections: Student's t-distribution and the Chi-square (χ^2) distribution, both widely used in modelling sampling distributions.

12.2.6.1 Student's t-Distribution

This is usually used when we have a small sample, $n < 30$, and also when the underlying population's standard deviation is unknown.

A random variable X is distributed normally with mean μ and variance σ^2, that is, $X \sim \mathcal{N}(\mu, \sigma^2)$. Suppose we take a sample $(x_1, \ldots, x_n)$ of size n. Let $\bar{x}$ be the sample mean and s^2 be the sample variance. From the central limit theorem, the random variable $Z = \frac{\bar{x}-\mu}{\sigma/\sqrt{n}}$ has a standard normal distribution. The random variable T, which has the sample standard deviation (a known quantity) rather than the population standard deviation, is defined as follows:

$$t = \frac{\bar{x} - \mu}{s/\sqrt{n}}. \tag{12.9}$$

This random variable has a distribution known as Student's t-distribution with $n-1$ degrees of freedom. This distribution is similar to a normal distribution but has wider tails. As n increases, or equivalently, the number of degrees of freedom increases, this distribution approaches a normal distribution.

This is illustrated in Fig. 12.5, which shows t distributions with different degrees of freedom (ν). As seen in the left panel, their density distributions all have a bell shape similar to normal distributions and are symmetrical about the mean. Student's t-distribution is mainly used for a small sample of less than 30, which is more likely to generate values far away from its mean (to have wider tails). As shown in the right panel, the cumulative probability approaches the value one faster as the value of degrees of freedom increases.

Similar to the normal distribution, people have constructed a mathematical table for the t statistic. The value in the table gives the number of standard deviations from the mean you need to capture a certain proportion of the data. For instance, we already know that we capture 95% of the data for a normal distribution if we move 1.96 standard deviations on either side of the mean. (See the properties of a standard normal distribution in Sect. 10.6.2.2 of Chap. 10.) There are different versions of t-tables, but a common one is shown in the footnote.[1] It has values such as $t_{.50}, t_{.75}, t_{.80}$ as headings for the columns. Suppose we want the column with $t_{.975}$

---

[1] The example table we have used is at https://www.tdistributiontable.com/.

12.2 Elementary Sampling Theory

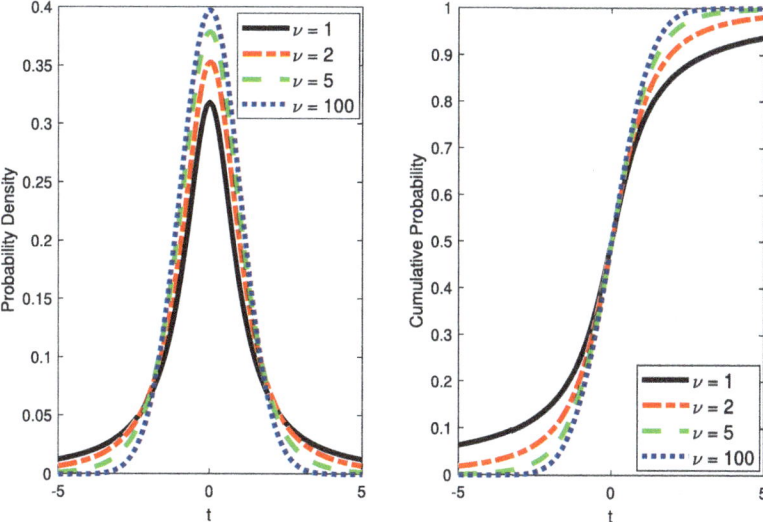

Fig. 12.5 Student's t-distributions with varying degrees of freedom. The left panel shows the density distributions and the right panel shows the cumulative probabilities

as the heading. This means that there are 2.5%(= 0.025) of the data to the right of the distribution and, therefore, 5%(= 0.05) of the data not in the centre (see second and third heading rows). If you look down the column, you will see that for $df = 10$ (degrees of freedom = 10) row, the value is 2.228. This means that for the t-distribution, you need to go 2.228 standard deviations on either side to capture 95% of the data. This is because there are more values in the tails (the tails are wider), so you need to go further out from the mean. If you continue down the $t_{.975}$ column, the values get closer to 1.96, and in fact, as you get to the ∞ value for df, you finally get 1.96.

Example 12.11 For Student's t-distribution with 8 degrees of freedom, find the value of t for which the probability on the right of t is 0.05.

Solution If the probability on the right is 0.05, then the probability to the left of t is $1-0.05 = 0.95$. Referring to the t statistics table, proceeding downward under the column noted with df (degrees of freedom) until reaching entry 8, and then proceeding right to the column headed $t_{.95}$, one can see the result is 1.860. That is the required value of t.

> **Exercise**
>
> **12.11** For Student's t-distribution with 15 degrees of freedom, find the value of t for which the probability on the right of t is (1) 0.05 and (2) 0.01.

12.2.6.2 The Chi-Square (χ^2) Distribution

Suppose $x_1, \ldots, x_n$ are independent and are drawn from a normal distribution with standard deviation σ. The sample mean is $\bar{x}$ and the standard deviation is s. We define the Chi-square (χ^2) statistic as follows:

$$\chi^2 = \sum_{i=1}^{n} \frac{(x_i - \bar{x})^2}{\sigma^2} \qquad (12.10)$$

with $n-1$ degrees of freedom. Each term of the summation is a squared z-score-like value. Therefore, all χ^2 values are equal to or greater than zero. From the definition of standard deviation (Sect. 4.2.1 of Chap. 4), the Chi-square statistic can also be written as

$$\chi^2 = \sum_{i=1}^{n} \frac{(x_i - \bar{x})^2}{\sigma^2} = \frac{(n-1)s^2}{\sigma^2}. \qquad (12.11)$$

It can be seen that the χ^2 statistic is a ratio measuring the deviation of the sample from the population.

If we consider many samples of size n drawn from the same normal population and compute χ^2 for each of them, then a sampling distribution for χ^2 can be obtained and is called the Chi-squared distribution with $n-1$ degrees of freedom.

Figure 12.6 shows Chi-square distributions with different degrees of freedom (ν). The left panel shows that Chi-square probability density functions are positively skewed for the lower values of the degrees of freedom. When $\nu = 1$ and $\nu = 2$, the curve starts high and then drops off. It shows a high probability that χ^2 is close to zero. When $\nu = 3$, 4, or 5, the distribution has a much longer tail on the right hand of the curve. As ν further increases, the distribution looks more similar to a normal distribution. The right panel shows that Chi-square cumulative distributions approach one faster when the value of degrees of freedom decreases.

Intuitively, if $\nu = 1$, then $n = 2$ and we are taking just two observations from a normal distribution. Recall that a normal distribution has a bell shape with a high probability for data being observed around the centre of the bell shape (see Fig. 10.9). Therefore, when taking only two observations, they are both likely to be close to the centre. So $(x_i - \bar{x})^2$ will be small, and hence χ^2 is small. This is why the $\nu = 1$ curve is so skewed to low values. As ν increases, you get more and

12.2 Elementary Sampling Theory

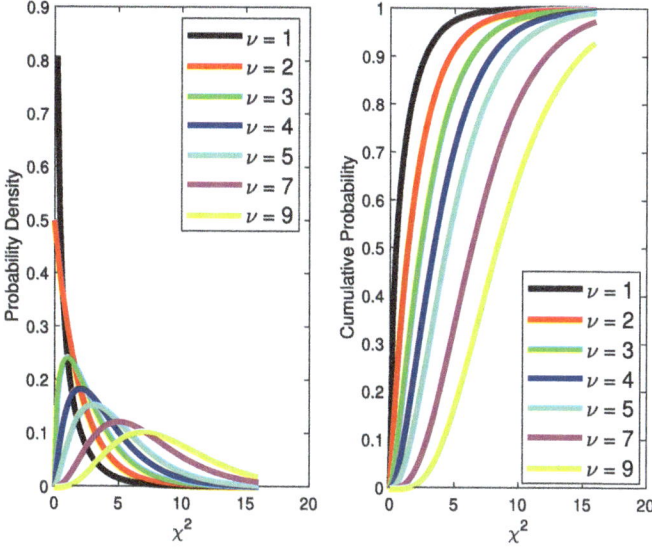

Fig. 12.6 Chi-square distributions with varying degrees of freedom. The left panel shows the density distributions, and the right panel shows the cumulative probabilities

more representative observations of the values from the original normal distribution, so the distribution of values for χ^2 begins to look more and more like a normal distribution, as is seen in Fig. 12.6.

The mean of a Chi-square distribution is its degrees of freedom, and the variance is two times its degrees of freedom.

People have constructed a mathematical table for the Chi-square statistic.[2] Each row of the χ^2 table represents a particular value of the degrees of freedom and, therefore, the sample size. For instance, if the sample size is eight, the number of degrees of freedom is 7, so that is the row to use. The columns represent the area you require under the relevant curve measured from the right, that is, the probability of exceeding a value of χ^2, since the χ^2 values are on the horizontal axis. So, for instance, if you wanted the probability of exceeding a certain value to be only 0.05 (so only 5% of the values are above it), then you use the column headed .05. The value in the table is the value of χ^2 that you must not exceed if you wish to have 95% of the area to the left and only 5% to the right. In the case of the $\nu = 7$ example, the value in the table is 14.07. If you look at the $\nu = 7$ curve in Fig. 12.6, then it looks reasonable that just 5% of the area under the curve is above (to the right of) 14.07 and that 95% is to the left of 14.07. If you continue to look along the 7 degrees of freedom row going left, the values of χ^2 get smaller (you are going to the left on the graph). So the next value in the table is 12.02, and for this value of χ^2 there is

---

[2] The example table we have used is at https://www.statisticshowto.com/tables/chi-squared-table-right-tail/.

10% of the area to the right, as it says at the top of the column. So the further you go to the left, the lower the values of χ^2 you get and therefore the larger percentage of the area under the curve there is to the right of that point. Hence, the values in the χ^2 table are critical values of χ^2, effectively telling you how far to the right of the area you are.

> **Example 12.12** For a Chi-square distribution with 4 degrees of freedom, find the value of χ^2 for which the probability on the right of χ^2 is 0.05.
>
> **Solution** If the probability on the right is 0.05, then the probability to the left of χ^2 is $1 - 0.05 = 0.95$. Referring to the χ^2 statistics table, proceeding downward under the column noted with df (degrees of freedom) until reaching entry 4, and then proceeding right to the column headed .05, one can see the result is 9.48773 or 9.49. That is the required value of χ^2.

> **Exercise**
>
> **12.12** For a Chi-square distribution with 12 degrees of freedom, find the value of χ^2 for which the probability on the right of χ^2 is (1) 0.05 and (2) 0.01.

12.3 Inference

People make inferences about entire populations based on certain samples of data. There are two main types of inference: classical and Bayesian inferences. We focus on classical inference in this book. Types of classical inference include:

- *Point estimation* computes a single number, an estimate of some parameter.
- *Interval estimation* provides a range of values in which the parameter is thought to lie based on the observed data.
- *Testing hypothesis* is to test a hypothesis about whether a parameter value is to be accepted or rejected based on the observed data.

12.3.1 Point Estimation

Point estimation uses sample data to calculate an unknown value for some parameter. For example, estimate a population mean, variance, or other statistics.

Example 12.13 Estimate the probability of heads-up for a coin. We have the following observations from 10 tosses of the coin:

$$H, T, T, T, T, H, T, H, T, H,$$

where H denotes heads-up and T tails-up.

Solution Among 10 observations, four are heads-up. We use p to denote the probability of having heads-ups. Let $\tilde{p}$ denote the estimator of p. We have

$$\tilde{p} = \frac{1+1+1+1}{10} = 0.4.$$

To approach the actual probability, more experiments (tosses) need to be done according to the law of large numbers (see Sect. 11.1.1 of Chap. 11). $\tilde{p} = 0.4$ represents our best estimate so far.

Exercise

12.13 The lifespan of a type of bulb follows $X \sim \mathcal{N}(\mu, \sigma^2)$. Both μ and σ are unknown. Randomly take five of this type of bulb and test their lifespan (in hours):

$$1400, 1520, 1368, 1600, 1544.$$

Estimate μ and σ.

More systematic methods of constructing estimators include the least-squares technique and the maximum likelihood (ML) method. We have introduced the least-squares technique in Sect. 8.2 of Chap. 8, which minimises the sum of the square of the differences between the observations and their expected values. We will show how ML works in Chap. 13.

12.3.2 Interval Estimation

A confidence interval is a range of values constructed using a sampling method based on a point estimate. With this method, many intervals can be constructed across repeated samples. The proportion of intervals constructed across many samples containing the true parameter is called the confidence level. Therefore, a confidence interval is constructed using a method that will capture the true parameter at the specified confidence level. Suppose the confidence level is at 68%. If we repeat the sampling process 100 times from a population, we expect 68 of the 100 confidence intervals around the found sample parameter will contain the true population parameter.

Calculating a confidence interval involves finding a point estimate and then incorporating a margin of error to create a range:

$$\text{Confidence interval} = \text{Point estimate} \pm \text{Margin of error}. \tag{12.12}$$

The margin of error is a value that represents the variability of the point estimate and is based on our desired confidence level, the variance of the data, and how big the sample is.

12.3.2.1 Confidence Intervals for Means

Let us consider estimating the population mean using sample means. As mentioned in Sect. 12.2.2 of this chapter, the distribution of sample means for a large number of samples approximates a normal distribution with a mean value around the population mean (see Eq. (12.3)). For simplicity, we consider a standard normal distribution, though it can be any normal distribution. Figure 12.7 shows the

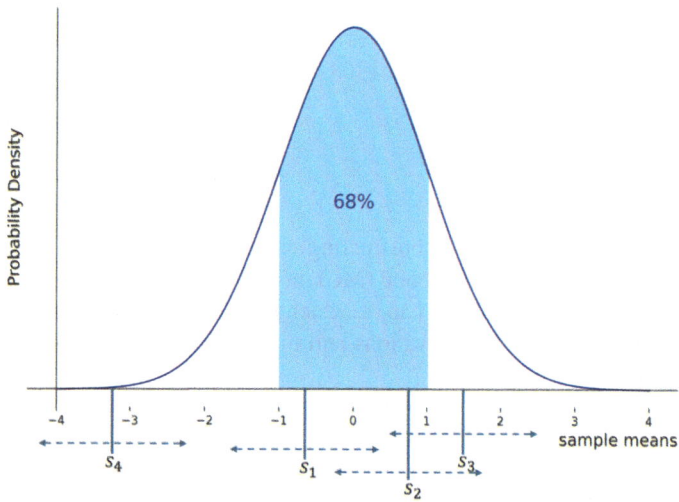

Fig. 12.7 An illustration of the relationship between the population mean and sample means. Four horizontal lines show the distance of 1 SD on either side of the specific sample mean

12.3 Inference

standard normal distribution of sample means with one standard deviation (SD)[3] marked off on either side of the population mean (0). That is 68% of all sample-means fall within this range. We have illustrated four sample means: s_1, s_2, s_3 and s_4. For each sample mean we have shown the range of the distance for 1 standard deviation each side as a horizontal bar below the horizontal axis. For s_1 and s_2, the range *sample mean* ± *1 SD* will include the real population mean, since both s_1 and s_2 are within the range of 1 standard deviation of the real mean, as shown in Fig. 12.7. However, s_3 and s_4 lie outside one standard deviation of the population mean, so their range *sample mean* ± *1 SD* will not contain the population mean.

If a normal distribution is a good fit to the sampling distribution, then the margin of error is usually estimated as

$$\text{Margin of error} = Z_c \times \text{SD}, \tag{12.13}$$

where Z_c was 1 in the example in the previous discussion. So the margin of error is a certain number of standard deviations.

Remark 12.3 Suppose we want a 68% confidence level, as we did above. We want the probability, the area under the standard normal distribution curve between $-Z_c$ and Z_c, to be 68%. We will check using a z-score table (the normal distribution table), where the cumulative probability is the area under the standard normal curve to the left of Z. We want approximately 16% above and 16% below the limits for an interval containing 68%. From the table, the cumulative probability for $-Z_c \leq -1$ is 15.87% (0.1587 in the table, being virtually as near to 0.16 as we can get). For the upper bound, we want to get as near as possible to $16\% + 68\% = 84\%$, and the cumulative probability for $Z_c \leq 1$ is 84.13% (0.84134 in the table). So we obtain $Z_c = 1$ and the margin of error $= 1 \times$ SD as we illustrated.

To construct a 95% confidence interval for the true population mean, we use the standard normal distribution. By checking the same z-score table, the probability for $-Z_c \leq -1.96$ is 2.5%, and the probability for $Z_c \leq 1.96$ is 97.5%. Since $97.5\% - 2.5\% = 95\%$, we have $Z_c = 1.96$ and the Margin of error $= 1.96 \times$ SD.

In the example shown in Fig. 12.7, the 95% confidence interval $s_3 \pm 1.96 \, SD$ will include the population mean, which is not captured by the 68% confidence interval $s_3 \pm 1 \, SD$. Thus, we can make a wide estimate with a high confidence level or a more accurate estimate with a low confidence level. The 95% is probably the most common value that is used.

♦

Consider the standard deviation of sample means as given by Eq. (12.4), we can compute the confidence interval at a certain confidence level for the population mean μ when σ^2 is known (if the sampling is either from an infinite population or with

---

[3] More precisely, it should be the standard error since we consider the sampling distribution of a statistic (see Sect. 12.2.4).

replacement from a finite population) using the following equation:

$$\left[\bar{x} - Z_c \frac{\sigma}{\sqrt{n}}, \bar{x} + Z_c \frac{\sigma}{\sqrt{n}}\right], \quad (12.14)$$

where n is the sample size, $\bar{x}$ is the sample mean, σ is the population standard deviation and $\frac{\sigma}{\sqrt{n}}$ is the sample standard deviation. Note that as n gets larger, that is, the sample size gets larger, then the quantity $Z_c \frac{\sigma}{\sqrt{n}}$ gets smaller. This makes the bounds tighter to the population mean. So, if Z_c is chosen as 1.96, giving a 95% confidence interval, then a larger sample size gives a tighter 95% interval.

Another way to construct a confidence interval is by using the left and right critical values. That is, we construct these intervals between the left critical value c_L and the right critical value c_R based on the observed statistic, s_{obs}. Under repeated sampling, approximately $(1 - \alpha) \times 100\%$ of such intervals will contain the true population parameter (p_{true}):

$$c_L \leq s_{obs} \leq c_R, \quad (12.15)$$

where both the left and right critical values can be obtained from the corresponding probability distribution table. To get 95% confidence, $(1 - \alpha)$ must be 0.95, so $\alpha = 0.05$. This means we must have $\frac{\alpha}{2} = 0.025$ above (to the right) and below (to the left) of the interval.

Figure 12.8 illustrates that an interval from the left critical value to the right critical value represents the confidence interval at $(1 - \alpha) \times 100\%$ confidence level.

We can obtain Eq. (12.14) by applying Eq. (12.15). If we consider the distribution of sample means and use the property that the left critical value is the negative of the right critical value Z_c for the standard normal distribution, then by converting an observed sample mean to its corresponding z-score, we have

$$-Z_c \leq \frac{\bar{x} - \mu_{\bar{x}}}{\frac{\sigma}{\sqrt{n}}} \leq Z_c,$$

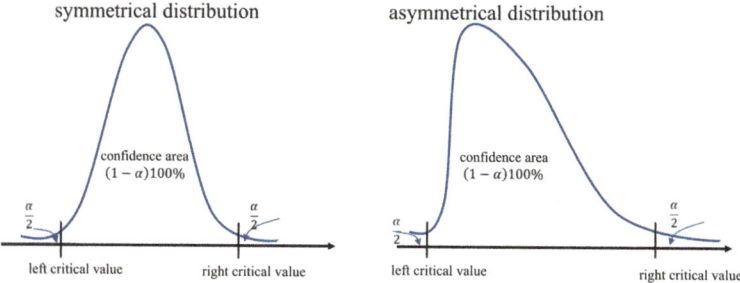

Fig. 12.8 An illustration of confidence intervals. The left panel shows a confidence interval for a symmetrical distribution, while the right panel shows one for an asymmetrical distribution

12.3 Inference

where $\mu_{\bar{x}}$ equals to the true population mean (see Sect. 12.2.2 of this chapter). Rearranging these inequalities, we get

$$\bar{x} - Z_c \frac{\sigma}{\sqrt{n}} \leq \mu_{\bar{x}} \leq \bar{x} + Z_c \frac{\sigma}{\sqrt{n}},$$

which is Eq. (12.14) for the bounds of the population mean.

If the sampling is without replacement from a population of finite size n_p, the standard error can be calculated using Eq. (12.5). However, it is usually approximated by Eq. (12.4) when $n_p \gg n$.

Example 12.14 A sample of size $n = 225$ produced the sample mean of $\bar{x} = 16$. Assuming the population standard deviation $\sigma = 3$, compute a 95% confidence interval for the population mean μ.

Solution We substitute $n = 225$, $\bar{x} = 16$, and $\sigma = 3$ in Eq. (12.14). Let us go through the working to find the bounds, Z_c, again.

We require 95% to be in the middle on either side of the mean. This means we need 2.5% to be at each end of the distribution or 0.025 of the total. Since the normal distribution table gives the cumulative totals from the left, we need to find the points with 2.5%, or 0.025 of the total on the left, and then 97.5%, or 0.975 of the total, which is nearly through to the right end.

So we look up 0.025 probability in the body of the table and find that it occurs when we look up a bound of -1.96. Similarly, we look up 0.975 in the table and get a bound of $+1.96$. So, as before, the probability is between $Z_{97.5\%} = 1.96$ and $Z_{2.5\%} = -1.96$, and this will give us 95% confidence limits. The interval is

$$\left[16 - 1.96 \times \frac{3}{\sqrt{225}}, 16 + 1.96 \times \frac{3}{\sqrt{225}}\right] = [15.608, 16.392].$$

The interval [15.608, 16.392] is the confidence interval at the 95% confidence level for μ.

Remark 12.4 Of course, if the standard deviation of the population is not known, then the standard deviation of the sample can be used. In this case, we use the values in the t-statistic table rather than the z-score table. For example, if we have 60 degrees of freedom (a sample size of 61) and want a 95% confidence interval, then in Eq. (12.14), we replace σ by the sample standard deviation, s, and $Z_c = 1.96$ with $t = 2.00$.

♦

Remark 12.5 Note that when you construct a 95% confidence interval for a population mean from a sample mean, it is not true to say that there is a 95% probability that the true mean lies within this interval. This is because the true mean is a fixed value, and it is either in or not in a confidence interval constructed for that sample. The confidence level applies to the method used to construct the interval and the sample from which it comes. Hence, the correct interpretation is that, if we were to repeat the sampling process 100 times, then about 95 of the resulting constructed intervals would contain the true mean [12].

♦

Exercise

12.14 Suppose that the heights of 2000 female students at a university are normally distributed with a standard deviation of 8 cm. Compute a 95% confidence interval and a 99% confidence interval for estimating the population mean (μ) height of the university's female students. The mean height of a sample of 100 female students is 163cm.

12.3.2.2 Confidence Intervals for Proportions

We are now back to using the binomial distribution since we are considering multiple examples of an event being true or not, that is, Bernoulli events. Consider sampling from an infinite population or a finite population with replacement. We can compute the confidence interval for the population proportion p at a certain confidence level using the following equation:

$$\left[\bar{p} - Z_c \sqrt{\frac{\bar{p}(1 - \bar{p})}{n}}, \bar{p} + Z_c \sqrt{\frac{\bar{p}(1 - \bar{p})}{n}} \right], \tag{12.16}$$

where $\sqrt{\frac{\bar{p}(1-\bar{p})}{n}}$ is the standard deviation of the sampling distribution of proportions (see Eq. 12.7) with $\bar{p}$ being the proportion from a sample and n the sample size.

For sampling from a finite population without replacement, the standard deviation of the sampling distribution of proportions should be calculated using Eq. (12.8).

12.3 Inference

Example 12.15 1000 randomly selected students were asked whether they preferred going swimming or doing athletics. The results show that 696 students prefer going swimming, and 304 prefer doing athletics. Compute a 99% confidence interval for the population proportion p.

Solution Let p be the true fraction of preferring going swimming in the population; $\bar{p} = \frac{696}{1000} = 0.696$ be the fraction in the sample. We substitute $n = 1000$, $\bar{p} = 0.696$ in Eq. (12.16). This time we want 99% confidence intervals, so we want 0.5% at either end. We therefore look up 0.005 and $1 - 0.005 = 0.995$ in the body of the table. This gives the bounds Z_c as -2.58 and $+2.58$. The interval is

$$\left[0.696 - 2.58 \times \sqrt{\frac{0.696 \times 0.304}{1000}}, 0.696 + 2.58 \times \sqrt{\frac{0.696 \times 0.304}{1000}}\right]$$

$$\approx [0.658, 0.734].$$

Exercise

12.15 A sample poll of 200 voters chosen randomly from all voters in a city showed that 59% of them favoured a particular candidate. Find the 95% confidence interval and the 99% confidence interval for the proportion of the voters in favour of this candidate.

12.3.2.3 Confidence Intervals for χ^2

By applying Eq. (12.15), we can define confidence limits for χ^2 by using a χ^2 distribution table.

The χ^2 distribution table has values for χ^2 with the column heading representing the area, α, to the right of the value. So, you use the column headed 0.05 if you want 5% above (to the right of) the value and, therefore, 95% below (to the left of) it. To get an interval of 95%, you need equal areas on either side. Hence, you use the column for 0.025, giving 2.5% to the right, and the column for 0.975, giving 97.5% to the right and 2.5% to the left. This gives 2.5% on either side of the interval (in both tails of the distribution).

Consider using a χ^2 distribution table for the areas α to the right of the critical value. Since $\alpha \leq 1$, then $\frac{\alpha}{2} \leq 1 - \frac{\alpha}{2}$. Thus, $\chi^2_{\frac{\alpha}{2}} \geq \chi^2_{1-\frac{\alpha}{2}}$, since χ^2 values get larger as you progress across the table to the right. Substituting Eq. (12.11) into Eq. (12.15), we have

$$\chi^2_{1-\frac{\alpha}{2}} \leq \frac{(n-1)s^2}{\sigma^2} \leq \chi^2_{\frac{\alpha}{2}},$$

where $\chi^2_{\frac{\alpha}{2}}$ and $\chi^2_{1-\frac{\alpha}{2}}$ are critical values for which $\frac{\alpha}{2}$ of the area lies in each tail of the χ^2 distribution.

Therefore, we can estimate the confidence interval of the population standard deviation σ in terms of an observed sample standard deviation by rearranging the above inequalities giving the following equation:

$$\frac{(n-1)s^2}{\chi^2_{\frac{\alpha}{2}}} \leq \sigma^2 \leq \frac{(n-1)s^2}{\chi^2_{1-\frac{\alpha}{2}}}. \tag{12.17}$$

Example 12.16 A sample of size $n = 30$ produced the sample standard deviation $s = 12.5$. Compute a 90% confidence interval for the population standard deviation σ.

Solution First, we need to find the critical values. The value of the degrees of freedom is given by $30 - 1 = 29$. A 90% level of confidence leaves 10% of the total area in the tails of the χ^2 distribution, with 5% in each tail. We have used a χ^2 distribution table showing the area to the right of the critical value:

$$\chi^2_{0.05} \approx 42.557$$

and

$$\chi^2_{0.95} \approx 17.708.$$

Substitute values to Eq. (12.17), we have

$$\frac{(30-1) \times 12.5^2}{42.557} \leq \sigma^2 \leq \frac{(30-1) \times 12.5^2}{17.708},$$

$$106.47 \leq \sigma^2 \leq 255.89.$$

Therefore, the confidence interval for the population standard deviation at the 90% confidence level is approximately $[\sqrt{106.47}, \sqrt{255.89}] \approx [10.32, 16.00]$.

12.3 Inference

Exercise

12.16 A sample of size $n = 25$ produced the sample standard deviation $s = 6.4$. Compute a 90% confidence interval and a 95% confidence interval for the population standard deviation σ.

12.3.2.4 Summary of Confidence Intervals

- Confidence intervals for means: involve calculating the confidence limits for the **population mean** μ given the **sample mean** $\bar{x}$. The sample size n is needed and should be ≥ 30. You also need the standard deviation σ for the population. The values are looked up in a standard normal distribution table, or z-score table, for whatever confidence level you require; this gives you the values of Z_c to use. The limits are calculated using Eq. (12.14).
- Confidence intervals for proportions: involve calculating the confidence limits for the **population proportion** p given the **sample proportion** $\bar{p}$. The sample size n is needed and should be ≥ 30 again. The values are again looked up in a standard normal distribution table, or z-score table, for whatever confidence level you require; this gives you the values of Z_c to use. The limits are calculated using Eq. (12.16).
- Confidence interval for χ^2: involves finding the confidence limits for the **population standard deviation** σ given the **sample standard deviation** s. The sample size n is needed. This value gives you the degrees of freedom to look up in a χ^2 table which gives you the values of $\chi^2_{\frac{\alpha}{2}}$ and $\chi^2_{1-\frac{\alpha}{2}}$ to use. The limits are calculated using Eq. (12.17).

Exercises

12.17 A random sample of size $n = 5000$ of people that are 55 years old and above in the UK found that 82% had a smartphone. Compute a 95% confidence interval for the proportion of the whole population of the UK that are 55 years old and above and that own a smartphone.

12.18 A random sample of size 20 of screw lengths that are supposed to be 5cm long had a standard deviation of 0.07cm. Find a 90% confidence interval for the actual standard deviation.

12.19 A random sample of 2000 7-year-old boys in the UK had a mean weight of 22.9kg. Assuming that the population standard deviation is 3.2kg, compute a 99% confidence interval for the population mean.

12.3.3 Testing Hypothesis

This is an important section and is, in a sense, what we have been leading up to. Hypothesis testing is used in lots of areas in science, social science, and business. A hypothesis test is a statistical test used to ascertain whether we can assume a specific condition is valid for the entire population, given a data sample. A hypothesis test generally looks at two opposing hypotheses about a population: the null and the alternative hypotheses.

- A null hypothesis is a statement being tested and is the default correct answer.
- An alternative hypothesis is a statement that is opposite to the null hypothesis.

Based on the sample data from a population, a hypothesis test determines whether or not to reject the null hypothesis.

If we reject a hypothesis when it should be accepted, the error is called *type I error*. On the other hand, if we accept a hypothesis when it should be rejected, the error is called *type II error*. The only way to reduce both types of error is to increase the sample size, which may or may not be possible.

Suppose that the sampling distribution of a statistic with a specific hypothesis is a normal distribution with mean $\mu_{\bar{x}}$ and standard deviation $\sigma_{\bar{x}}$. The distribution of the standardised variable is the standardised normal distribution. If the null hypothesis is true, there is a 95% probability that the z-score of the observed sample statistic will lie within $[-1.96, 1.96]$. If we observe a sample statistic whose z-score lies outside the $[-1.96, 1.96]$ range, we conclude that such an event could only happen with a probability of only 0.05 or 5% if the null hypothesis were true. Therefore, we would be inclined to reject the null hypothesis.

In hypothesis testing, the test's significance level is the maximum probability with which we would be willing to risk a type I error. This significance level is a predefined threshold chosen by the user, such as 0.05. For example, at a significance level of 0.05 in a two-tailed test, the chance of getting a result outside the $[-1.96, 1.96]$ range is only 5% or less. Suppose the probability computed based on the observed test statistic is less than the significance level. In that case, it is reasonable to reject the null hypothesis since the result is unlikely to have happened by chance under the null hypothesis.

This range, ± 1.96, is almost 2 standard deviations on either side of the mean in a standard normal distribution and is commonly used for hypothesis tests at the 5% significance level in social sciences or with subject areas where only small sample sizes are available. This can be referred to as a 2σ result. In particle physics, a 5σ result is usually needed. For example, to reject the null hypothesis that the Higgs boson did not exist and that the results found were just chance, a one-sided 5σ result was needed. This corresponds to approximately a 0.00003% chance that the observed data could happen by chance, with a significance level of 0.0000003 for a one-sided test.

Remark 12.6 Distributions like the normal distribution and t-distribution are symmetrical and have two tails. If actions following the result from a statistical test

12.3 Inference

are the same regardless of which tail is considered, then we can use *two-tailed* or *two-sided* tests. We usually consider half of the significance level for each side when given a significance level for such tests. *One-tailed* or *one-sided* tests may be used when we may be interested in only extreme values to one side of the distribution, or actions following the result from a statistical test are different if different tails are considered. That is, the choice of a two-sided or one-sided test is determined by the actual problem. Researchers should decide whether they want to use a one-tailed or two-tailed test before collecting the data. If in doubt it is usual to do a two-tailed test.

♦

The following shows the general procedure for conducting a hypothesis test:

- Specify the hypotheses:
  H_0: represents the null hypothesis. H_1: represents the alternative hypothesis.
- Determine (1) which test to use, one-tailed or two-tailed with the significance level, and (2) the sample size for the test sample.
- Collect the data.
- Decide whether to reject or fail to reject the null hypothesis.
  - Compute the sample statistic. This step varies based on the test used, for example, t-statistic or χ^2 statistic.
  - Check the corresponding distribution table:
    · If the probability of the computed statistic occurring is less than the significance level, then we reject the null hypothesis;
    · If the probability of the computed statistic occurring is greater than the significance level, we fail to reject the null hypothesis.

In this chapter we consider two types of test: the t-test and the Chi-squared test. They are each illustrated by considering two kinds of test.

The t-test is used to:

- Investigate the likely mean of a normally distributed population from a single small sample
- Investigate whether two small samples come from the same underlying normally distributed population

The Chi-squared test is used to:

- To compare an observed distribution from an expected, or desired, distribution (one-way classification)
- To compare two distributions to test for any significant differences (two-way classification)

12.3.3.1 The t-Tests

The t−tests are used for small samples, usually less than 30. We assume that observations are independent, the population distribution should be normal, and the population size should be at least ten times larger than the sample size.

- Means
  To test the null hypothesis that a normal population has the mean μ_0, that is, $H_0 : \mu = \mu_0$, while $H_1 : \mu \neq \mu_0$, we use the t statistic, that is, to apply Eq. (12.9) to check if the observed sample mean is significantly different from the population mean.

Example 12.17 A machine has produced mugs with a bottom thickness of 4 millimetres (mm). To determine whether the machine is in proper working order, a sample of 15 mugs is chosen, for which the mean thickness is 4.2 mm and the standard deviation is 0.2 mm. Test the hypothesis that the machine is in proper working order using a significance level of 0.05.

Solution

- Step 1: Write the hypotheses as follows:
  - $H_0 : \mu = 4$ mm and the machine is properly working.
  - $H_1 : \mu \neq 4$ mm and the machine is not properly working.
- Step 2: Which test should be used: a two-tailed or one-tailed test?
  Since the thickness can be greater or less than the mean value and both are treated as not working properly, we do a two-tailed test. The sample size is $n = 15$ and the significance level is set to 0.05 in the question.
- Step 3: The sample information includes the sample mean $\bar{x} = 4.2$ and the sample standard deviation $s = 0.2$.
- Step 4: Compute the t-statistic using Eq. (12.9) as follows:

$$t = \frac{4.2 - 4}{0.2/\sqrt{15}} \approx 3.87.$$

We want a significance of 0.05 on a two-tailed test, so we need to look in the t-table column with its third row saying 0.05 (the top row says $t_{.975}$). Two critical values for which 2.5% of the area lies in each tail of the t distribution with $15 - 1 = 14$ degrees of freedom are found in the column specified as -2.145 and 2.145 (the value in the table being 2.145). Since 3.87 is not covered in $[-2.145, 2.145]$, indicating the probability of observing a t-statistic value like this is less than 5%, we reject H_0 at the 0.05 significance level.

12.3 Inference

Exercises

12.20 The lifespan of a type of product should be at least 1000 hours. Randomly take 16 products from this type and test their lifespans. The average lifespan of this sample is 920 hours. It is known that the lifespan of this type of product follows a normal distribution with a standard deviation of 100 hours. Test the hypothesis that this product type is non-defective using a significance level of 0.05.

12.21 A packet of washing powder is labelled as having 235 grams in it. In order to adhere to EU packaging regulations, it must not be too far lower than this. We do not care if it is higher. We randomly take 5 samples and note the actual weights; they are 237, 230, 232, 234, and 232 grams. Determine whether this test shows that the washing powder is likely to conform to the regulations with a 0.05 significance level. [Hint: Use a one-tailed t-test.]

12.22 A certain brand of phone is advertised as having a 23-hour video play back capability. 12 phones are taken at random and tested to give the following times: 20, 22, 23, 23, 28, 23, 20, 22, 22, 20, 20, and 21. Use a two-tailed t-test to determine if it satisfies this claim with a 0.1 significance level.

- Difference between means of two samples
  Suppose that two random samples of sizes n_1 and n_2 are drawn from normal populations whose standard deviations are equal, that is, $\sigma_1 = \sigma_2$. The sample means are $\bar{x}_1$ and $\bar{x}_2$, and the sample standard deviations are s_1 and s_2, respectively. To test the null hypothesis that the samples come from the same population, that is, $H_0 : \mu_1 = \mu_2$ and $\sigma_1 = \sigma_2$, we use the t score given by

$$t = \frac{\bar{x}_1 - \bar{x}_2}{\sigma \sqrt{\frac{1}{n_1} + \frac{1}{n_2}}}, \quad (12.18)$$

where

$$\sigma = \sqrt{\frac{(n_1 - 1)s_1^2 + (n_2 - 1)s_2^2}{n_1 + n_2 - 2}}. \quad (12.19)$$

The distribution of t is Student's distribution with $n_1 + n_2 - 2$ degrees of freedom.

Example 12.18 The history GCSE scores of 16 students from Year 11 of a secondary school have a mean of 65 and a standard deviation of 7, while the history GCSE scores of 15 students from Year 11 of another secondary school have a mean of 60 and a standard deviation of 13. Is there a significant difference between the history GCSE scores of the two groups at a significance level of 0.05?

Solution

- Step 1: Denote two schools' mean scores of history GCSE scores as μ_1 and μ_2, separately. Write the hypotheses as follows:
  - $H_0 : \mu_1 = \mu_2$,
  - $H_1 : \mu_1 \neq \mu_2$.
- Step 2: Which test should be used: a two-tailed or one-tailed test? Since one school's mean score can be higher or lower than the other school's mean value and both indicate that the two schools' mean scores differ, we do a two-tailed test. The significance level is set to 0.05, and the sample means are $\bar{x}_1 = 65$ and $\bar{x}_2 = 60$ in the question.
- Step 3: Under the null hypothesis H_0, applying Eq. (12.19), we have

$$\sigma = \sqrt{\frac{(16-1)7^2 + (15-1)13^2}{16+15-2}} \approx 10.3.$$

- Step 4: Compute the t-statistic by applying Eq. (12.18):

$$t = \frac{65 - 60}{10.3\sqrt{\frac{1}{16} + \frac{1}{15}}} \approx 1.35.$$

We want a significance of 0.05 on a two-tailed test, so we need to look in the t-table column with its third row saying 0.05 (the top row says $t_{.975}$). So, two critical values for which 2.5% of the area lies in each tail of the t distribution with $16 + 15 - 2 = 29$ degrees of freedom are -2.045 and 2.045. Since 1.35 is covered in $[-2.045, 2.045]$, indicating the probability of observing a t-statistic value like this is at least 95%, we do not reject H_0 at the 0.05 significance level.

Remark 12.7 Let us change the scenario a bit. One of the two schools has adopted a new practice in teaching history and performs better in the GCSE exam than the other. We are interested only in differences where the 'new practice' students' mean score is greater than the 'normal' teaching students' mean score.

12.3 Inference

In this case we use a one-tailed test. Instead of taking the 2.5% differences in each tail, we take the whole 5% from one tail, the tail on the side where the 'new practice' samples would outperform the 'normal' samples. In this case, it happens to be the right-hand tail. Now we want a significance of 0.05 on a one-tailed test, so we need to look in the t-table column with its **second** row saying 0.05 (the top row says $t_{.95}$).

♦

Exercise

12.23 The weight of 10 students from Year 1 of a primary school has a mean of 18.5kg and a standard deviation of 2kg, while the weight of 20 students from Year 1 of another primary school has a mean of 17.5kg and a standard deviation of 3.5kg. Is there a significant difference between the weight of the two groups at a significance level of 0.01?

12.24 The length of 8 newborn baby boys in one area of the UK has a mean of 55cm and a standard deviation of 5cm, while the length of 12 newborn baby boys in another area of the UK has a mean of 50cm and a standard deviation of 4cm. Is there a significant difference between the lengths of the two sets of babies at a significance level of 0.05?

12.25 The weight of 10 newborn baby girls in one area of the UK has a mean of 3.5kg and a standard deviation of 0.5kg, while the weight of 15 newborn baby girls in another area of the UK has a mean of 3.2kg and a standard deviation of 0.3kg. Is there a significant difference between the weight of the two sets of babies at a significance level of 0.01?

12.3.3.2 Chi-Square Tests

The Chi-square test can be used to determine how well a theoretical distribution, for example, the normal and binomial distributions, fits an empirical distribution which is obtained from the sample data. The key is to estimate the expected frequency.

Since the Chi-squared test is a 'goodness of fit' test, people usually use it as a one-tailed test. Using it as a two-sided test means we are also concerned about whether the fit may be far too good, which usually is not a problem in real-world applications.

- One-way classification tables
  The Chi-square test can test whether the observed frequencies differ significantly from the expected frequencies. To do a Chi-square test, we may set up a table like Table 12.2. It is called a one-way classification table, in which the observed frequencies occupy a single row and n columns. The number of degrees of

Table 12.2 Observed and theoretical frequencies

Event	E_1	E_2	$\cdots$	E_n
Observed frequency	o_1	o_2	$\cdots$	o_n
Expected frequency	e_1	e_2	$\cdots$	e_n

freedom is $n - 1$ for a one-way classification table. Based on the theoretical work showing the relationships among the normal, binomial, and Chi-square distributions [13], rather than using Eq. (12.11), people compute the χ^2 statistic as follows:

$$\chi^2 = \frac{(o_1 - e_1)^2}{e_1} + \frac{(o_2 - e_2)^2}{e_2} + \cdots + \frac{(o_n - e_n)^2}{e_n} = \sum_{i=1}^{n} \frac{(o_i - e_i)^2}{e_i}, \quad (12.20)$$

where o_i are the n observed frequencies and e_i are the n expected frequencies.

Remark 12.8 As a rule of thumb, to use the Chi-square test, the expected number of counts in each cell should be at least 5.

♦

Example 12.19 We have a die with 10 faces. Suppose there is an equally likely chance that the die can land on any face. The die has the following numbers on its faces: 0, 1, 2, 3, 4, 5, 6, 7, 8, 9, respectively. We throw the die 400 times. Let X be the number shown on the landed face. Table 12.3 shows X values and the respective frequencies. Use the Chi-square test at a significance level of 0.05 to test the hypothesis that the die is fair.

Solution

- Specify the hypotheses:
    H_0: The die is fair. H_1: The die is not fair.
- We consider a one-tailed test since it evaluates the right-hand tail area, indicating a significant disagreement between two distributions. The sample size is 400 and the significance level is set to 0.05 in the question.
- The observed data is shown in Table 12.3.
- Decide whether to reject or fail to reject the null hypothesis.
    · Compute the χ^2 statistic. First, we need to calculate the expected frequencies. Under H_0 that the die is fair, we expect the frequency value to be $\frac{400}{10} = 40$ for each face value. Therefore, we can set up Table 12.4.

(continued)

Example 12.19 (continued)

From Table 12.4, we can compute χ^2 using Eq. (12.20) as follows:
$$\chi^2 = \frac{(37-40)^2}{40} + \frac{(46-40)^2}{40} + \frac{(41-40)^2}{40} + \frac{(40-40)^2}{40} + \frac{(40-40)^2}{40} + \frac{(36-40)^2}{40} + \frac{(39-40)^2}{40} + \frac{(37-40)^2}{40} + \frac{(38-40)^2}{40} + \frac{(46-40)^2}{40} = 2.8.$$

- Check the critical value from the corresponding distribution table based on the significance level and the degrees of freedom.
  The degree of freedom of this one-way classification is $10 - 1 = 9$. Therefore, $\chi^2_{0.05} = 16.919$ obtained from a χ^2 table showing the area to the right of the critical value. So for the value of $\chi^2 = 16.919$ the area to the right is 5%. For any larger value of χ^2, the area is less than 5% because we are further to the right hand end of the curve. For any smaller value, the area is larger than 5%, and we are further left on the curve.
- Compare the computed statistic to the critical value.
  Our value of χ^2 is 2.8, and this is less than 16.919 and so further left. This indicates that the area to the right of our observed statistic value (2.8) is greater than 5%. Remember, we are testing how close our observed frequencies are from the expected frequencies, that is, we are testing if the die appears to be fair. A larger value of χ^2 indicates that the observed values are further from the expected values and, in this case, the value of $\chi^2 = 2.8$ does not go past the significance level of 0.05. We therefore conclude that the result is not significant at the 0.05 level. Thus, we do not reject H_0 at the significance level of 5%, and we either conclude that the die is fair or pending further tests.
  Only if we had got a much higher value for χ^2, one above 16.919, would we feel able to reject the null hypothesis. In this case, the chance of getting such a result from a fair dice would have been less than 5%, and so we could conclude that the die was unfair.

Table 12.3 The distribution of X used in Example 12.19

X	0	1	2	3	4	5	6	7	8	9
Frequency	37	46	41	40	40	36	39	37	38	46

Table 12.4 The observer and expected distribution of X used in Example 12.19

X	0	1	2	3	4	5	6	7	8	9
Observed frequency	37	46	41	40	40	36	39	37	38	46
Expected frequency	40	40	40	40	40	40	40	40	40	40

Table 12.5 The theoretical distribution of Z in Exercise 12.26

z_i	1	2	3	4	5	6
p_i	$\frac{11}{36}$	$\frac{9}{36}$	$\frac{7}{36}$	$\frac{5}{36}$	$\frac{3}{36}$	$\frac{1}{36}$

Table 12.6 The distribution of Z of 360 throws in Exercise 12.26

z_i	1	2	3	4	5	6
Observed frequency	90	145	35	50	15	25
Expected frequency						

Table 12.7 The actual distribution of chocolates X in Exercise 12.27

x_i	1	2	3	4	5	6	7	8
Observed frequency	24	34	25	35	36	26	36	24

Exercises

12.26 A pair of fair dice is thrown. Let Z be the smaller value of the two numbers. The distribution of Z consists of its values with their respective probabilities. This has been worked out for you and is given in Table 12.5.

Table 12.6 shows the observed frequencies of Z of 360 throws. Fill in their expected frequencies.

Calculate the Chi-square (χ^2) value. Does the observed distribution differ significantly from the expected distribution, using a significance level of 0.05?

12.27 A box of chocolates of a certain make has 8 types of chocolates: hazelnuts, caramel, orange, fudge, almond, raspberry, coffee, and truffle. These are labelled types $1, 2, \cdots, 8$, respectively. Each box contains 24 chocolates, and 10 boxes were selected, and the types counted to get a distribution X as shown in Table 12.7.

It is expected that each type of chocolate has an equal chance of appearing in a box. Calculate the Chi-square (χ^2) value. Does the observed distribution differ significantly from the expected distribution, using a significance level of 0.05?

12.28 A certain university has percentage targets for its overall degree classification and a percentage distribution Z of actual results from a recent year. Both the expected frequency and the observed frequency, Z, are shown in Table 12.8.

Calculate the Chi-square (χ^2) value. Does the observed distribution differ significantly from the expected distribution, using a significance level of 0.05?

12.3 Inference

Table 12.8 The distribution of observed and expected grades at a university in Exercise 12.28

z_i	1st	2i)	2ii)	3	Pass	Fail
Observed frequency	20	25	29	14	11	1
Expected frequency	15	20	25	20	15	5

- Two-way classification tables
  A two-way classification table, also called a contingency table, concerns two variables. The observed frequencies occupy m rows (that is, m specified values for one variable) and n columns (that is, n specified values for another variable). As mentioned in Sect. 12.2.5 of this chapter, the number of degrees of freedom of a two-way classification table is $(m-1) \times (n-1)$.
  The Chi-square test can test the association between the two variables in a two-way table. The null hypothesis H_0 assumes no association between the two variables in the rows and columns, while the alternative hypothesis H_1 states that some association exists.

Example 12.20 Two groups, A and B, consist of 150 people, each with type 2 diabetes. The two groups are treated identically, except that group A must do supervised aerobic activities for two hours each week. After half a year, it is found that in groups A and B, 120 and 100 people, respectively, put diabetes into remission. Use the Chi-square test at a significance level of 0.05 to test the hypothesis that the supervised aerobic activities help diabetes remission.

Solution

– Specify the hypotheses:
  H_0: There is no difference between the results of the two groups of people—that is, doing the supervised aerobic activities has no effect. In other words that the diabetes remission is independent of, or has no association with, doing the supervised aerobic activities.
  H_1: Represents alternative hypothesis.
– We consider a one-tailed test. The sample size is $150 + 150 = 300$, and the significance level is set to 0.05 in the question.
– We set up Table 12.9 using the data in the question.
– Decide whether to reject or fail to reject the null hypothesis.
  · Compute the statistic. First, we need to calculate the expected frequencies. Under H_0 that doing the supervised aerobic activities has no effect, we expect $\frac{(120+100)}{2} = 110$ people in each of the groups to put them into diabetes remission and $\frac{30+50}{2} = 40$ in each group not to put them into remission. We can set up Table 12.10.
  We can compute χ^2 using Eq. (12.20) as follows:
  $$\chi^2 = \frac{(120-110)^2}{110} + \frac{(100-110)^2}{110} + \frac{(30-40)^2}{40} + \frac{(50-40)^2}{40} = 6.82.$$

(continued)

Example 12.20 (continued)

- Check the critical value from the corresponding distribution table based on the significance level and the degrees of freedom.
  The degrees of freedom of this two-way classification is $(2-1) \times (2-1) = 1$. Therefore, $\chi^2_{0.05} = 3.84$ obtained from a χ^2 table showing the area to the right of the critical value.
- Comparing our computed statistics to the critical value.
  Since $\chi^2 = 6.82 > 3.84$, showing that our value is further to the right and indicating the area to the right of our observed statistic value (6.82) is less than 5%. We therefore conclude that the result is significant at the 0.05 level, that is, our value of χ^2 is too big. Thus, we reject H_0; that is, there is a difference between the two groups of people, and we can conclude that doing supervised aerobic activities helps diabetes remission.

Exercises

12.29 Two universities have percentage distributions of actual results from a recent year. Both results are shown in Table 12.11.

The null hypothesis H_0 is that there is no difference between the two universities in exam performance. A student chosen randomly has the same probability of obtaining any grade from either university. Use the Chi-square test at a significance level of 0.05 to test this hypothesis.

12.30 Two factories are making the same product and both produce a certain number of defective items. The percentage of good and defective items is given in Table 12.12.

The null hypothesis H_0 is that there is no difference between the two factories. Use the Chi-square test at a significance level of 0.05 to test this hypothesis.

Table 12.9 The observed frequencies used in Example 12.20 testing diabetes and aerobic exercise

	Remission	No remission	Total
Group A	120	30	150
Group B	100	50	150
Total	220	80	300

Table 12.10 The expected frequencies used in Example 12.20 testing diabetes and aerobic exercise

	Remission	No remission	Total
Group A	110	40	150
Group B	110	40	150
Total	220	80	300

12.3 Inference

Table 12.11 The distribution of grades in two universities in Exercise 12.29

	1st	2i)	2ii)	3	Pass	Fail
University 1	20	25	29	14	11	5
University 2	12	25	27	20	14	6

Table 12.12 The distribution of defective products in two factories in Exercise 12.30

	Good	Defective
Factory 1	95	5
Factory 2	87	13

Table 12.13 The observed frequencies

	Remission	No remission	Total
Group A	120	40	160
Group B	120	40	160
Total	240	80	320

Table 12.14 The observed frequencies

	Remission	No remission	Total
Group A	24	8	32
Group B	120	40	160
Total	144	48	192

In the examples and exercises in this section so far, both row totals were the same. But what happens if the row totals are different?

First, consider a slightly silly and unrealistic example with data similar to that in Example 12.20, as shown in Table 12.13. Here the data in groups A and B are identical! Hence, doing supervised aerobic activities obviously has no effect, and so we know we will not reject the hypothesis H_0 (H_0 was that doing the supervised aerobic activities has no effect). In fact, if we tried to calculate the χ^2 statistic, we would immediately find that the expected values are all the same as the actual values. This is because the expected values are found in each case by dividing the column sum in the observed frequencies table by two. All the terms in the χ^2 calculation would be zero. The value from the χ^2 table would be the same as in Example 12.20, so our χ^2 value of 0 is less than the 3.84 value in the χ^2 table, and we get the correct conclusion that we do not reject the hypothesis H_0.

Now suppose each of the values in Group A were 5 times smaller since Group A was a small group. This is shown in Table 12.14.

Table 12.15 The observed frequencies in Example 12.21

	Distinction	Pass	Fail	Total
F	10	60	10	80
M	5	29	6	40
Total	15	89	16	120

The proportions of people in group A are the same as in Group B, so again, we do not reject the hypothesis H_0, that doing the supervised aerobic activities has no effect. But if we blindly, and incorrectly, tried to calculate the χ^2 statistic using the method used before involving column totals, that is, the expected values are $\frac{24+120}{2} = 72$ and $\frac{8+40}{2} = 24$, respectively, we would get a very large number for the χ^2 statistic. In fact, we would get

$$\frac{(24-72)^2}{72} + \frac{(120-72)^2}{72} + \frac{(8-24)^2}{24} + \frac{(40-24)^2}{24} \approx 91.3.$$

This is obviously totally wrong, and the problem is we have not considered the different row sums.

The correct method to use to find the expected values for each cell is to use the formula:

$$\text{expected value} = ((\text{row total}) * (\text{column total}))/(\text{grand total}). \tag{12.21}$$

So, for instance, Group A remission cell would be $\frac{32 \times 144}{192} = 24$, which is the same as the observed value. Check the other cells all give the same expected value as the observed value. Now we would again get all zeros in the χ^2 calculation, and all would be well.

Remark 12.9 In fact, Eq. (12.21) reduces to the previous method if the row totals are equal. That is because each row total is half the grand total, so $\frac{\text{row total}}{\text{grand total}} = \frac{1}{2}$ and hence expected value $= \frac{\text{column total}}{2}$ as before.

♦

We will now do a proper example.

Example 12.21 The results for an MSc exam for 80 female students and 40 male students are shown in Table 12.15. Use the Chi-square test at a significance level of 0.05 to test the hypothesis that the there is no difference between the females and males in their performance.

(continued)

Example 12.21 (continued)
Solution

- Specify the hypotheses:
  H_0: There is no difference between the females and males in their performance.
  H_1: Represents alternative hypothesis.
- We consider a one-tailed test. The sample size is $80 + 40 = 120$, and the significance level is set to 0.05 in the question.
- Decide whether to reject or fail to reject the null hypothesis.
  · Compute the statistic. First, we need to calculate the expected frequencies. Here, since the row sums are different, we use Eq. (12.21). These are shown in Table 12.16. For instance, in the cell for females that pass, we have that the expected value $= \frac{80 \times 89}{120} = 59.33$.
  We can compute χ^2 using Eq. (12.20) as follows:
  $\chi^2 = \frac{(10-10)^2}{10} + \frac{(5-5)^2}{5} + \frac{(60-59.33)^2}{59.33} + \frac{(29-29.67)^2}{29.67} + \frac{(10-10.67)^2}{10.67} + \frac{(6-5.33)^2}{5.33} \approx 0.149$.
  · Check the critical value from the corresponding distribution table based on the significance level and the degrees of freedom.
  The degrees of freedom of this two-way classification is $(2-1) \times (3-1) = 2$. Therefore, $\chi^2_{0.05} = 5.99$ obtained from a χ^2 table showing the area to the right of the critical value.
  · Comparing our computed statistics to the critical value.
  Since $\chi^2 = 0.149 < 5.99$, indicating the area to the right of the observed statistic value (0.149) is greater than 5%, we conclude that the result is not significant at the 0.05 level. Thus, we do not reject H_0 at the 5% level.

Table 12.16 The expected frequencies in Example 12.21

	Distinction	Pass	Fail	Total
F	10	59.33	10.67	80
M	5	29.67	5.33	40
Total	15	89	16	120

Table 12.17 The observed frequencies in Exercise 12.31

	Distinction	Pass	Fail	Total
F	9	64	7	80
M	6	26	8	40
Total	15	90	15	120

Table 12.18 Observed frequencies in Exercise 12.32

	A	B	C	D
Girls	20	50	60	15
Boys	25	60	50	15

Exercises

12.31 Do the same calculation as in Example 12.21 but with the new values given in Table 12.17. That is, use the Chi-square test at a significance level of 0.05 to test the hypothesis that there is no difference between the females and males in their performance.

12.32 Conduct a standard Chi-squared test using a significance level of 0.05. Maths exam grades are compared between 150 boys and 145 girls in the same year group in a secondary school (see Table 12.18). The null hypothesis H_0 is that there is no difference between girls and boys in the maths exam performance. A girl and boy chosen randomly have the same probability of obtaining any grade. Use the Chi-square test at a significance level of 0.05 to test this hypothesis.

Remark 12.10 The Chi-square test relies on the Chi-square distribution, which is a continuous probability distribution. However, data in contingency tables consist of discrete counts. This mismatch can cause inconsistencies between the observed (discrete) test statistic and the theoretical (continuous) Chi-square distribution.

To address this, a continuity correction is sometimes applied to adjust the test statistic and better approximate the continuous distribution, especially when the degrees of freedom are equal to one or when sample sizes are small, such as when the expected frequencies in some cells are less than 5.

For example, Yates' continuity correction can be applied when the number of degrees of freedom is one. Yates' correction adjusts the test statistic as follows:

$$\chi^2(correct) = \sum_{i=1}^{n} \frac{(|o_i - e_i| - 0.5)^2}{e_i}.$$

Please note that in this book, we use the original Chi-square test formula without applying Yates' continuity correction, even when the degrees of freedom equals 1.

♦

Chapter 13
Algorithms 4: Maximum Likelihood Estimation and Its Application to Regression

In this chapter, we first introduce the maximum likelihood estimation method. Then we show how it can be applied in enhancing the linear regression algorithm introduced in Chap. 8. Moreover, since the algorithm is now configured in a proper probability and statistical framework, we can set up confidence intervals for estimators using the methods presented in Chap. 12. Finally, we use the maximum likelihood estimation technique to introduce logistic regression, which is actually a classification algorithm.

13.1 Maximum Likelihood Estimation

Before we start to introduce the maximum likelihood estimation method, let us have a look at Example 13.1 first.

Example 13.1 Suppose data are generated independently from an identical Gaussian distribution, but we do not know which Gaussian distribution it was. That is, we do not know the parameters μ and σ for the Gaussian distribution. Suppose that two of the data points are $x_1 = 0$ and $x_2 = 1$. Different Gaussian distributions will give a different probability (density) for these two points being generated. That is, each Gaussian distribution will give a different **likelihood** for these two points being generated. We can illustrate this by considering two particular Gaussian distributions. So we want to know the probability of obtaining $x_1 = 0$ and $x_2 = 1$ from the distribution, that is, to compute $p(x_1 = 0, x_2 = 1 | \mu, \sigma)$. Let us consider the two Gaussian distributions:

(continued)

Example 13.1 (continued)
1. $X \sim \mathcal{N}(2, 2^2)$.
2. $X \sim \mathcal{N}(1, 1^2)$.

Figure 13.1 shows these two Gaussian distributions using solid and dash-dotted lines, respectively, and presents the point $x = 0$ with a cross and $x = 1$ with a square.

We can apply Eq. (10.19) from Chap. 10 to compute the probability (density).

If $X \sim \mathcal{N}(2, 2^2)$, we have

$$p(x = 0 | \mu = 2, \sigma = 2) = \frac{1}{\sqrt{2\pi \times 2^2}} e^{-\frac{(0-2)^2}{2 \times 2^2}} \approx 0.12,$$

and

$$p(x = 1 | \mu = 2, \sigma = 2) = \frac{1}{\sqrt{2\pi \times 2^2}} e^{-\frac{(1-2)^2}{2 \times 2^2}} \approx 0.18.$$

It gives $p(x_1 = 0, x_2 = 1 | \mu = 2, \sigma = 2) = 0.12 \times 0.18 = 0.0216$.
If $X \sim \mathcal{N}(1, 1)$, we have

$$p(x = 0 | \mu = 1, \sigma = 1) = \frac{1}{\sqrt{2\pi \times 1^2}} e^{-\frac{(0-1)^2}{2 \times 1^2}} \approx 0.24,$$

$$p(x = 1 | \mu = 1, \sigma = 1) = \frac{1}{\sqrt{2\pi \times 1^2}} e^{-\frac{(1-1)^2}{2 \times 1^2}} \approx 0.40.$$

It gives $p(x_1 = 0, x_2 = 1 | \mu = 1, \sigma = 1) = 0.24 \times 0.40 = 0.096$. Intersections of the horizontal dashed lines in Fig. 13.1 indicate the corresponding $p(x|\mu, \sigma)$ values of a specified Gaussian distribution for each data observation.

Therefore, the likelihood that we observe data $x_1 = 0$ and $x_2 = 1$ generated from the Gaussian distribution with a mean value of 1 and a standard deviation of 1 is higher than observing them from the Gaussian distribution with a mean of 2 and a standard deviation of 2.

The above example shows that the likelihood of generating 0 and 1 can differ with different Gaussian distributions. The maximum likelihood (ML) method finds the estimate that gives the observed data the highest likelihood, the highest probability (density) that an event has occurred. That is, it finds the values of the parameters μ and σ that maximise the probability (density). As was shown in Chap. 5,

13.1 Maximum Likelihood Estimation

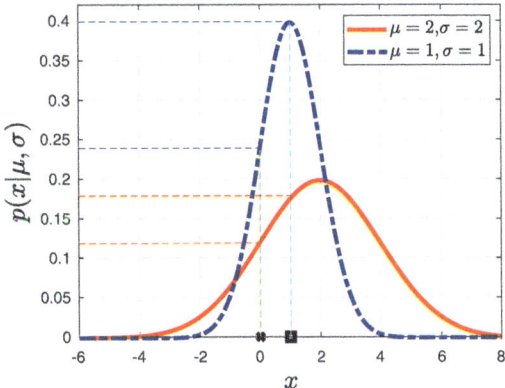

Fig. 13.1 Illustration for Example 13.1

finding maximums involves differentiation and, in particular, since there are usually multiple parameters, it involves partial differentiation as discussed in Chap. 6.

Definition 13.1 (Maximum Likelihood Estimation) Suppose X is a random variable with a probability density function $f(x; \theta_1, \cdots, \theta_k)$, where $\theta_1, \cdots, \theta_k$ are k unknown parameters. Let $x_1, \cdots, x_N$ be observations of X. The joint probability of observations is given by

$$L(\theta_1, \cdots, \theta_k) = \prod_{i=1}^{N} f(x_i; \theta_1, \cdots, \theta_k).$$

This function is called the likelihood function and is a function of unknown parameters $(\theta_1, \cdots, \theta_k)$ [14]. We wish to maximise it. So if there exists $\hat{\theta}_1, \cdots, \hat{\theta}_k$ where the following equation holds:

$$L(\hat{\theta}_1, \cdots, \hat{\theta}_k) = \max_{(\theta_1, \cdots, \theta_k) \in \Theta} \{L(\theta_1, \cdots, \theta_k)\}, \tag{13.1}$$

then $\hat{\theta}_1, \cdots, \hat{\theta}_k$ are the maximum likelihood estimation of θ_j, where $j = 1, \ldots, k$.

The notation used in $f(x; \theta_1, \cdots, \theta_k)$ just reminds you that the values used after the semicolon are parameters of the particular probability density function being considered. So, for example, in the Gaussian distribution used in Example 13.1, the probability density function would be written as $f(x; \mu, \sigma)$. In Example 13.1 we just took the product of two values of x, namely, $x_1 = 0$ and $x_2 = 1$. So

$$L(\mu, \sigma) = \prod_{i=1}^{2} f(x_i; \mu, \sigma).$$

Example 13.2 Suppose X is a random variable $X \sim \mathcal{N}(\mu, \sigma^2)$. Show that $\hat{\mu} = \frac{1}{N} \sum_i^N x_i$ and $(\hat{\sigma})^2 = \frac{1}{N} \sum_i^N (x_i - \mu)^2$ using the maximum likelihood estimate method.

Solution Since $X \sim \mathcal{N}(\mu, \sigma^2)$, its probability density function (Eq. 10.19) is given by

$$f(x; \mu, \sigma) = \frac{1}{\sqrt{2\pi\sigma^2}} e^{-\frac{(x-\mu)^2}{2\sigma^2}}.$$

The likelihood function over n observations is

$$L(\mu, \sigma) = \prod_{i=1}^{N} f(x_i; \mu, \sigma)$$

$$= \prod_{i=1}^{N} \left\{ \frac{1}{\sqrt{2\pi\sigma^2}} e^{-\frac{(x_i-\mu)^2}{2\sigma^2}} \right\}$$

$$= \frac{1}{(\sqrt{2\pi\sigma^2})^N} e^{-\frac{1}{2\sigma^2} \sum_{i=1}^{N}(x_i-\mu)^2}.$$

It is more convenient to maximise the logarithm of the likelihood function, that is, $\max_{(\theta_1, \cdots, \theta_k) \in \Theta} \{\ln L(\theta_1, \cdots, \theta_k)\}$. When taking logarithms of both sides, things that are multiplied become added, and $\ln(e^{g(x)}) = g(x)$ for any $g(x)$. Hence, we get the following:

$$\ln L(\mu, \sigma) = \ln\left(\frac{1}{(\sqrt{2\pi\sigma^2})^N}\right) - \frac{1}{2\sigma^2} \sum_{i=1}^{N}(x_i - \mu)^2$$

$$= -N \ln(\sigma) - N \ln(\sqrt{2\pi}) - \frac{1}{2\sigma^2} \sum_{i=1}^{N}(x_i - \mu)^2.$$

To find $\hat{\mu}$ and $(\hat{\sigma})^2$ so that $\ln L(\hat{\mu}, \hat{\sigma})$ is maximised, we can do the partial derivatives with respect to each parameter and then set them to zero:

$$\begin{cases} \frac{\partial \ln L(\mu,\sigma)}{\partial \mu} = \frac{1}{\sigma^2} \sum_{i=1}^{N}(x_i - \mu) = 0 \\ \frac{\partial \ln L(\mu,\sigma)}{\partial \sigma} = -\frac{N}{\sigma} + \frac{1}{\sigma^3} \sum_{i=1}^{N}(x_i - \mu)^2 = 0. \end{cases}$$

Rearranging these, the solution is

(continued)

13.1 Maximum Likelihood Estimation

Example 13.2 (continued)

$$\begin{cases} \hat{\mu} = \frac{1}{N} \sum_i^N x_i, \\ (\hat{\sigma})^2 = \frac{1}{N} \sum_{i=1}^N (x_i - \mu)^2. \end{cases}$$

We can now apply the result from Example 13.2 to Example 13.1.

Example 13.3 We now want to maximise the parameters of a Gaussian distribution (Gaussian probability density function) to get the most likely Gaussian distribution that contains just the two points $x_1 = 0$ and $x_2 = 1$, as in Example 13.1. From Example 13.2 we have

$$\begin{cases} \hat{\mu} = \frac{1}{N} \sum_i^N x_i, \\ (\hat{\sigma})^2 = \frac{1}{N} \sum_{i=1}^N (x_i - \mu)^2. \end{cases}$$

So for $N = 2$, $x_1 = 0$ and $x_2 = 1$, this gives $\hat{\mu} = 0.5$ and $(\hat{\sigma})^2 = 0.25$ and so the standard deviation, $\hat{\sigma}$, is 0.5.

To show that this gives a higher probability than the previous Gaussian distributions used in Example 13.1, we will do the same calculations as we did in Example 13.1:

$$p(x = 0 | \mu = 0.5, \sigma = 0.5) = \frac{1}{\sqrt{2\pi \times 0.5^2}} e^{-\frac{(0-0.5)^2}{2 \times 0.5^2}} \approx 0.48,$$

and

$$p(x = 1 | \mu = 0.5, \sigma = 0.5) = \frac{1}{\sqrt{2\pi \times 0.5^2}} e^{-\frac{(1-0.5)^2}{2 \times 0.5^2}} \approx 0.48.$$

This gives $p(x_1 = 0, x_2 = 1 | \mu = 0.5, \sigma = 0.5) = 0.48 \times 0.48 = 0.23$ which is higher than both the values in Example 13.1.

Intuitively, looking at Fig. 13.2, for the values we get for $\hat{\mu}$ and $\hat{\sigma}$ (the dotted curve), then a value for the mean of halfway between the two points $x_1 = 0$ and $x_2 = 1$ and a standard deviation of 0.5 seems right for the best fitting Gaussian.

Fig. 13.2 Illustration for Example 13.3

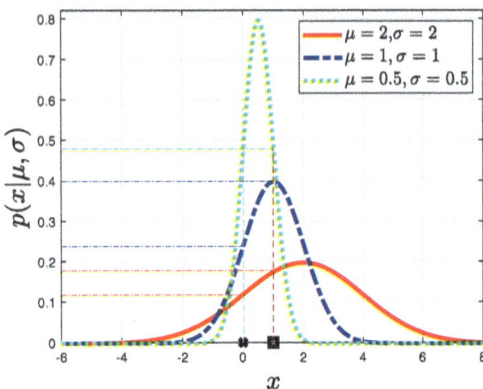

Exercise

13.1 Suppose X is a random variable, and its probability density function is

$$f(x) = \begin{cases} \lambda e^{-\lambda x}, & x > 0 \\ 0, & x < 0, \end{cases}$$

where $\lambda > 0$. If $x_1, \ldots, x_n$ are observations of X, find the maximum likelihood estimate of λ.

Now suppose we have just two points in our distribution, namely, $x_1 = 1$ and $x_2 = 2$.

(1) Suppose $\lambda = 1$, find $p(x_1 = 1, x_2 = 2 | \lambda = 1)$.
(2) Suppose $\lambda = 2$, find $p(x_1 = 1, x_2 = 2 | \lambda = 2)$.
(3) Now calculate the value you would get for λ when you have the two points $x_1 = 1$ and $x_2 = 2$ using the maximum likelihood estimate of $\lambda = \lambda_{best}$ you found in the first part of this Exercise. Using this value of λ calculate $p(x_1 = 1, x_2 = 2 | \lambda = \lambda_{best})$.

13.2 Revisiting Linear Regression

13.2.1 Linear Regression with Maximum Likelihood Estimation

Suppose Y is the dependent variable, and X is the independent variable. We use y and x to denote the sample values of these two variables separately. We want to understand the relationship between these two variables from the given data and to be able to predict y when a new sample value x is observed.

13.2 Revisiting Linear Regression

Given a sample value x from X, there could be a set of different values found for y, and those values of y would form some sort of distribution. We can model this conditional distribution of y given the sample value of X. This distribution has some unknown parameters. For example, if we use a normal distribution (see Eq. 10.19 in Chap. 10) to model the given data, we need to estimate the mean μ and the standard deviation σ from the data.

When introducing the least-squares method in Chap. 8, the regression line of y on x is written as

$$f_\mathbf{a}(x) = a_0 + a_1 x,$$

where x is the observed input value, and $f_\mathbf{a}(x)$ gives the estimated value of y. This equation is referred to as noise-free and gives a fixed value for y. However, we can include a noise or error element and can write the linear regression model as follows:

$$y = a_0 + a_1 x + \epsilon, \tag{13.2}$$

where ϵ is an unobserved error term. This model has two parts: the deterministic part, $a_0 + a_1 x$, and the stochastic part, ϵ. There are three assumptions made for this model:

1. For each observed x value, there is a normal distribution of Y from which the sample value y is drawn at random.
2. The mean μ of the normal distribution of Y to the corresponding x value lies on the straight line $a_0 + a_1 x$.
3. The standard deviation σ of each normal distribution of Y is a constant as x varies. That is, the noise is assumed to be the same at all points.

Figure 13.3 shows the normal distribution of Y about the regression line $a_0 + a_1 x$ for three selected values of x.

Based on these assumptions, we have the probability density distribution of y given by

$$f(y|x; a_0, a_1, \sigma^2) = \frac{1}{\sqrt{2\pi\sigma^2}} e^{\frac{(y-(a_0+a_1 x))^2}{2\sigma^2}}, \tag{13.3}$$

with the expected y value lying along the solid line shown in Fig. 13.3, that is, $E(y) = a_0 + a_1 x$, and the variance of y is σ^2. So the deterministic part of y is given by its mean lying on the line $y = a_0 + a_1 x$, and the stochastic part is the normally

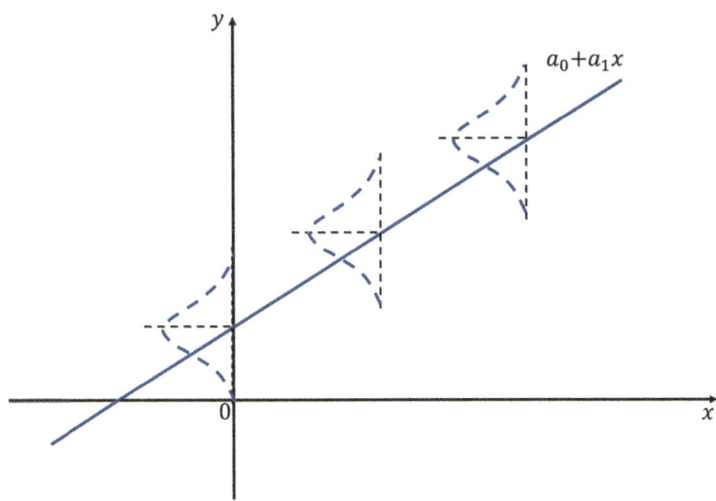

Fig. 13.3 A schematic diagram illustrating the three assumptions of the linear regression model

distributed error ϵ. In fact, we can see that given Eq. (13.2), Eq. (13.3) can be written as

$$f(\epsilon|\sigma) = \frac{1}{\sqrt{2\pi\sigma^2}} e^{\frac{\epsilon^2}{2\sigma^2}}, \qquad (13.4)$$

which says the unobserved error, ϵ, has a Gaussian distribution with a zero mean and standard deviation of σ.

Example 13.4 Consider a simple linear regression model given by $y = 2 + 1.5x + \epsilon$, where $\epsilon \sim \mathcal{N}(\mu = 0, \sigma^2 = 9)$. What is the mean of y given that $x = 2$, that is, what is $E(y|x = 2)$?

Solution Substituting $x = 2$ into $y = 2 + 1.5x + \epsilon$, we have $y = 2 + 1.5 \times 2 + \epsilon = 5 + \epsilon$ and $E(y|x = 2) = E(5 + \epsilon)$. Applying the third property (3) in Sect. 10.5.1.1 of Chap. 10, that is, $E(a + X) = E(X) + a$, where a is a constant, we obtain $E(y|x = 2) = 5 + E(\epsilon) = 5$, since ϵ has zero mean.
This, of course, is the value of y found by substituting $x = 2$ into $y = 2 + 1.5x$, using the second assumption above that says the mean value lies on the straight line $a_0 + a_1 x$, namely, the deterministic part of the model.

Exercise

13.2 Consider a simple linear regression model given by $y = -4 + 3x + \epsilon$, where $\epsilon \sim \mathcal{N}(\mu = 0, \sigma^2 = 0.01)$.

(1) What is $E(y|x = 1)$?
(2) What is the standard deviation (std) of y given that $x = 1$, that is, what is $std(y|x = 1)$?

Suppose there are N data points (x_i, y_i) for $i = 1, \cdots, N$. If the y_i's are independently normal distributed, that is, $y_i \sim \mathcal{N}(a_0 + a_1 x_i, \sigma^2)$, then the likelihood function is again formed by a product of probabilities:

$$L(a_0, a_1, \sigma) = \prod_{i=1}^{N} p(y_i | x_i; a_0, a_1, \sigma)$$

$$= \prod_{i=1}^{N} \frac{1}{\sqrt{2\pi\sigma^2}} e^{-\frac{(y_i - (a_0 + a_1 x_i))^2}{2\sigma^2}} \qquad (13.5)$$

$$= \frac{1}{(\sqrt{2\pi\sigma^2})^N} e^{-\frac{1}{2\sigma^2} \sum_{i=1}^{N}(y_i - (a_0 + a_1 x_i))^2}.$$

Equation (13.5) is the likelihood function for data that can be plotted in two dimensions, one independent variable X and one dependent variable Y. This can be extended by considering multiple independent variables. Here we consider d independent variables. This analysis closely follows the exposition given in Sect. 8.3 of Chap. 8. The main difference between this and Chap. 8 is that Chap. 8 does not have any error or noise terms, and so has no reference to ϵ or to σ^2. In fact, the result we now obtain for the coefficients of the line of best fit, or, since we are now in multiple dimensions, the coefficients of hyperplane of best fit, $\mathbf{a} = (a_0, a_1, \ldots, a_d)^T$, will be the same as we got in Chap. 8 since it is the deterministic part of the model.

Let the data be a matrix of the size of $N \times (d + 1)$, $\mathbf{X}$, where the first column is a column vector including N ones. Let $\mathbf{x}_i$ be the ith data item of a row vector including $d + 1$ elements, $\mathbf{a}$ be a $d + 1$ column vector including d coefficients $a_1, \ldots, a_d$ plus a_0, and $\mathbf{y}$ and $\boldsymbol{\epsilon}$ be column vectors including y_i and ϵ_i, $i = 1, \ldots, N$, respectively. Then we have

$$\mathbf{y} = \mathbf{X}\mathbf{a} + \boldsymbol{\epsilon}. \qquad (13.6)$$

The assumptions are $\epsilon \sim \mathcal{N}(0, \sigma^2)$, and the error terms are uncorrelated. That is,

$$E[\epsilon_i \epsilon_j | \mathbf{X}] = 0. \tag{13.7}$$

Therefore, the likelihood function is

$$L(\mathbf{a}, \sigma^2) = \prod_{i=1}^{N} p(y_i | \mathbf{x}_i; \mathbf{a}, \sigma^2)$$

$$= \prod_{i=1}^{N} \frac{1}{\sqrt{2\pi\sigma^2}} e^{-\frac{(y_i - \mathbf{x}_i \mathbf{a})^T (y_i - \mathbf{x}_i \mathbf{a})}{2\sigma^2}} \tag{13.8}$$

$$= (2\pi\sigma^2)^{-N/2} e^{-\frac{1}{2\sigma^2}(\mathbf{y} - \mathbf{Xa})^T (\mathbf{y} - \mathbf{Xa})}.$$

The logarithm of $L(\mathbf{a}, \sigma^2)$, denoted as $\mathcal{L}(\mathbf{a}, \sigma^2)$, is

$$\mathcal{L}(\mathbf{a}, \sigma^2) = -\frac{N}{2} \ln(2\pi) - \frac{N}{2} \ln(\sigma^2) - \frac{1}{2\sigma^2} (\mathbf{y} - \mathbf{Xa})^T (\mathbf{y} - \mathbf{Xa}). \tag{13.9}$$

In order to find the maximum-likelihood estimations, we need to find the first-order derivative of $\mathcal{L}$ with respect to $\mathbf{a}$ and σ^2 as follows:

$$\frac{\partial \mathcal{L}}{\partial \mathbf{a}} = -\frac{1}{2\sigma^2} 2(-\mathbf{X})^T (\mathbf{y} - \mathbf{Xa}) = \frac{1}{\sigma^2} \mathbf{X}^T (\mathbf{y} - \mathbf{Xa}), \tag{13.10}$$

and

$$\frac{\partial \mathcal{L}}{\partial \sigma} = -\frac{N}{\sigma} + \frac{1}{\sigma^3} (\mathbf{y} - \mathbf{Xa})^T (\mathbf{y} - \mathbf{Xa}). \tag{13.11}$$

The detailed derivative of $\frac{\partial (\mathbf{y} - \mathbf{Xa})^T (\mathbf{y} - \mathbf{Xa})}{\partial \mathbf{a}}$ can be viewed in Eq. (8.16) in Chap. 8. We set the derivatives of Eqs. (13.10) and (13.11) to zero to obtain $\mathbf{a}$ and σ^2, and the solution is the estimator

$$\tilde{\mathbf{a}} = (\mathbf{X}^T \mathbf{X})^{-1} \mathbf{X}^T \mathbf{y},$$

the same as Eq. (8.17) in Chap. 8, and

$$\tilde{\sigma}^2 = \frac{(\mathbf{y} - \mathbf{Xa})^T (\mathbf{y} - \mathbf{Xa})}{N}.$$

13.2 Revisiting Linear Regression

Remember that **a** is a $d+1$ length vector since it contains a_0. So, we can obtain an unbiased estimator for $\tilde{\sigma}^2$, by dividing $(\mathbf{y} - \mathbf{Xa})^T(\mathbf{y} - \mathbf{Xa})$ by $N - (d+1)$ instead of N, that is,

$$\tilde{\sigma}^2 = \frac{(\mathbf{y} - \mathbf{Xa})^T(\mathbf{y} - \mathbf{Xa})}{N - (d+1)}. \tag{13.12}$$

We have yet to introduce the concept of *unbiased estimator*. Readers may refer to [15].

We use the notation $\tilde{\mathbf{a}}$ and $\tilde{\sigma}^2$ to denote they are estimators to distinguish them from the true parameters **a** and σ^2.

Example 13.5 Suppose there are a set of data as shown in Table 13.1, where x_i are the values of an independent variable X, and y_i are the corresponding values of the dependent variable Y. Note that here $d+1 = 2$ since there is just one independent variable. This is picked to make the calculation easier. Also we only have three data points, so $N = 3$. Realistic examples would have both d and N larger.

Fit a linear regression model for the data using the maximum likelihood estimate method and round results to two decimal places.

Solution Let $\mathbf{y} = \begin{bmatrix} 1 \\ 1.8 \\ 4.2 \end{bmatrix}$ and **X** be the inputs including x_i and one column of 1s corresponding to a_0. We have $\mathbf{X}^T = \begin{bmatrix} 1 & 1 & 1 \\ 1 & 2 & 4 \end{bmatrix}$.

Since

$$\mathbf{X}^T\mathbf{X} = \begin{bmatrix} 1 & 1 & 1 \\ 1 & 2 & 4 \end{bmatrix} \begin{bmatrix} 1 & 1 \\ 1 & 2 \\ 1 & 4 \end{bmatrix} = \begin{bmatrix} 3 & 7 \\ 7 & 21 \end{bmatrix},$$

then

$$(\mathbf{X}^T\mathbf{X})^{-1} \approx \begin{bmatrix} 1.5 & -0.5 \\ -0.5 & 0.214 \end{bmatrix}.$$

Substituting $(\mathbf{X}^T\mathbf{X})^{-1}$, $\mathbf{X}^T$, and **y** into Eq. (8.17), we have

$$\tilde{\mathbf{a}} = \begin{bmatrix} 1.5 & -0.5 \\ -0.5 & 0.214 \end{bmatrix} \begin{bmatrix} 1 & 1 & 1 \\ 1 & 2 & 4 \end{bmatrix} \begin{bmatrix} 1 \\ 1.8 \\ 4.2 \end{bmatrix} \approx \begin{bmatrix} -0.2 \\ 1.086 \end{bmatrix} \approx \begin{bmatrix} -0.2 \\ 1.09 \end{bmatrix} = \begin{bmatrix} a_0 \\ a_1 \end{bmatrix}.$$

(continued)

Example 13.5 (continued)

Furthermore, substituting $\mathbf{y}, \mathbf{X}, \mathbf{a}, N = 3$, and $d = 1$ into Eq. (13.12), we have

$$\tilde{\sigma}^2 = \frac{1}{(3-2)}\left(\begin{bmatrix} 1 \\ 1.8 \\ 4.2 \end{bmatrix} - \begin{bmatrix} 1 & 1 \\ 1 & 2 \\ 1 & 4 \end{bmatrix}\begin{bmatrix} -0.2 \\ 1.086 \end{bmatrix}\right)^T \left(\begin{bmatrix} 1 \\ 1.8 \\ 4.2 \end{bmatrix} - \begin{bmatrix} 1 & 1 \\ 1 & 2 \\ 1 & 4 \end{bmatrix}\begin{bmatrix} -0.2 \\ 1.086 \end{bmatrix}\right) \approx 0.05.$$

Therefore, the fitted linear regression is

$$\tilde{y} = -0.2 + 1.09x + \epsilon,$$

where $\epsilon \sim \mathcal{N}(\mu = 0, \tilde{\sigma}^2 = 0.05)$.

The fitted linear regression line is shown in Fig. 13.4.

Suppose now there is new data at $x = 3$. Estimate its expected $\tilde{y}$ value using the above fitted linear regression model, that is, to compute $E(\tilde{y}|x = 3, \mu = 0, \tilde{\sigma}^2 = 0.05)$.

Solution $E(\tilde{y}|x = 3; \mu = 0, \sigma^2 = 0.05) = -0.2 + 1.086 \times 3 \approx 3.06$.

Table 13.1 The data for Example 13.5

x_i	y_i
1	1
2	1.8
4	4.2

Fig. 13.4 Visualisation of the fitted linear regression line in Example 13.5

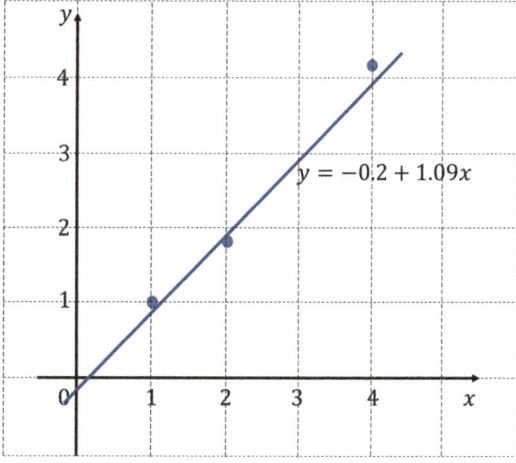

13.2 Revisiting Linear Regression

To illustrate this a bit further, we will do another example using the same X values as in Example 13.5 and Table 13.1 but two different sets of Y values. This is just to save having to redo the calculation of $\mathbf{X}^T\mathbf{X}$!

Example 13.6 The new values for the first set of data are shown in the left two columns of Table 13.2. Since the x_i values are the same as in Example 13.5, we get the same matrix for $\mathbf{X}$ and the same matrix for $(\mathbf{X}^T\mathbf{X})^{-1}$, namely,

$$(\mathbf{X}^T\mathbf{X})^{-1} \approx \begin{bmatrix} 1.5 & -0.5 \\ -0.5 & 0.214 \end{bmatrix}.$$

However, $\mathbf{y}$ now is $\mathbf{y} = \begin{bmatrix} 1 \\ 3 \\ 3 \end{bmatrix}$.

To find $\tilde{\mathbf{a}}$ we again use Eq. (8.17):

$$\tilde{\mathbf{a}} = \begin{bmatrix} 1.5 & -0.5 \\ -0.5 & 0.214 \end{bmatrix} \begin{bmatrix} 1 & 1 & 1 \\ 1 & 2 & 4 \end{bmatrix} \begin{bmatrix} 1 \\ 3 \\ 3 \end{bmatrix} \approx \begin{bmatrix} 1 \\ 0.566 \end{bmatrix} \approx \begin{bmatrix} 1 \\ 0.57 \end{bmatrix}.$$

Finally, we again use Eq. (13.12) to find $\tilde{\sigma}^2$:

$$\tilde{\sigma}^2 = \frac{1}{(3-2)} \left(\begin{bmatrix} 1 \\ 3 \\ 3 \end{bmatrix} - \begin{bmatrix} 1 & 1 \\ 1 & 2 \\ 1 & 4 \end{bmatrix} \begin{bmatrix} 1 \\ 0.566 \end{bmatrix} \right)^T \left(\begin{bmatrix} 1 \\ 3 \\ 3 \end{bmatrix} - \begin{bmatrix} 1 & 1 \\ 1 & 2 \\ 1 & 4 \end{bmatrix} \begin{bmatrix} 1 \\ 0.566 \end{bmatrix} \right) \approx 1.14.$$

Therefore, the fitted linear regression is

$$\tilde{y} = 1 + 0.57x + \epsilon,$$

where $\epsilon \sim \mathcal{N}(\mu = 0, \tilde{\sigma}^2 = 1.14)$.

This fitted linear regression line is shown in Fig. 13.5. We can see by comparing Figs. 13.4 and 13.5 that the line is closer to the points in Fig. 13.4 and hence we need a smaller value for $\tilde{\sigma}^2$ in Example 13.5 than in Example 13.6.

Finally, we will use the two right-hand columns in Table 13.2. Again $\mathbf{X}$ is the same, but now $\mathbf{y} = \begin{bmatrix} 1 \\ 2 \\ 4 \end{bmatrix}$.

(continued)

Example 13.6 (continued)

Using the same method we get $\tilde{\mathbf{a}} = \begin{bmatrix} 0 \\ 1 \end{bmatrix}$. However, when we come to find $\tilde{\sigma}^2$, we need to calculate $(\mathbf{y} - \mathbf{Xa})$ which is as follows:

$$\left(\begin{bmatrix} 1 \\ 2 \\ 4 \end{bmatrix} - \begin{bmatrix} 1 & 1 \\ 1 & 2 \\ 1 & 4 \end{bmatrix} \begin{bmatrix} 0 \\ 1 \end{bmatrix} \right) = \begin{bmatrix} 0 \\ 0 \\ 0 \end{bmatrix}.$$

Hence, $\tilde{\sigma}^2 = 0$.
This is because these values of X and Y actually lie on a straight line.

Finally, we will do an example with more data points.

Table 13.2 The data for Example 13.6

x_i	y_i	x_i	y_i
1	1	1	1
2	3	2	2
4	3	4	4

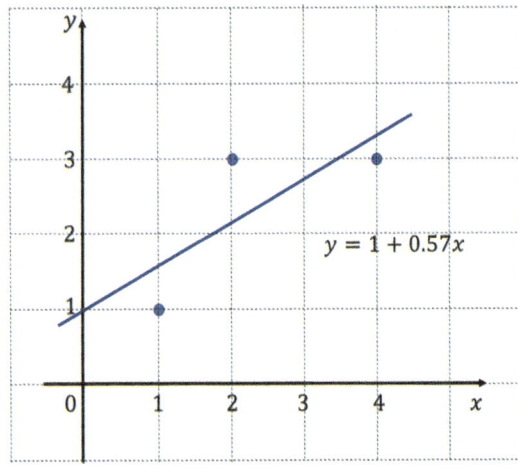

Fig. 13.5 Visualisation of the fitted linear regression line in Example 13.6

13.2 Revisiting Linear Regression

Example 13.7 Suppose there are a set of data as shown in Table 13.3, where x_i are the values of an independent variable X, and y_i are the corresponding values of the dependent variable Y. Again this data can be plotted in two dimensions since it has just one independent variable, so $d + 1 = 2$, but now we have five data points, that is $N = 5$.

Fit a linear regression model for the data using the maximum likelihood estimate method and round results to two decimal places.

Solution

Let $\mathbf{y} = \begin{bmatrix} 2.4 \\ 2 \\ 1.6 \\ 1 \\ 0.4 \end{bmatrix}$ and $\mathbf{X}^T = \begin{bmatrix} 1 & 1 & 1 & 1 & 1 \\ 1 & 2 & 3 & 4 & 5 \end{bmatrix}.$

So

$$\mathbf{X}^T \mathbf{X} = \begin{bmatrix} 5 & 15 \\ 15 & 55 \end{bmatrix},$$

and

$$(\mathbf{X}^T \mathbf{X})^{-1} = \begin{bmatrix} 1.1 & -0.3 \\ -0.3 & 0.1 \end{bmatrix}.$$

Substituting $(\mathbf{X}^T \mathbf{X})^{-1}$, $\mathbf{X}^T$, and $\mathbf{y}$ into Eq. (8.17), we have

$$\tilde{\mathbf{a}} = \begin{bmatrix} 2.98 \\ -0.5 \end{bmatrix}.$$

Furthermore, substituting $\mathbf{y}, \mathbf{X}, \mathbf{a}, N = 5$, and $d = 1$ into Eq. (13.12), we have

$$\tilde{\sigma}^2 \approx \frac{1}{(5-2)} \times 0.028 \approx 0.0093 \approx 0.01.$$

Therefore, the fitted linear regression is

$$\tilde{y} = 2.98 - 0.5x + \epsilon,$$

where $\epsilon \sim \mathcal{N}(\mu = 0, \tilde{\sigma}^2 = 0.01)$.

Suppose now there is new data at $x = 3.5$. Estimate its expected $\tilde{y}$ value using the above fitted linear regression model, that is, to compute $E(\tilde{y}|x = 3.5, \mu = 0, \tilde{\sigma}^2 = 0.01)$.

Solution
$E(\tilde{y}|x = 3.5; \mu = 0, \sigma^2 = 0.01) = 2.98 - 0.5 \times 3.5 = 1.23.$

Exercise

13.3 Given the two sets of data as shown in Table 13.4, where x_i are the values of an independent variable X and y_i are the corresponding values of the dependent variable Y, fit a linear regression model for each dataset using the maximum likelihood estimate method and round results to two decimal places.

13.4 Given the data as shown in Table 13.5, where x_i are the values of an independent variable X and y_i are the corresponding values of the dependent variable Y, fit a linear regression model for each set of data using the maximum likelihood estimate method and round results to two decimal places, except $\tilde{\sigma}^2$, which should be given to 3 decimal places.

13.2.2 Sampling Distribution of the Linear Regression Estimators

We now know how to estimate values for $\tilde{a}$, the mean value or the hyperplane (line in 2-dimensional space) of best fit, and $\tilde{\sigma}^2$, the variance, and to get an estimated value for a new data point. The question now is: How accurate are those estimates?

Table 13.3 The data for Example 13.7

x_i	y_i
1	2.4
2	2
3	1.6
4	1
5	0.4

Table 13.4 The data for Exercise 13.3

x_i	y_i	x_i	y_i
1	7	1	7
3	4	3	5
4	1	4	0

Table 13.5 The data for Exercise 13.4

x_i	y_i
1	0.6
2	1
3	1.6
4	2

13.2 Revisiting Linear Regression

That is the topic of this section—finding confidence limits for these estimates with a specified level of confidence.

Values of estimators $\tilde{a}$ and $\tilde{\sigma}^2$ obtained from Eqs. (8.17) and (13.12), respectively, may change when the observations of y and X change. Just like sampling distributions of means or sampling distributions of proportions (covered in Sects. 12.2.2 and 12.2.3 of Chap. 12), the $\tilde{a}$ and $\tilde{\sigma}^2$ of linear regression models have their sampling distributions, which approximate a normal distribution.

To get to the equations for the sampling distributions of $\tilde{a}$ and $\tilde{\sigma}^2$ and derive confidence limits, we need some fairly detailed preliminary mathematical results. These mathematical results are separated out and collected in the next subsection. They are needed to justify the equations on the sampling distributions that will be covered in the subsections afterwards, namely, Sects. 13.2.2.2, 13.2.2.3, and 13.2.2.4. All of this is leading up to the key results on confidence limits of the sampling distributions, namely, Eqs. (13.19), (13.20), and (13.24).

13.2.2.1 Preliminary Knowledge

- Expectation and Variance of Random Vectors
  If $X_1, X_2, \ldots, X_t$ are random variables, then the vector $\mathbf{x} = (X_1, X_2, \ldots, X_t)^T$ is a vector of random variables.

Definition 13.2 The expected value of a vector of random variables is defined as the vector of the expected values of the component parts. That is,

$$E[\mathbf{x}] = \Big(E[X_1], E[X_2], \ldots, E[X_t]\Big)^T.$$

The variance-covariance matrix of $\mathbf{x}$ is

$$Var[\mathbf{x}] = E\Big[(\mathbf{x} - E[\mathbf{x}])(\mathbf{x} - E[\mathbf{x}])^T\Big]$$

$$= \begin{bmatrix} Var(X_1) & cov(X_1, X_2) & \ldots & cov(X_1, X_t) \\ cov(X_2, X_1) & Var(X_2) & \ldots & cov(X_2, X_t) \\ \vdots & \vdots & \ddots & \vdots \\ cov(X_t, X_1) & cov(X_t, X_2) & \ldots & Var(X_t) \end{bmatrix}. \quad (13.13)$$

We know some properties of mean and variance in Sects. 10.5.1.1 and 10.5.2.1 of Chap. 10 for random variables. Here are a couple of vector generalisations of these properties.

Suppose **A** is a $s \times t$ matrix of constants, **b** is a $s \times 1$ vector of constants, and **x** is a $t \times 1$ vector of random variables.

$$E[\mathbf{Ax} + \mathbf{b}] = \mathbf{A}E[\mathbf{x}] + \mathbf{b}, \qquad (13.14)$$

$$Var(\mathbf{Ax} + \mathbf{b}) = \mathbf{A}Var(\mathbf{x})\mathbf{A}^T. \qquad (13.15)$$

The first of these two properties is illustrated in Example 13.8 since the expected value of a vector of random variables was easily defined above. It also looks like the properties of mean given in Sect. 10.5.1.1 of Chap. 10 and so an example is easy to understand. The second is more complicated since the variance of a vector of random variables is a matrix involving variances on the main diagonal and covariances in the other places, as shown in Eq. (13.13) above. So, we give proof of it instead of giving an example so that you can apply it with trust. If you are happy to accept the result, you can skip the proof. First, we give an example on the first property, which is the one involving the mean.

Example 13.8 Suppose **A** is the matrix of constants: $\mathbf{A} = \begin{bmatrix} 1 & 2 \\ 3 & 4 \\ 5 & 6 \end{bmatrix}$; **b** is a vector of constants, $\mathbf{b} = (1, 2, 3)^T$; and **x** is a vector of random variables, $\mathbf{x} = (X_1, X_2)^T$.
Now

$$\mathbf{Ax} + \mathbf{b} = \begin{bmatrix} 1 & 2 \\ 3 & 4 \\ 5 & 6 \end{bmatrix} \begin{bmatrix} X_1 \\ X_2 \end{bmatrix} + \begin{bmatrix} 1 \\ 2 \\ 3 \end{bmatrix} = \begin{bmatrix} X_1 + 2X_2 + 1 \\ 3X_1 + 4X_2 + 2 \\ 5X_1 + 6X_2 + 3 \end{bmatrix}.$$

So

$$E[\mathbf{Ax} + \mathbf{b}] = \begin{bmatrix} E[X_1 + 2X_2 + 1] \\ E[3X_1 + 4X_2 + 2] \\ E[5X_1 + 6X_2 + 3] \end{bmatrix}$$

$$= \begin{bmatrix} E[X_1] + 2E[X_2] + 1 \\ 3E[X_1] + 4E[X_2] + 2 \\ 5E[X_1] + 6E[X_2] + 3 \end{bmatrix}$$

$$= \mathbf{A}E[\mathbf{x}] + \mathbf{b},$$

as required. The second line uses the first three properties of mean from Sect. 10.5.1.1 of Chap. 10.

13.2 Revisiting Linear Regression

Now we prepare for the proof of the second property by developing a useful variation on the definition of $Var[\mathbf{x}]$.

Example 13.9 Prove the following useful variation on the definition of $Var[\mathbf{x}]$.

$$Var[\mathbf{x}] = E[\mathbf{x}\mathbf{x}^T] - E[\mathbf{x}](E[\mathbf{x}])^T. \quad (13.16)$$

Solution Before we start, let us recall the second property in Sect. 3.3.11.1 of Chap. 3: $(\mathbf{A} + \mathbf{B})^T = \mathbf{A}^T + \mathbf{B}^T$. Therefore, we have $(\mathbf{x} - E[\mathbf{x}])^T = \mathbf{x}^T - (E[\mathbf{x}])^T$. Recall the second property in Sect. 10.5.1.1 of Chap. 10: $E(aX) = aE(X)$, where a is a constant. Thus, we have $E[E[\mathbf{x}]\mathbf{x}^T] = E[\mathbf{x}]E[\mathbf{x}^T]$ since each element of $E[\mathbf{x}]$ is a constant. In addition, by the definition (13.2) above, of the expected random vector of random variables, we have $E[\mathbf{x}^T] = (E[\mathbf{x}])^T$.

Proof Applying Eq. (13.13), we have

$$Var[\mathbf{x}] = E\left[(\mathbf{x} - E[\mathbf{x}])(\mathbf{x} - E[\mathbf{x}])^T\right]$$
$$= E\left[(\mathbf{x} - E[\mathbf{x}])(\mathbf{x}^T - (E[\mathbf{x}])^T)\right]$$
$$= E\left[\mathbf{x}\mathbf{x}^T - E[\mathbf{x}]\mathbf{x}^T - \mathbf{x}(E[\mathbf{x}])^T + E[\mathbf{x}](E[\mathbf{x}])^T\right]$$
$$= E[\mathbf{x}\mathbf{x}^T] - E[\mathbf{x}]E[\mathbf{x}^T] - E[\mathbf{x}](E[\mathbf{x}])^T + E[\mathbf{x}](E[\mathbf{x}])^T$$
$$= E[\mathbf{x}\mathbf{x}^T] - E[\mathbf{x}](E[\mathbf{x}])^T.$$
□

Now we can give the proof of the second property as promised:

Proof $Var(\mathbf{A}\mathbf{x} + \mathbf{b}) = \mathbf{A}Var(\mathbf{x})\mathbf{A}^T$.

Applying Eqs. (13.16), (13.14), and the third property in Sect. 3.3.11.1 of Chap. 3, that is, $(\mathbf{A}\mathbf{B})^T = \mathbf{B}^T\mathbf{A}^T$, we have

$$Var(\mathbf{A}\mathbf{x} + \mathbf{b}) = E\left[(\mathbf{A}\mathbf{x} + \mathbf{b})(\mathbf{A}\mathbf{x} + \mathbf{b})^T\right] - E[\mathbf{A}\mathbf{x} + \mathbf{b}](E[\mathbf{A}\mathbf{x} + \mathbf{b}])^T$$
$$= E[\mathbf{A}\mathbf{x}\mathbf{x}^T\mathbf{A}^T + \mathbf{b}\mathbf{x}^T\mathbf{A}^T + \mathbf{A}\mathbf{x}\mathbf{b}^T + \mathbf{b}\mathbf{b}^T]$$
$$- (\mathbf{A}E[\mathbf{x}] + \mathbf{b})(\mathbf{A}E[\mathbf{x}] + \mathbf{b})^T$$

$$= AE[\mathbf{xx}^T]\mathbf{A}^T + \mathbf{b}E(\mathbf{x}^T)\mathbf{A}^T + AE(\mathbf{x})\mathbf{b}^T + \mathbf{bb}^T$$
$$- AE[\mathbf{x}](E[\mathbf{x}])^T\mathbf{A}^T - \mathbf{b}(E[\mathbf{x}])^T\mathbf{A}^T - AE(\mathbf{x})\mathbf{b}^T - \mathbf{bb}^T$$
$$= AE[\mathbf{xx}^T]\mathbf{A}^T - AE[\mathbf{x}](E[\mathbf{x}])^T\mathbf{A}^T$$
$$= A Var(\mathbf{x})\mathbf{A}^T.$$

□

The last step of the above has used the result shown in Example 3.26 in Chap. 3.
- Suppose **C** and **D** are two matrices with a size of $m \times n$ and $n \times m$, respectively; then

$$tr(\mathbf{CD}) = tr(\mathbf{DC}),$$

as illustrated in Example 13.10.

Example 13.10 Let $\mathbf{U} = \begin{bmatrix} 2 & -1 & 3 \\ 0 & 5 & 1 \end{bmatrix}$, and $\mathbf{V} = \begin{bmatrix} 4 & 1 \\ 0 & 2 \\ 2 & 1 \end{bmatrix}$.

$$\mathbf{UV} = \begin{bmatrix} 2 & -1 & 3 \\ 0 & 5 & 1 \end{bmatrix} \begin{bmatrix} 4 & 1 \\ 0 & 2 \\ 2 & 1 \end{bmatrix} = \begin{bmatrix} 14 & 3 \\ 2 & 11 \end{bmatrix}.$$

$$\mathbf{VU} = \begin{bmatrix} 4 & 1 \\ 0 & 2 \\ 2 & 1 \end{bmatrix} \begin{bmatrix} 2 & -1 & 3 \\ 0 & 5 & 1 \end{bmatrix} = \begin{bmatrix} 8 & 1 & 13 \\ 0 & 10 & 2 \\ 4 & 3 & 7 \end{bmatrix}.$$

$$tr(\mathbf{UV}) = tr(\mathbf{VU}) = 25.$$

- A square matrix **M** is defined to be idempotent if and only if $\mathbf{M}^2 = \mathbf{M}$.

Example 13.11 Suppose **M** is idempotent. We use **I** to denote an identity matrix whose size is the same as **M**.

$$(\mathbf{I} - \mathbf{M})(\mathbf{I} - \mathbf{M}) = \mathbf{I}^2 - \mathbf{MI} - \mathbf{IM} + \mathbf{M}^2 = \mathbf{I} - 2\mathbf{M} + \mathbf{M} = \mathbf{I} - \mathbf{M}.$$

Therefore, $\mathbf{I} - \mathbf{M}$ is also idempotent.

Every idempotent matrix, except the identity matrix, is singular. Its rank is equal to its trace.

Example 13.12 Let the *hat* matrix, $\mathbf{H} = \mathbf{X}(\mathbf{X}^T\mathbf{X})^{-1}\mathbf{X}^T$, where $\mathbf{X}$ is a $N \times (d+1)$ matrix with $N \geq d+1$. $\mathbf{H}$ is therefore an $N \times N$ matrix. Since

$$\mathbf{H}^2 = \mathbf{X}(\mathbf{X}^T\mathbf{X})^{-1}\mathbf{X}^T\mathbf{X}(\mathbf{X}^T\mathbf{X})^{-1}\mathbf{X}^T$$
$$= \mathbf{X}[(\mathbf{X}^T\mathbf{X})^{-1}(\mathbf{X}^T\mathbf{X})](\mathbf{X}^T\mathbf{X})^{-1}\mathbf{X}^T$$
$$= \mathbf{X}\mathbf{I}(\mathbf{X}^T\mathbf{X})^{-1}\mathbf{X}^T$$
$$= \mathbf{X}(\mathbf{X}^T\mathbf{X})^{-1}\mathbf{X}^T$$
$$= \mathbf{H},$$

$\mathbf{H}$ is therefore idempotent, and the rank of it equals

$$tr(\mathbf{H}) = tr(\mathbf{X}(\mathbf{X}^T\mathbf{X})^{-1}\mathbf{X}^T)$$
$$= tr(\mathbf{X}^T\mathbf{X}(\mathbf{X}^T\mathbf{X})^{-1})$$
$$= tr(I_{(d+1)\times(d+1)})$$
$$= d+1.$$

Inside the trace calculation, between the first and second line, we reversed two matrices to put $\mathbf{X}^T$ at the front as we illustrated in Example 13.10.

- The simplified version of Cochran's theorem
  Let $\mathbf{Y} \sim \mathcal{N}(\mathbf{0}, \sigma^2\mathbf{I})$ and $\mathbf{H}$ is an idempotent matrix of rank $d+1$. Then $\mathbf{Y}^T\mathbf{H}\mathbf{Y}$ is a Chi-square distribution, that is, $\mathbf{Y}^T\mathbf{H}\mathbf{Y} \sim \sigma^2\chi^2_{d+1}$; and $\mathbf{Y}^T(\mathbf{I}-\mathbf{H})\mathbf{Y}$ is also a Chi-square distribution, that is, $\mathbf{Y}^T(\mathbf{I}-\mathbf{H})\mathbf{Y} \sim \sigma^2\chi^2_{N-(d+1)}$.

That completes the set of preliminary results we need.

13.2.2.2 Sampling Distribution of Estimators ã

What form of distribution should we consider for $\tilde{\mathbf{a}}$? Recall the central limit theorem (see Sect. 11.1.2 of Chap. 11), which states that samples of sums or means of a random variable tend to be normally distributed in large samples. Consider $\tilde{\mathbf{a}}$ as a weighted mean value of $\mathbf{y}$. We can use the normal distribution for $\tilde{\mathbf{a}}$.

Consider the data generated from Eq. (13.6):

$$\mathbf{y} = \mathbf{Xa} + \boldsymbol{\epsilon}.$$

Based on the three assumptions of the ordinary linear regression model presented in Sect. 13.2.1, we have $E[\mathbf{y}] = \mathbf{Xa}$ and $Var(\mathbf{y}) = \sigma^2 \mathbf{I}_N$, where $\mathbf{I}_N$ is a $N \times N$ identity matrix. This last result is because the matrix of $Var(\boldsymbol{\epsilon})$ is just a diagonal matrix of variances of the components of $\boldsymbol{\epsilon}$ (all equal to σ^2) with the other values in the matrix equal to zero since the error terms are uncorrelated and therefore all covariance terms are zero.

Now using Eq. (8.17) for $\tilde{\mathbf{a}}$, namely,

$$\tilde{\mathbf{a}} = (\mathbf{X}^T \mathbf{X})^{-1} \mathbf{X}^T \mathbf{y},$$

the expected value of $\tilde{\mathbf{a}}$ is

$$\begin{aligned} E[\tilde{\mathbf{a}}] &= E[(\mathbf{X}^T \mathbf{X})^{-1} \mathbf{X}^T \mathbf{y}] \\ &= (\mathbf{X}^T \mathbf{X})^{-1} \mathbf{X}^T \mathbf{Xa} \quad \text{since } E[\mathbf{y}] = \mathbf{Xa} \\ &= \mathbf{a}, \end{aligned} \quad (13.17)$$

and the variance of $\tilde{\mathbf{a}}$ is

$$\begin{aligned} Var(\tilde{\mathbf{a}}) &= Var((\mathbf{X}^T \mathbf{X})^{-1} \mathbf{X}^T \mathbf{y}) \\ &= (\mathbf{X}^T \mathbf{X})^{-1} \mathbf{X}^T Var(\mathbf{y})((\mathbf{X}^T \mathbf{X})^{-1} \mathbf{X}^T)^T \\ &= (\mathbf{X}^T \mathbf{X})^{-1} \mathbf{X}^T \sigma^2 \mathbf{I} \mathbf{X} (\mathbf{X}^T \mathbf{X})^{-1} \\ &= \sigma^2 (\mathbf{X}^T \mathbf{X})^{-1}. \end{aligned} \quad (13.18)$$

The third line of the last result relies on three results from previous work: namely, that $(\mathbf{A}^T)^T = \mathbf{A}$, $(\mathbf{AB})^T = \mathbf{B}^T \mathbf{A}^T$, and $(\mathbf{A}^{-1})^T = (\mathbf{A}^T)^{-1}$. That is,

$$((\mathbf{X}^T \mathbf{X})^{-1} \mathbf{X}^T)^T = \mathbf{X}((\mathbf{X}^T \mathbf{X})^{-1})^T = \mathbf{X}((\mathbf{X}^T \mathbf{X})^T)^{-1} = \mathbf{X}(\mathbf{X}^T \mathbf{X})^{-1}.$$

Therefore, the distribution of $\tilde{\mathbf{a}}$ is $\mathcal{N}(\mathbf{a}, \sigma^2 (\mathbf{X}^T \mathbf{X})^{-1})$.

One of the properties of the multivariate normal distribution is that every single variable has a univariate normal distribution (see Sect. 11.2.4 of Chap. 11). If we let $\tilde{a}_i$ be the ith element of $\tilde{\mathbf{a}}$, then

$$\tilde{a}_i \sim \mathcal{N}(a_i, \sigma^2 m_{ii}),$$

where a_i is the ith element of $\mathbf{a}$ and m_{ii} is the ith diagonal element of $M = (\mathbf{X}^T \mathbf{X})^{-1}$.

13.2 Revisiting Linear Regression

If we replace the actual standard deviation with the estimated standard deviation, then the distribution is Student's t-distribution (see Sect. 12.2.6.1 of Chap. 12). The confidence interval can be constructed by applying Eq. (12.13) on the confidence intervals of means that is found in Sect. 12.3.2.1 of Chap. 12. We need to change Z_c to the critical value from a t-distribution table, so that margin of error $= t_c \times$ SD. Since the standard deviation is the square root of the variance, we get the following confidence limits for α_i:

$$\tilde{a}_i \pm t_c \sqrt{\tilde{\sigma}^2 m_{ii}}, \tag{13.19}$$

where t_c is the critical value with degrees of freedom of $N - (d + 1)$ if there are N observations and $d + 1$ elements in **a** (there are d independent variables) and the required confidence level is c.

In this subsection and the following two subsections, we will continue with Example 13.5 to quickly illustrate finding the appropriate confidence limits. At the end of all three subsections we will do a full example involving all three subsections. You need to bear in mind that these examples are really simple ones in the sense that there is only a small amount of data. Hence, all the confidence limits come out very wide. Realistic examples have much larger amounts of data, but would be done using an appropriate computer program.

Example 13.13 Continue with Example 13.5—Part 1: constructing the 95% confidence interval for **a**. Since this is a two-tailed test, we need $t_{0.975}$ in the t-table.

We substitute $\tilde{\mathbf{a}} = \begin{bmatrix} -0.2 \\ 1.09 \end{bmatrix}$, $\tilde{\sigma}^2 = 0.05$, the elements of main diagonal of $(\mathbf{X}^T\mathbf{X})^{-1} = \begin{bmatrix} 1.5 \\ 0.214 \end{bmatrix}$, and $t_{0.975} = 12.71$ with degrees of freedom of $N - (d + 1) = 1$ into Eq. (13.19) and obtain

$$\begin{bmatrix} -0.2 \\ 1.09 \end{bmatrix} \pm 12.71 \times \sqrt{0.05 \times \left(\begin{bmatrix} 1.5 \\ 0.214 \end{bmatrix} \right)} = \begin{bmatrix} -0.2 \\ 1.09 \end{bmatrix} \pm \begin{bmatrix} 3.48 \\ 1.31 \end{bmatrix}.$$

Therefore, the confidence interval for the intercept, a_0, is $[-0.2 - 3.48, -0.2 + 3.48] = [-3.68, 3.28]$ and similarly the confidence interval for the gradient, a_1, is $[-0.22, 2.40]$.

As indicated before, these values give a very wide confidence interval since we only had three data points. With lots more data points, we would be much further down the appropriate t-table column and would get a much smaller t_c value.

13.2.2.3 Sampling Distribution of Variance $\tilde{\sigma}^2$

The residual vector is given by

$$e = y - \tilde{y}$$
$$= (I_N - X(X^TX)^{-1}X^T)y$$
$$= (I_N - X(X^TX)^{-1}X^T)(Xa + \epsilon)$$
$$= (I_N - X(X^TX)^{-1}X^T)Xa + (I_N - X(X^TX)^{-1}X^T)\epsilon$$
$$= 0 + (I_N - X(X^TX)^{-1}X^T)\epsilon$$
$$= (I_N - H)\epsilon,$$

where H is the *hat* matrix introduced in Example 13.12. Note that we substituted $\tilde{y} = X\tilde{a} = X(X^TX)^{-1}X^Ty$, using the equation for $\tilde{a}$, into the second equation line. Also, the first term of the fourth equation line is equal to

$$I_NXa - X(X^TX)^{-1}X^TXa$$
$$= I_NXa - X(X^TX)^{-1}(X^TX)a$$
$$= Xa - XI_{d+1}a$$
$$= Xa - Xa$$
$$= 0.$$

As shown in Sect. 13.2.2.1, if H is idempotent, then $I_N - H$ is also idempotent. Also since $H^T = (X(X^TX)^{-1}X^T)^T = H$, therefore H is symmetric. This means that $I_N - H$ is also symmetric, that is, $(I_N - H)^T = (I_N - H)$.

So the sum of the squares of the residuals is

$$e^Te = \epsilon^T(I_N - H)^T(I_N - H)\epsilon$$
$$= \epsilon^T(I_N - H)^2\epsilon$$
$$= \epsilon^T(I_N - H)\epsilon,$$

where we have used the idempotent property, that is, if matrix $I_N - H$ is idempotent, $(I_N - H)^2 = (I_N - H)$.

Since $\tilde{\sigma}^2 \propto e^Te$ (see Eq. (13.12)), we have

$$\tilde{\sigma}^2 \propto \epsilon^T(I_N - H)\epsilon,$$

13.2 Revisiting Linear Regression

and by applying Cochran's theorem, we have

$$\tilde{\sigma}^2 \propto \sigma^2 \chi^2_{N-(d+1)}.$$

Recall that we constructed the confidence interval for the population standard deviation in Sect. 12.3.2.3 of Chap. 12 by applying Eq. (12.17). The confidence interval for σ^2 can be constructed in the same way with degrees of freedom of $N - (d + 1)$ in this case, that is,

$$\frac{(N-(d+1))\tilde{\sigma}^2}{\chi^2_{\frac{\alpha}{2}}} \leq \sigma^2 \leq \frac{(N-(d+1))\tilde{\sigma}^2}{\chi^2_{1-\frac{\alpha}{2}}}. \tag{13.20}$$

Example 13.14 Continue with Example 13.5, part 2: constructing the 95% confidence interval for σ^2.

By applying Eq. (13.20), where $\chi^2_{0.025} = 5.024$ and $\chi^2_{0.975} = 0.001$ with degrees of freedom of 1 (we have used a χ^2 distribution table showing the area to the right of critical value), then we have

$$\left[\frac{(3-2) \times 0.05}{5.024}, \frac{(3-2) \times 0.05}{0.001}\right] = [0.01, 50].$$

Again we have very large confidence limits.

13.2.2.4 Prediction

Finally, assume that we use the fitted model to make a prediction y_{new} for a new data point $\mathbf{x}_{new}$. We have

$$E[y_{new}] = E[\mathbf{x}_{new}\tilde{\mathbf{a}} + \epsilon] = \mathbf{x}_{new}\tilde{\mathbf{a}}, \tag{13.21}$$

and

$$Var(\mathbf{x}_{new}\tilde{\mathbf{a}}) = \mathbf{x}_{new} Var(\tilde{\mathbf{a}})(\mathbf{x}_{new})^T. \tag{13.22}$$

Substituting Eq. (13.18) into Eq. (13.22), we have

$$Var(\mathbf{x}_{new}\tilde{\mathbf{a}}) = \sigma^2 \mathbf{x}_{new}(\mathbf{X}^T\mathbf{X})^{-1}(\mathbf{x}_{new})^T. \tag{13.23}$$

If we replace the true standard deviation with the estimated standard deviation, then the distribution is Student's t-distribution. Hence, we can again use Eq. (12.13) on the confidence intervals of means that is found in Sect. 12.3.2.1 of Chap. 12. Again, we change Z_c to the critical value from a t-distribution table as we did in Sect. 13.2.2.2 of this chapter; then the margin of error $= t_c \times$ SD. Hence, the confidence interval can be easily constructed as follows:

$$\mathbf{x}_{new}\tilde{\mathbf{a}} \pm t_c \sqrt{\tilde{\sigma}^2 \mathbf{x}_{new}(\mathbf{X}^T\mathbf{X})^{-1}(\mathbf{x}_{new})^T}. \tag{13.24}$$

Example 13.15 Continue with Example 13.5, part 3: We use the fitted model to make a mean prediction y_{new} for $x_{new} = 3$. As before, we have that

$$y_{new} = a_0 + a_1 \times x_{new} = -0.2 + 1.086 \times 3 = 3.06.$$

Or, in vector form, using Eq. (13.21), we have

$$y_{new} = \mathbf{x}_{new}\tilde{\mathbf{a}} = \begin{bmatrix} 1 & 3 \end{bmatrix} \begin{bmatrix} -0.2 \\ 1.086 \end{bmatrix} = 3.06.$$

By applying Eq. (13.24) with $\mathbf{x}_{new} = [1, 3]$, the 95% confidence interval of y_{new} is computed as follows:

$$3.06 \pm 12.71 \sqrt{0.05 \times [1, 3] \times \begin{bmatrix} 1.5 & -0.5 \\ -0.5 & 0.214 \end{bmatrix} \times \begin{bmatrix} 1 \\ 3 \end{bmatrix}}$$

$$\approx 3.06 \pm 1.85$$

$$= [1.21, 4.91].$$

We will now do a full example.

Example 13.16 Continue with Example 13.7. Here $N = 5$ and $d = 1$, so the degrees of freedom is $N - (d+1) = 3$. First, let us construct the 95% confidence interval for **a**.

We substitute $\tilde{\mathbf{a}} = \begin{bmatrix} 2.98 \\ -0.5 \end{bmatrix}$, $\tilde{\sigma}^2 = 0.0093$, the elements of main diagonal of $(\mathbf{X}^T\mathbf{X})^{-1} = \begin{bmatrix} 1.1 \\ 0.1 \end{bmatrix}$, and $t_{0.975} = 3.182$ with degrees of freedom of 3 into Eq. (13.19) and obtain

$$\begin{bmatrix} 2.98 \\ -0.5 \end{bmatrix} \pm 3.182 \times \sqrt{0.0093 \times \left(\begin{bmatrix} 1.1 \\ 0.1 \end{bmatrix} \right)} = \begin{bmatrix} 2.98 \\ -0.5 \end{bmatrix} \pm \begin{bmatrix} 0.32 \\ 0.097 \end{bmatrix}.$$

(continued)

13.2 Revisiting Linear Regression

Example 13.16 (continued)
Therefore, the confidence interval for the intercept, a_0, is [2.66, 3.30] and similarly the confidence interval for the gradient, a_1, is [−0.60, −0.40].

These confidence limits are tighter than in the previous example since we now have five data points.

Second, we construct 95% confidence interval for σ^2.

By applying the Eq. (13.20), where $\chi^2_{0.025} = 9.348$ and $\chi^2_{0.975} = 0.216$ with degrees of freedom of 3 (we have again used a χ^2 distribution table showing the area to the right of critical value), we have

$$\left[\frac{(5-2) \times 0.0093}{9.348}, \frac{(5-2) \times 0.0093}{0.216} \right] = [0.003, 0.13].$$

Finally, use the fitted model to make a mean prediction y_{new} for $x_{new} = 3.5$. That is, by applying Eq. (13.21), we have

$$y_{new} = a_0 + a_1 \times x_{new} = 2.98 - 0.5 \times 3.5 = 1.23.$$

By applying Eq. (13.24) with $\mathbf{x}_{new} = [1, 3.5]$, the 95% confidence interval of y_{new} is computed as follows:

$$1.23 \pm 3.182 \sqrt{0.0093 \times [1, 3.5] \begin{bmatrix} 1.1 & -0.3 \\ -0.3 & 0.1 \end{bmatrix} \times \begin{bmatrix} 1 \\ 3.5 \end{bmatrix}}$$

$$= 1.23 \pm 0.15$$

$$= [1.08, 1.38].$$

Table 13.6 The data for Exercise 13.5

x_i	y_i
−0.5	2.5
1	0.5
2	0

> **Exercise**
>
> **13.5** Suppose a set of data is shown in Table 13.6, where x_i is the value of an independent variable X, and y_i is the corresponding value of the dependent variable Y. Fit a linear regression model for the data and construct 95% confidence intervals for **a** and σ^2. Round results to two decimal places.
>   A new data point is given at $x = 1.5$, find the expected $\tilde{y}$ value, and construct 95% confidence limits for this value.
>
> **13.6** Continuation of Exercise 13.4. The data is given in Table 13.5, where x_i is the value of an independent variable X, and y_i is the corresponding value of the dependent variable Y. Construct 95% confidence intervals for the values of **a** and σ^2 that you have already found in Exercise 13.4.
>   A new data point is given at $x = 2.5$, find the expected $\tilde{y}$ value, and construct 95% confidence limits for this value.

13.3 The Logistic Regression Algorithm

As mentioned in Chap. 1, there are two categories of supervised learning: regression and classification. In this section, we briefly show how to formulate a two-class classification algorithm, logistic regression, also called logit regression, using the maximum likelihood technique. This is a classification method despite its name. A classification problem could actually be a classification into multiple classes (a picture is a flower, a dog, a human, or a tank) or a two-class problem (with these symptoms, you have or have not got a particular medical condition). The most common type is two classes, so we only consider that here.

Recall a sigmoid function defined as $\sigma(z) = \frac{1}{1+e^{-z}}$ is bounded between 0 and 1 (see Fig. 5.5 in Chap. 5). Since it only has values between 0 and 1, it is a suitable function for converting a real number into a probability. So, logistic regression uses a sigmoid function to estimate the probability of $P(y_i = c_j | \mathbf{x}_i)$, that is, the probability that an instance i belongs to a specific class c_j given its features $\mathbf{x}_i$. Here, each of the $i = N$ data points $\mathbf{x}_i$ is a vector of d features.

Let z denote a linear combination of features, $z = \mathbf{x}_i \mathbf{a}$, where $\mathbf{a}$ includes coefficients. That is, $z = a_0 x_{i0} + a_1 x_{i1} + \cdots + a_d x_{id}$, where x_{i0} is 1. For instance, with only one feature, we have $z = a_0 + a_1 x_{i1}$. Also, let Y be a discrete random variable that only takes two values, $c_1 = 1$ and $c_2 = 0$, giving a two-class classification. We have

$$P(y_i = 1 | \mathbf{x}_i) = \frac{1}{1+e^{-z}}, \quad P(y_i = 0 | \mathbf{x}_i) = 1 - \frac{1}{1+e^{-z}}, \quad \text{where } z = \mathbf{x}_i \mathbf{a}.$$
(13.25)

13.3 The Logistic Regression Algorithm

Remark 13.1 We are finding the estimated value for Y (which is either 0 and 1 using a threshold on the sigmoid function) from some linear combination of the features. For example, with only one feature, we have for the probability in the $y_i = 1$ case as

$$\frac{1}{1 + e^{-(a_0 + a_1 x_{i1})}}.$$

Note that in the general sigmoid function

$$\frac{1}{1 + e^{-(a_0 + a_1 x)}},$$

the a_1 gives the steepness of the curve. Figure 13.6 shows that the curve with $a_1 = 2$ (black) is steeper than the one with $a_1 = 1$ (red), and the curve with $a_1 = 2$ (black) is symmetrical with the one with $a_1 = -2$ (blue) about the line of $x = 0$.

For a given value of a_1, a_0 is related to where the sigmoid gets to the half-height value (0.5), that is, when $e^{-(a_0 + a_1 x)} = 1$ or when $a_0 + a_1 x = 0$. So the sigmoid reaches half height when $x = -\frac{a_0}{a_1}$. For example, as illustrated in Fig. 13.7, when a_1 is fixed as 1 and $y = 0.5$ is indicated using the green dashed line, we can see how the curve changes as a_0 varies from -1 (red), 0 (black), and 1 (blue). Thus, varying the value of a_0 is related to how far the curve moves to the left or right.

Hence, adjusting the coefficients in **a** varies where and how the sigmoid comes. The same sort of thing is true in the multiple features case, but it is not possible to draw.

♦

We can rewrite Eq. (13.25) using Eq. (10.14) from Chap. 10 as follows:

$$P(y_i | \mathbf{x}_i) = \sigma(\mathbf{x}_i \mathbf{a})^{y_i} (1 - \sigma(\mathbf{x}_i \mathbf{a}))^{1 - y_i}. \tag{13.26}$$

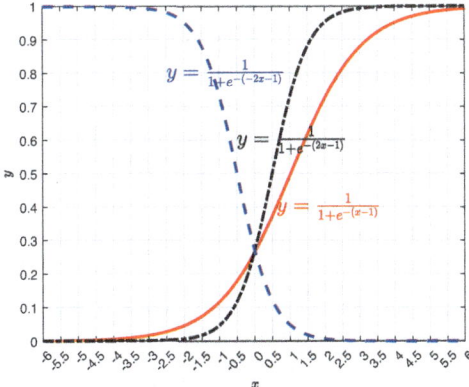

Fig. 13.6 An illustration showing how varying a_1 in the sigmoid function $\frac{1}{1 + e^{-(a_1 x + a_0)}}$ affects the steepness of the curve

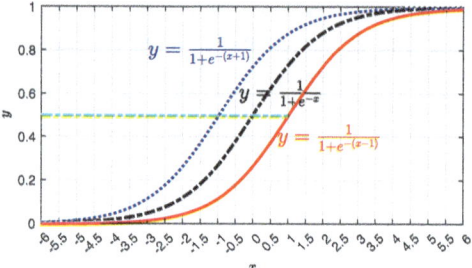

Fig. 13.7 An illustration of the effect of varying a_0 in the sigmoid function $\frac{1}{1+e^{-(a_1x+a_0)}}$: varying a_0 is related to how far the curve moves to the left or right

If we treat y_i's as independent, then the likelihood function is

$$L = \prod_{i=1}^{N} \sigma(\mathbf{x}_i\mathbf{a})^{y_i}(1-\sigma(\mathbf{x}_i\mathbf{a}))^{1-y_i}.$$

It is more convenient to minimise the negative logarithm of the likelihood:

$$\mathcal{L} = -\sum_{i}^{N}\Big(y_i \ln \sigma(\mathbf{x}_i\mathbf{a}) + (1-y_i)\ln(1-\sigma(\mathbf{x}_i\mathbf{a}))\Big). \tag{13.27}$$

Unlike simple linear regression, there is no closed form (that is, no explicit formula) to calculate $\mathbf{a}$. However, we can find a suitable estimated, $\tilde{\mathbf{a}}$, where $\tilde{\mathbf{a}} = [\tilde{a}_0, \ldots, \tilde{a}_d]^T$, using an optimisation method. Let us apply the gradient descent algorithm. To do so, we need to compute the partial derivatives $\frac{\partial \mathcal{L}}{\partial a_j}$, where $j = 0, \ldots, d$.

Let us take each of the summation terms in $\mathcal{L}$ in turn. So let

$$S\mathcal{L} = -\Big(y_i \ln \sigma(\mathbf{x}_i\mathbf{a}) + (1-y_i)\ln(1-\sigma(\mathbf{x}_i\mathbf{a}))\Big).$$

Because the derivative of a sum is the sum of derivatives (see Addition Rule in Sect. 5.2.2 of Chap. 5), the partial derivative $\frac{\partial \mathcal{L}}{\partial a_j}$ is simply the sum of the $\frac{\partial S\mathcal{L}}{\partial a_j}$, that is, the sum of the following for each data item $(\mathbf{x}_i, y_i)$ for each i from $1 \cdots N$:

$$\begin{aligned}
\frac{\partial S\mathcal{L}}{\partial a_j} &= -\left(\frac{\partial}{\partial a_j}y_i \ln \sigma(\mathbf{x}_i\mathbf{a}) + \frac{\partial}{\partial a_j}(1-y_i)\ln(1-\sigma(\mathbf{x}_i\mathbf{a}))\right) \\
&= -\left(\frac{y_i}{\sigma(\mathbf{x}_i\mathbf{a})} - \frac{1-y_i}{1-\sigma(\mathbf{x}_i\mathbf{a})}\right)\frac{\partial}{\partial a_j}\sigma(\mathbf{x}_i\mathbf{a}) \quad \text{derivative of a log function} \\
&= -\left(\frac{y_i}{\sigma(\mathbf{x}_i\mathbf{a})} - \frac{1-y_i}{1-\sigma(\mathbf{x}_i\mathbf{a})}\right)\sigma(\mathbf{x}_i\mathbf{a})(1-\sigma(\mathbf{x}_i\mathbf{a}))x_{ij} \quad \text{chain rule} \\
&= -\left(\frac{y_i - \sigma(\mathbf{x}_i\mathbf{a})}{\sigma(\mathbf{x}_i\mathbf{a})(1-\sigma(\mathbf{x}_i\mathbf{a}))}\right)\sigma(\mathbf{x}_i\mathbf{a})(1-\sigma(\mathbf{x}_i\mathbf{a}))x_{ij} \\
&= (\sigma(\mathbf{x}_i\mathbf{a}) - y_i)x_{ij}.
\end{aligned} \tag{13.28}$$

13.3 The Logistic Regression Algorithm

Note that for the third equation line above, we have used Example 5.13 from Sect. 5.2.2 in Chap. 5 to differentiate the sigmoid function and have applied the following:

$$\mathbf{x}_i \mathbf{a} = \sum_{j=0}^{d} a_j x_{ij} = a_0 x_{i0} + a_1 x_{i1} + \cdots + a_d x_{id},$$

and so

$$\frac{\partial \mathbf{x}_i \mathbf{a}}{\partial a_j} = x_{ij}.$$

Therefore, we have

$$\frac{\partial \mathcal{L}}{\partial a_j} = \sum_{i=1}^{N} (\sigma(\mathbf{x}_i \mathbf{a}) - y_i) x_{ij}.$$

To update a_j over all data items, we apply Eq. (6.7) from Chap. 6 and have

$$a_j^{new} = a_j^{old} - \epsilon \left(\sum_{i=1}^{N} (\sigma(\mathbf{x_{ia}}) - y_i) x_{ij} \right), \tag{13.29}$$

where ϵ is the learning rate.

Example 13.17 Suppose we have a set of data items $[-1.5, -1, 0, 0.3]$, so we only have one feature. We apply the sigmoid function where $z_i = 2 + 4x_i$ to the data, that is, $a_0 = 2$ and $a_1 = 4$. Figure 13.8 shows the inputs against their sigmoid function values. Since the sigmoid function values of -1.5 and -1 are less than the threshold 0.5, we set the class labels for these two data items to 0 (circle signs) and for the other two data items to 1 (square signs). Therefore, the corresponding estimated class labels are $[0, 0, 1, 1]$.

More usually, of course, we already have the class for the data items and are using this technique as a supervision technique to determine the model (finding the values of a_0 and a_1). Having found the model, we can use it to determine the class for a new, unknown data item. So now assume that we do not know a_0 and a_1, and we want to fit a logistic regression model for the data using the gradient descent algorithm with one iteration and a learning rate of 0.1. We already know that the first two data items are in the class labelled 0, and the others are in the class labelled 1. So the class labels are $[0, 0, 1, 1]$.

(continued)

Example 13.17 (continued)
Solution Let us set initial values for a_0 and a_1 as 0.5 and 0.4, respectively. Substituting each data value to $z_i = 0.5 + 0.4x_i$, we have

$$z_1 = 0.5 + 0.4 \times (-1.5) = -0.1,$$

$$z_2 = 0.5 + 0.4 \times (-1) = 0.1,$$

$$z_3 = 0.5 + 0.4 \times 0 = 0.5,$$

$$z_4 = 0.5 + 0.4 \times 0.3 = 0.62.$$

Substituting z_i to Eq. (13.25), we have

$$P(y_1 = 1|x_1) = \frac{1}{1 + e^{0.1}} \approx 0.475,$$

$$P(y_2 = 1|x_2) = \frac{1}{1 + e^{-0.1}} \approx 0.525,$$

$$P(y_3 = 1|x_3) = \frac{1}{1 + e^{-0.5}} \approx 0.622,$$

$$P(y_4 = 1|x_4) = \frac{1}{1 + e^{-0.62}} \approx 0.650.$$

Considering the threshold is 0.5, then the estimated class label y_i is 0, 1, 1, 1, respectively. So this setting gives a misclassification to the second data item. Note that the place where the sigmoid gets to the halfway value is $-\frac{a_0}{a_1} = -\frac{0.5}{0.4} = -1.25$ which is to the left of the second point (see Fig. 13.8). So not surprisingly it does not classify this one correctly. This sort of analysis would be impossible visually with multiple points and many dimensions.

We apply Eq. (13.29) to update the a_j. When updating a_0, we use $x_{i0} = 1$ and have

$$a_0^{new} = 0.5 - 0.1 \times ((0.475 - 0)$$
$$+ (0.525 - 0) + (0.622 - 1) + (0.650 - 1))1 \approx 0.473.$$

$$a_1^{new} = 0.4 - 0.1 \times ((0.475 - 0) \times (-1.5) + (0.525 - 0)$$
$$\times (-1) + (0.622 - 1)0 + (0.650 - 1) \times 0.3) \approx 0.534.$$

(continued)

Example 13.17 (continued)
With the initial **a** values, the second data item was misclassified. After the first iteration,

$$z_1 = 0.473 + 0.534 \times (-1.5) = -0.328,$$

$$z_2 = 0.473 + 0.534 \times (-1) = -0.061,$$

$$z_3 = 0.473 + 0.534 \times 0 = 0.473,$$

$$z_4 = 0.473 + 0.534 \times 0.3 = 0.633.$$

Substituting z_i to Eq. (13.25), we have

$$P(y_1 = 1|x_1) = \frac{1}{1 + e^{0.328}} \approx 0.419 \approx 0.42,$$

$$P(y_2 = 1|x_2) = \frac{1}{1 + e^{0.061}} \approx 0.485 \approx 0.49,$$

$$P(y_3 = 1|x_3) = \frac{1}{1 + e^{-0.473}} \approx 0.616 \approx 0.62,$$

$$P(y_4 = 1|x_4) = \frac{1}{1 + e^{-0.633}} \approx 0.653 \approx 0.65.$$

So using the threshold of 0.5, all 4 data points are now classified correctly. Again looking at where the sigmoid gets to the halfway value, that is, $-\frac{a_0}{a_1} = -\frac{0.473}{0.534} = -0.88$, which is to the right of the second point, it is reasonable that it gets this point classified correctly now. However, it is only just below the 0.5 probability, so perhaps more iterations are needed to make it a better predictor of new points. Getting a result in one iteration is not realistic. In this example, data are classified correctly with one iteration partly due to having an unrealistically high value for the learning rate (and choosing good values for the example).

Fig. 13.8 Scatter plot showing data and their sigmoid values in Example 13.17. The dashed line marks the threshold at 0.5

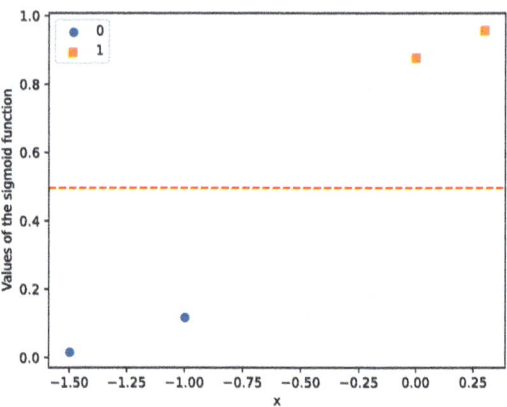

Example 13.18 Suppose we have a new set of data items $[-1, 0, 1, 4]$, so we only have one feature again. Suppose the class labels are $[0, 0, 1, 1]$. Fit a logistic regression model for the data using the gradient descent algorithm with two iterations and a learning rate of 0.05.

Solution Let us set initial values for a_0 and a_1 as -0.3 and 0.1, respectively. Substituting each data value into $z_i = -0.3 + 0.1 x_i$ gives

$$z_1 = -0.3 + 0.1 \times (-1) = -0.4,$$

and similarly,

$$z_2 = -0.3, \quad z_3 = -0.2, \text{ and } z_4 = 0.1.$$

Substituting each z_i into Eq. (13.25), we have

$$P(y_1 = 1 | x_1) = \frac{1}{1 + e^{0.4}} \approx 0.401,$$

and similarly,

$P(y_1 = 1 | x_2) \approx 0.426$, $P(y_1 = 1 | x_3) \approx 0.450$, and $P(y_1 = 1 | x_4) \approx 0.525$.

Using the threshold of 0.5, then the estimated class label y_i is $0, 0, 0, 1$. So this setting gives a misclassification to the third data item.

To update the a_j we apply Eq. (13.29). When updating a_0, we again use $x_{i0} = 1$ and have

$$a_0^{new} = -0.3 - 0.05 \times ((0.401 - 0) + (0.426 - 0) + (0.450 - 1))$$

(continued)

13.3 The Logistic Regression Algorithm

Example 13.18 (continued)
$$+(0.525-1)\times 1 \approx -0.290.$$

$$a_1^{new} = 0.1 - 0.05 \times ((0.401-0)\times(-1) + (0.426-0)\times 0 + (0.450-1)$$
$$\times 1 + (0.525-1)\times 4) \approx 0.243.$$

So after the first iteration we can recalculate the z_i, using the new values of a_0 and a_1, and then use Eq. (13.25) to get the probabilities:

$$z_1 = -0.543, \ z_2 = -0.290, \ z_3 = -0.047, \text{ and } z_4 = 0.682.$$

So

$$P(y_1 = 1|x_1) \approx 0.367, \ P(y_1 = 1|x_2) \approx 0.428, \ P(y_1 = 1|x_3) \approx 0.488, \text{ and}$$

$$P(y_1 = 1|x_4) \approx 0.664.$$

With the same threshold of 0.5, we get estimated class labels of [0, 0, 0, 1], so the third data item is still misclassified.

For the next iteration we again update the a_j giving

$$a_0 = -0.287, \text{ and } a_1 = 0.354.$$

Finally, to see how we are doing, we calculate the z_is and the probabilities to get

$$z_1 = -0.641, \ z_2 = -0.287, \ z_3 = 0.067, \text{ and } z_4 = 1.13,$$

and

$$P(y_1 = 1|x_1) \approx 0.35, \ P(y_1 = 1|x_2) \approx 0.43, \ P(y_1 = 1|x_3) \approx 0.51, \text{ and}$$

$$P(y_1 = 1|x_4) \approx 0.76.$$

Now we have the correct classification of [0, 0, 1, 1].

Exercise

13.7 Continue Example 13.17. Do the second iteration of the gradient descent algorithm and see if it gets all the points correctly classified more definitely, that is, it might make a better predictor.

13.8 Suppose we have a new set of data items $[-1, 0, 1, 3]$ and suppose the class labels are $[0, 0, 1, 1]$. Fit a logistic regression model for the data using the gradient descent algorithm with one iteration and a learning rate of 0.05, starting with values of a_0 and a_1 as -0.3 and 0.2, respectively.

Chapter 14
Data Modelling in Practice

In previous chapters, we have introduced some fundamental mathematical and statistical knowledge needed to understand algorithms and create new approaches. This chapter will deal with some of the important issues surrounding data analysis. The fields of machine learning and data science have developed rapidly recently with many new versions of algorithms being presented and evaluated, each suited for different tasks. There are too many to describe here and specialised literature is needed to introduce you to the ones in any area that you wish to study.

However, there are some really fundamental issues that need mentioning in this book, such as data pre-processing, model selection, model evaluation, and understanding the bias-variance trade-off in model design. In this chapter, all of these will be discussed and we will use these issues to motivate the detailed discussion of two particular algorithms that can improve model generalisation, namely, ridge regression and early stopping.

14.1 Data Pre-Processing

Chapter 1 mentions that data scientists need to explore data to understand the relationships among the data better after obtaining some new raw data. To do that, one should spend some time learning some essential knowledge in the problem domain, for example, understanding the meaning of each feature or attribute and how they relate to the target of the problem going to be solved.

14.1.1 Questions to Ask When Pre-Processing the Data

As mentioned before, we need to check whether the data is organised. If it is unorganised, we need to convert it into a table-like structure. Then we need to understand what each row and column represents and whether each attribute is quantitative or qualitative. Apart from these, some common issues must be considered when exploring the data.

- Is the dataset balanced or imbalanced?
  This question is vital to a classification problem, though it has also been drawn to attention when dealing with the regression problem in recent years [16]. Let us consider a two-class application. If the ratio of the sizes of the two classes is much lower than 1, that is, the number of patterns in one class is much higher than in the other, then the trained model will tend to predict any unseen data belonging to the majority class. Therefore, it is helpful to balance the training dataset. Methods for balancing can be categorised into two groups: undersampling the majority class and oversampling the minority class. Readers interested in this topic may start by reading [17] and [18].
- Are there any inconsistent[1] data points?
  We briefly discuss two types of inconsistency. First, by inconsistent, we mean data items that have the same feature values but different target values or class labels. For example, we have a dataset of customer profiles, and we want to use the data to train a model to predict whether a customer would like to buy a newly published book about cooking. Consider two attributes we use: the amount of money each customer spent and the number of books the customer bought in the last six months. With only two attributes available, there may be many inconsistent data items. That is, customers who spent the same amount of money and bought the same number of books in the last six months may or may not buy the newly published book as shown in Customer 1 and Customer 3 in Table 14.1. To cope with this problem, we can add more features to the dataset. For example, the gender and age of each customer and the types of book each customer prefers to buy as shown in Table 14.2. In this way, we can alleviate many inconsistent data items.
  The second type of inconsistency refers to data that violate general observations in the training set. Consider two attributes of a dataset: a book title and its author. The data displayed in Table 14.3 would be a data inconsistency. When

Table 14.1 An example of data inconsistency

ID	The amount spent (£)	The number of books	Preference on the new book
1	3100	1	Yes
2	4060	2	Yes
3	3100	1	No

---

[1] This book focuses on addressing data quality issues such as label inconsistency caused by duplicates or measurement errors. In contrast, modelling input-dependent noise variance (heteroscedasticity), where noise levels vary systematically with the inputs, requires probabilistic approaches such as heteroscedastic Gaussian Processes or quantile-based regression. These methods explicitly account for structured noise and are beyond the scope of this book.

14.1 Data Pre-Processing

Table 14.2 Adding more features for dealing with inconsistent data

ID	Gender	Age	The amount spent (£)	The number of books	Types of purchased book	Preference on the new book
1	Female	35	3100	1	Cooking	Yes
2	Female	56	4060	2	Fiction	Yes
3	Male	41	3100	1	Finance	No

Table 14.3 Another example of data inconsistency

Book Title	Author
Harry Potter and the Philosopher's Stone	Chris Columbus

the book is *Harry Potter and the Philosopher's Stone*, we expect the author to be *J. K. Rowling*. Chris Columbus is the director of the film *Harry Potter and the Philosopher's Stone*. So there is something wrong here. To deal with this inconsistency, we need to check data accuracy by identifying and removing the causes of errors. We can compare data from different sources to identify and resolve any discrepancies.

- Are there any missing data?

It is common to see missing values in the collected dataset. It may be caused by errors during collection or by the fact that data are not available. For example, some people are likely to want to avoid answering specific questions in a survey. Data may be missing completely at random. That is, missing values can be observed in all features, and all data items have the same probability of having missing values. Data may be missing randomly for a specific feature or a set of features, or not all data items have the same chance of being missing. Alternatively, missing data may not be at random. That is, the probability of being missing is entirely different for different values of the same feature.

It is crucial to identify missing values and determine why they are generated. When we have enough data representing the underlying distribution, removing observations involving missing values from the dataset is the easiest way to deal with missing values. Many methods have been proposed to deal with missing values—for example, replacing with the mean value of the corresponding attribute. However, replacing missing values may introduce bias. Therefore, extra care should be taken to check whether it still makes sense with the new filled-in values. Readers may find more details in [19].

- Are there repeated data?

By repeated data, we mean those duplications among observations. How to deal with replicated data depends on the algorithm being used and the size of the dataset. It is generally a good idea to remove duplicated data items. Alternatively, we may consider whether it is necessary to add more features, as discussed in dealing with inconsistent data.

Example 14.1 illustrates whether involving repeated data may affect results. Here we consider applying the principal component analysis technique on a small dataset with repeated data items. This material was covered in Sect. 4.2.3 of Chap. 4.

Example 14.1 Suppose we have two datasets as follows:

$$X1 = \begin{bmatrix} 5 & 4 \\ -2 & 2 \\ -4 & -4 \\ 4 & 4 \\ 2 & -2 \\ 0 & 0 \end{bmatrix},$$

and

$$X2 = \begin{bmatrix} 5 & 4 \\ -2 & 2 \\ -4 & -4 \\ 4 & 4 \\ 2 & -2 \\ 0 & 0 \\ 5 & 4 \\ 5 & 4 \end{bmatrix}.$$

As seen, $X1$ includes six unique data items, while $X2$ includes the same data except that the data point [5, 4] is duplicated three times. The following analysis is done with the aid of suitable programs on a computer. The mean of $X1$ is $[0.8\dot{3}, 0.\dot{6}]$ and the mean of $X2$ is $[1.875, 1.5]$. After removing the mean values from each data matrix, we obtain the following:

$$\mathbf{newX1} = \begin{bmatrix} 4.1\dot{6} & 3.\dot{3} \\ -2.8\dot{3} & 1.\dot{3} \\ -4.8\dot{3} & -4.\dot{6} \\ 3.1\dot{6} & 3.\dot{3} \\ 1.1\dot{6} & -2.\dot{6} \\ -0.8\dot{3} & -0.\dot{6} \end{bmatrix},$$

(continued)

Example 14.1 (continued)
and

$$\mathbf{newX2} = \begin{bmatrix} 3.125 & 2.5 \\ -3.875 & 0.5 \\ -5.875 & -5.5 \\ 2.125 & 2.5 \\ 0.125 & -3.5 \\ -1.875 & -1.5 \\ 3.125 & 2.5 \\ 3.125 & 2.5 \end{bmatrix}.$$

The covariance matrix of **newX1** is shown as follows:

$$\mathbf{cov\_newX1} = \begin{bmatrix} 12.1\dot{6} & 8.1\dot{3} \\ 8.1\dot{3} & 10.\dot{6} \end{bmatrix},$$

and the covariance matrix of **newX2** is given by

$$\mathbf{cov\_newX2} = \begin{bmatrix} 12.41 & 8.79 \\ 8.79 & 10.00 \end{bmatrix},$$

where two decimal places are kept for each value. The results of eigendecomposition on both covariance matrices are displayed in Table 14.4. The eigenvectors are the two columns of the matrices in each case. Again, two decimal places are kept in all results.

As can be seen, the PCA results of the two datasets are not identical, though they are close to each other in this example.

Table 14.4 The results of eigendecomposition on **cov_newX1** and **cov_newX2**

The covariance matrix	Eigenvalues	Eigenvectors
cov_newX1	[19.58, 3.25]	$\begin{bmatrix} 0.74 & -0.67 \\ 0.67 & 0.74 \end{bmatrix}$
cov_newX2	[20.07, 2.34]	$\begin{bmatrix} 0.75 & -0.66 \\ 0.66 & 0.75 \end{bmatrix}$

Exercise

14.1 If you fancy reminding yourself about performing PCA, or if you want to try a suitable program on a computer, try the following exercise. Given the two datasets in transpose form as follows:

$$\mathbf{X1}^T = \begin{bmatrix} 1 & 0 & 0 & 1 \\ 2 & 0 & 1 & 1 \end{bmatrix},$$

and

$$\mathbf{X2}^T = \begin{bmatrix} 1 & 0 & 0 & 1 & 0 & 0 & 0 & 0 \\ 2 & 0 & 1 & 1 & 1 & 1 & 1 & 1 \end{bmatrix}.$$

The second dataset has the same values as the first except that the data point [0, 1] is duplicated five times. Apply principal component analysis to both datasets, having first removed the mean values from the datasets. Show that the eigenvalues and eigenvectors are different.

Solving the above issues is also called data cleaning. After addressing these issues, data scientists need to focus on understanding the statistics of each predictor and the relations among predictors. The aim is to find more information about the data than was available when we initially saw it. This helps us to identify suitable models to apply, whether to adapt appropriate approaches or to create a new method to solve the problem.

Applying descriptive statistics is helpful at this stage. In addition, employing unsupervised learning methods, for example, principal component analysis, can help to visualise the data and extract features. The difference between extracted and selected features is that extracted features differ from the original data attributes. For example, features extracted from the principal component analysis are linear combinations of original features. However, selected features are a subset of the features in the original feature set. In the following subsection, we will follow [20] to present a simple approach to carrying out feature selection.

14.1.2 A Simple Feature Selection Method

Suppose we have a structured dataset. The procedure used to select features considers the correlation between attributes as follows:

14.1 Data Pre-Processing

Table 14.5 The correlation coefficient matrix of the dataset with six features

	f1	f2	f3	f4	f5	f6
f1	1	0.80	0.61	0.91	0.20	−0.45
f2	0.80	1	0.75	0.37	0.85	−0.30
f3	0.61	0.75	1	0.18	0.39	0.52
f4	0.91	0.37	0.18	1	0.21	−0.15
f5	0.20	0.85	0.39	0.21	1	0.52
f6	−0.45	−0.30	0.52	−0.15	0.52	1

1. Calculate the correlation matrix of the features.
2. Determine the two features A and B, associated with the largest absolute pairwise correlation.
3. Determine the average of the absolute correlations between A and the other features and the average between B and the other features.
4. Remove the feature whose average correlation is the biggest.
5. Repeat Steps 2–4 until no absolute correlations are above the threshold.

Example 14.2 Consider a structured dataset with six features: from $f1$ to $f6$. Table 14.5 shows its correlation coefficients matrix. Select five features from the original six features using the method introduced above.

Solution

1. Determine the two features associated with the largest absolute pairwise correlation:
   As seen in Table 14.5, features $f1$ and $f4$ have the largest correlation: 0.91.
2. Determine the average absolute correlation between $f1$ and the other features and the average between $f4$ and the other features.
   $\frac{0.80+0.61+0.20+0.45}{4} = 0.515$ is the average of the absolute correlations between $f1$ and the other four features.
   $\frac{0.37+0.18+0.21+0.15}{4} = 0.2275$ is the average of the absolute correlations between $f4$ and the other four features.
3. Remove the feature whose average correlation is the biggest.
   We remove $f1$ since its average correlation is bigger than $f4$'s.
   Therefore, the five selected features are $f2, f3, f4, f5,$ and $f6$.

> **Exercise**
>
> **14.2** Continue Example 14.2 and further remove one feature using the method introduced in Sect. 14.1.2 of this chapter.

Remark 14.1 Why do we remove highly correlated features?

A set of highly correlated features usually provides little or no additional information but increases the model complexity. The model complexity usually refers to the number of features, the number of terms included in a given predictive model, and whether the model is linear or non-linear.

♦

14.2 Model Selection

14.2.1 Data Splitting

When training a model using a supervised learning algorithm, we usually separate the whole dataset into a training, validation, and test set.

The training set provides examples for the model to learn the mapping from inputs to the corresponding targets. The validation set helps search for the most suitable hyperparameters (user pre-set parameters). The test set determines how well the trained model performs on data it has never seen before, which is crucial: The model must never see the test data in the training phase.

The goal of model training is not to learn an exact representation of the training data itself but rather to build a statistical model of the process that generates the data. That is, to find the model having the best performance on new data, this is known as the model's generalisation ability.

14.2.2 Model Evaluation

Performance metrics are usually calculated for the validation and test sets when assessing how well a model fits data.

14.2.2.1 Regression Models

Let us use n to denote the number of data items in the validation or test set, whichever we are evaluating at the time, $\tilde{y}_i$ denotes the estimated value for data item i, y_i is the actual target value for data value i, and $\overline{y}$ is the mean of all target values in the validation or test set.

- The mean squared and root mean squared errors.
  The mean squared error (MSE) is defined as follows:

$$MSE = \frac{\sum_i^n (\tilde{y}_i - y_i)^2}{n}. \tag{14.1}$$

The root mean squared error (RMSE) is given by

$$RMSE = \sqrt{\frac{\sum_i^n (\tilde{y}_i - y_i)^2}{n}}. \tag{14.2}$$

The lower the MSE or RMSE, the better a model fits a dataset. In practice, we use the RMSE more often since it is measured in the same units as the target value of the dependent variable.

- Methods used for linear regression models

  1. The coefficient of determination
     The coefficient of determination was defined in Sect. 8.5.2 of Chap. 8, and is denoted as R^2. It is a value between 0 and 1. It is defined based on assumptions underlying the linear regression algorithm. Readers are referred to [15] for more details about the linear regression method. Here, we simply repeat its definition:

$$R^2 = 1 - \frac{\sum_i^n (y_i - \tilde{y}_i)^2}{\sum_i^n (y_i - \overline{y})^2}. \tag{14.3}$$

  If all estimated values are equal to their target values, then the numerator in Eq. (14.3) is zero and R^2 equals one, indicating the model fits the data perfectly. If the ratio of the numerator and denominator is one, then R^2 equals zero, suggesting the model cannot fit the data, and all estimated values are equal to the mean of the actual target values.

  2. Scatter plots of residual against predictions
     It is common to plot residuals against features to look for extra regression structures. Residuals were defined in Sect. 8.5.1 of Chap. 8 as $e_i = y_i - \tilde{y}_i$, or in vector form as $\mathbf{e} = \mathbf{y} - \tilde{\mathbf{y}}$, where $\tilde{\mathbf{y}}$ is the expected or estimated value.
     Recall (using Eq. (8.17) in Chap. 8) the expected prediction from a linear regression model is given by

$$\mathbf{X}\tilde{\mathbf{a}} = \mathbf{X}(\mathbf{X}^T\mathbf{X})^{-1}\mathbf{X}^T\mathbf{y}.$$

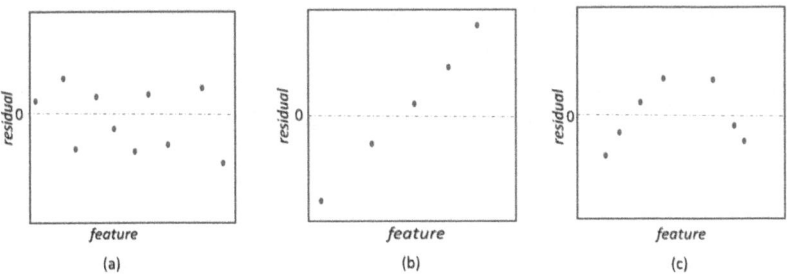

Fig. 14.1 Scatter plots of residuals versus a feature

The residual is calculated as follows:

$$e = y - X(X^T X)^{-1} X^T y.$$

Therefore, we have

$$X^T e = X^T y - X^T X(X^T X)^{-1} X^T y = 0. \tag{14.4}$$

Equation (14.4) shows that if we calculate the dot product of residuals and any feature in the data matrix X, the result must always be equal to zero. So the residuals and the features are always independent. In Sect. 11.2.2.3 of Chap. 11, we show if two random variables are independent, they are also uncorrelated. Thus, we expect to observe a scatter plot similar to Panel (a) in Fig. 14.1, where the residuals and feature values are uncorrelated. We do not expect a straight line, a positive relationship as shown in Panel (b), or a negative relationship between the residuals and feature values. If we see a quadratic curve, such as in Panel (c), it suggests we need a quadratic term in the regression.

14.2.2.2 Classification Models

First, let us define the confusion matrix shown in Table 14.6.

Table 14.6 A confusion matrix: where TN is the number of true-negative samples, FP false-positive samples, FN false-negative samples, and TP true-positive samples

	Predicted negative	Predicted positive
Actual negative	TN	FP
Actual positive	FN	TP

14.2 Model Selection

The classification accuracy rate is given by

$$\text{accuracy rate} = \frac{TN + TP}{TN + FP + FN + TP}. \quad (14.5)$$

For a problem domain with an imbalanced dataset, that is, a dataset where one class is much bigger than the other, the classification accuracy rate is not sufficient as a standard performance measure. This is because you can get good accuracy by always predicting the majority class. So, if you use accuracy as your sole training criterion of success you, are likely to get a model that just predicts the majority class. Several common performance metrics, such as *recall*, *precision*, and *F-score*, which are calculated to fairly quantify the performance of the classification algorithm on the minority class, can be defined as follows:

$$\text{Recall} = \frac{\text{TP}}{(\text{TP} + \text{FN})}, \quad (14.6)$$

$$\text{Precision} = \frac{\text{TP}}{(\text{TP} + \text{FP})}, \quad (14.7)$$

$$\text{F-score} = \frac{2 \cdot \text{Recall} \cdot \text{Precision}}{\text{Recall} + \text{Precision}}, \quad (14.8)$$

$$\text{FP rate} = \frac{\text{FP}}{\text{FP} + \text{TN}}, \quad (14.9)$$

$$\text{True-negative rate} = \frac{\text{TN}}{\text{TN} + \text{FP}}. \quad (14.10)$$

Recall, also called *sensitivity*, measures the true-positive rate, that is, the number of actual positives you get right. Precision, also called *positive predictive value*, measures the accuracy rate of predicted positive values, that is, the number of predicted positives you have got right. Usually, a trade-off between precision and recall is integrated into the metrics, such as the F-score. The false-positive rate, or FP rate, is the number of actual negatives that you get wrong. A high F-score and low FP rate are generally seen as the preferred criterion of success. The true-negative rate is also called *specificity*, which is equal to $1 - \text{FP rate}$.

Table 14.7 The confusion matrix of the balanced dataset used in Example 14.3

	Predicted negative	Predicted positive
Actual negative	496	4
Actual positive	0	500

Example 14.3 Consider a perfectly balanced dataset with 1000 data points, 500 in each class. Table 14.7 shows the confusion matrix. Compute accuracy rate, recall, precision, F-score, FP rate, and true-negative rate.

Solution We have $TP = 500$, $TN = 496$, $FP = 4$, and $FN = 0$ from Table 14.7.

$$\text{accuracy rate} = \frac{996}{1000} = 0.996.$$

$$\text{Recall} = \frac{500}{500} = 1,$$

$$\text{Precision} = \frac{500}{504} \approx 0.992,$$

$$\text{F-score} = \frac{2 \times 1 \times 0.992}{1 + 0.992} \approx 0.996,$$

$$\text{FP rate} = \frac{4}{500} = 0.008.$$

$$\text{True-negative rate} = \frac{496}{496 + 4} = 0.992.$$

Exercises

14.3 Let us consider a dataset with a highly imbalanced class distribution with 1000 data points in total, but with only 10 data points in the positive class and the rest in the negative class. Table 14.8 shows a confusion matrix that could have been produced by training the algorithm on accuracy alone— it has predicted most of the data as being negative since that was the majority class. Compute accuracy rate, recall, precision, F-score, FP rate, and true-negative rate.

(continued)

14.2 Model Selection

Now suppose it was trained on the F-score, and the possible results are shown in Table 14.9. Compute accuracy rate, recall, precision, F-score, FP rate, and true-negative rate for this confusion matrix.

14.4 Consider a reasonably balanced dataset, with 1000 data points in total. Table 14.10 shows the confusion matrix. Compute accuracy rate, recall, precision, F-score, FP rate, and true-negative rate.

14.5 Let us consider a dataset with a reasonably imbalanced class distribution with 1000 data points in total but with only 100 data points in the positive class and the rest in the negative class. Table 14.11 shows the confusion matrix. The algorithm has predicted most of the data as being negative since that was the majority class. Compute accuracy rate, recall, precision, F-score, FP rate, and true-negative rate. A perfect predictor for this dataset would get the results shown in Table 14.12. Compute accuracy rate, recall, precision, F-score, FP rate, and true-negative rate for this confusion matrix.

Table 14.8 The confusion matrix of the imbalanced dataset trained on accuracy in Exercise 14.3

	Predicted negative	Predicted positive
Actual negative	982	8
Actual Positive	8	2

Table 14.9 The confusion matrix of the imbalanced dataset trained on F-score in Exercise 14.3

	Predicted negative	Predicted positive
Actual negative	982	8
Actual positive	0	10

Table 14.10 The confusion matrix of the dataset in Exercise 14.4

	Predicted negative	Predicted positive
Actual negative	475	15
Actual positive	10	500

Table 14.11 The confusion matrix of the imbalanced dataset in Exercise 14.5

	Predicted negative	Predicted positive
Actual negative	855	45
Actual positive	80	20

Table 14.12 The perfect confusion matrix of the imbalanced dataset in Exercise 14.5

	Predicted negative	Predicted positive
Actual negative	900	0
Actual positive	0	100

14.2.3 Understanding Bias-Variance Trade-Off

Definition 14.1 (Bias) Bias means that an estimator is calculated in a way that is systematically different from the quantity that it is supposed to estimate.
Let $\tilde{f}(\mathbf{x})$ be a point estimator and $f(\mathbf{x})$ the ground truth. The bias of the estimator is defined as follows:

$$B(\tilde{f}(\mathbf{x})) = E\{\tilde{f}(\mathbf{x})\} - f(\mathbf{x}). \tag{14.11}$$

It says the bias measures how far the average estimate of a model is from the ground truth. A positive bias means that the model value is overestimated, and a negative bias means that the model value is underestimated. The bias may be caused by making wrong assumptions when choosing a model.

Example 14.4 Recall that the distribution of estimated $\tilde{\mathbf{a}}$ for a linear regression approximates the normal distribution given by $\mathcal{N}(\mathbf{a}, \sigma^2(\mathbf{X}^T\mathbf{X})^{-1})$ (see Sect. 13.2.2.2 in Chap. 13). That is, $E[\tilde{\mathbf{a}}] = (\mathbf{X}^T\mathbf{X})^{-1}\mathbf{X}^T\mathbf{X}\mathbf{a} = \mathbf{a}$ (see Eq. 13.17). Applying Eq. (14.11), we have $B(\tilde{\mathbf{a}}) = E[\tilde{\mathbf{a}}] - \mathbf{a} = \mathbf{0}$. That is, under the assumptions mentioned in Sect. 13.2.1 of Chap. 13, estimates of the ordinary linear regression coefficients are unbiased. This is an important result for ordinary linear regression.

For example, if the true underlying relationship between the independent variable and the dependent variable is $f(x) = a_0 + a_1 x$, then $E[\tilde{a}_0] = a_0$ and $E[\tilde{a}_1] = a_1$ if we estimate $\tilde{\mathbf{a}}$ from the ordinary linear regression method.

Definition 14.2 (Variance) Variance is due to the model's excessive sensitivity to small variations in the training data. Let $\tilde{f}(\mathbf{x})$ be a point estimator (see Sect. 12.3.1 of Chap. 12). The variance of the estimator is defined as follows:

$$var\{\tilde{f}(\mathbf{x})\} = E\left(\left(\tilde{f}(\mathbf{x}) - E\{\tilde{f}(\mathbf{x})\}\right)^2\right). \tag{14.12}$$

The variance measures the variability of a model estimate when changing the training examples.

Remark 14.2 If the model does not change much between samples, the model would be considered a low-variance model. On the other hand, if the model changes drastically between samples, then that model would be considered a high-variance model.

♦

14.2 Model Selection

Fig. 14.2 Scatter plot of the data from Example 14.5, with the solid line showing the estimated linear regression line

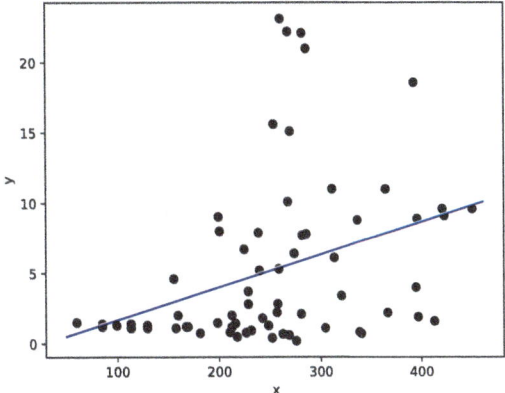

Example 14.5 Figure 14.2 shows 66 data points with two variables. Values of the independent variable X are in the interval of [50, 460], and values of the dependent variable Y are in the interval of [0, 24]. We have used simple linear regression to estimate the relationship between these two variables. The estimated line is shown as a solid line.

We divide the whole data set randomly into two samples, including 33 data points for each. Then simple linear regression is applied for each sample. Figure 14.3 shows the results. Data in sample 1 are denoted as plus signs, and data in sample 2 are denoted as square signs. The estimated line from sample 1 is shown as the solid line, while the line from sample 2 is depicted as the dash-dotted line. As observed, the estimated model remains largely consistent between the two samples. So, the model is considered a low-variance model.

Now, suppose we had employed a polynomial model with a degree of 5 for each of the two samples. The estimated curves for each sample are shown in Fig. 14.4 as solid and dash-dotted curves. As can be seen, this estimated model changes between samples, especially with x values less than 100 and between 300 and 400. So, this model is considered a high-variance model.

The generalisation error, measured on the test set, can be shown to be composed of the sum of the bias squared, the variance, and the irreducible error. We cannot do anything about the irreducible error, or noise, but it is important that the full generalisation error includes both the bias squared and the variance. We will show that the generalisation error, or mean squared error, is composed of these three factors by breaking it down, or decomposing it, as shown in the following example. This process is quite complicated and can be skipped if needed.

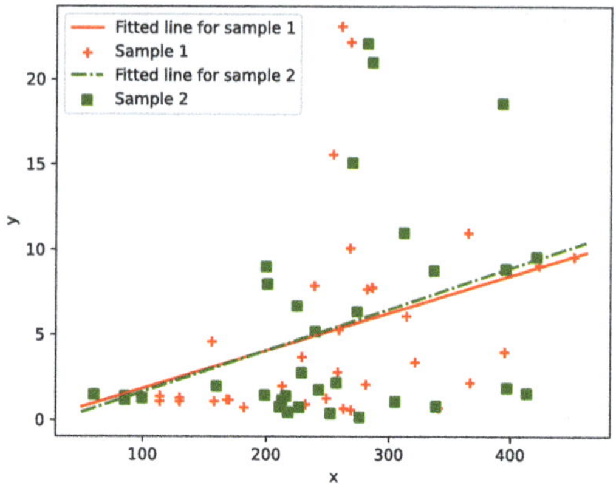

Fig. 14.3 The solid line is fitted to sample 1, while the dash-dotted line is fitted to sample 2, using the data in Example 14.5

Fig. 14.4 The solid curve is fitted to sample 1 and the dash-dotted curve to sample 2, using data from Example 14.5. Both models are fitted with a polynomial of degree 5

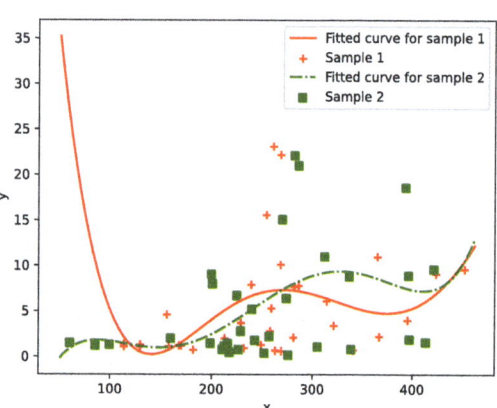

Example 14.6 Let us decompose the mean squared error in the ordinary linear regression model.

First, recall the equation for variance, namely, Eq. (10.11) of Chap. 10, showed the following equality:

$$E((X - E(X))^2) = E(X^2) - (E(X))^2.$$

If we add $(E(X))^2$ to both sides of the above equation, it gives us

$$E(X^2) = E((X - E(X))^2) + (E(X))^2. \tag{14.13}$$

Now suppose X and Y are two variables, and the underlying function is $y = f(x) + \epsilon$, where $\epsilon \sim \mathcal{N}(0, \sigma^2)$. So σ^2 is the variance of the error term ϵ, which has a Gaussian distribution (see Sect. 13.2.1 of Chap. 13).

(continued)

14.2 Model Selection

Example 14.6 (continued)

Since $E(y) = E(f(x) + \epsilon)$, $E(\epsilon) = 0$, and $E(f(x)) = f(x)$ for a fixed function f at the point x, we have $E(y) = f(x)$.

We fit a linear regression line $\tilde{f}(x) = ax + a_0$ using N training examples, (x_i, y_i), $i = 1, \ldots, N$, to minimise the square error:

$$\sum_i^N (y_i - \tilde{f}(x_i))^2.$$

Given any new data point (x_{new}, y_{new}) from the same distribution, we can estimate the expected square error, $E\left(\left(y_{new} - \tilde{f}(x_{new})\right)^2\right)$, as follows:

$$E\left(\left(y_{new} - \tilde{f}(x_{new})\right)^2\right) = E\left((y_{new})^2 - 2y_{new}\tilde{f}(x_{new}) + \left(\tilde{f}(x_{new})\right)^2\right)$$

$$= E\left((y_{new})^2\right) - 2E(y_{new})E\left(\tilde{f}(x_{new})\right)$$

$$+ E\left(\left(\tilde{f}(x_{new})\right)^2\right)$$

$$= E\left((y_{new})^2\right) - 2f(x_{new})E\left(\tilde{f}(x_{new})\right)$$

$$+ E\left(\left(\tilde{f}(x_{new})\right)^2\right), \tag{14.14}$$

where $E(y_{new}) = f(x_{new})$ for any x value as noted above.

Consider the first term on the right-hand side of the last equal sign of Eq. (14.14). We can rewrite it using Eq. (14.13) with $X = y_{new}$ and then can use $E(y_{new}) = f(x_{new})$ to obtain

$$E\left((y_{new})^2\right) = E\left(\left(y_{new} - E\left(y_{new}\right)\right)^2\right) + \left(E\left(y_{new}\right)\right)^2$$

$$= E\left(\left(y_{new} - f(x_{new})\right)^2\right) + \left(f(x_{new})\right)^2. \tag{14.15}$$

(continued)

Example 14.6 (continued)
Now consider the third term on the right-hand side of the last equal sign of Eq. (14.14). Again we can rewrite it using Eq. (14.13) with $X = \tilde{f}(x_{new})$. We obtain

$$E\left(\left(\tilde{f}(x_{new})\right)^2\right) = E\left(\left(\tilde{f}(x_{new}) - E\left(\tilde{f}(x_{new})\right)\right)^2\right) + \left(E\left(\tilde{f}(x_{new})\right)\right)^2. \tag{14.16}$$

Now we can substitute Equations (14.15) and (14.16) for the first and last terms of the last equal sign of Eq. (14.14) to get a new version of Eq. (14.14):

$$E\left(\left(y_{new} - \tilde{f}(x_{new})\right)^2\right) = E\left(\left(y_{new} - f(x_{new})\right)^2\right) + \left(f(x_{new})\right)^2$$

$$- 2f(x_{new})E\left(\tilde{f}(x_{new})\right) + E\left(\left(\tilde{f}(x_{new})\right.\right.$$

$$\left.\left. - E\left(\tilde{f}(x_{new})\right)\right)^2\right) + \left(E\left(\tilde{f}(x_{new})\right)\right)^2$$

$$= E\left(\left(y_{new} - f(x_{new})\right)^2\right) + E\left(\left(\tilde{f}(x_{new})\right.\right.$$

$$\left.\left. - E\left(\tilde{f}(x_{new})\right)\right)^2\right) + \left(f(x_{new}) - E\left(\tilde{f}(x_{new})\right)\right)^2. \tag{14.17}$$

The last line of Eq. (14.17) includes three terms. From left to right:

- The first term is the noise $E\left(\left(y_{new} - f(x_{new})\right)^2\right) = E(\epsilon^2)$. Now from Eq. (10.11) of Chap. 10 we have $E(\epsilon^2) = Var(\epsilon) + E(\epsilon)^2 = \sigma^2$ using Sect. 13.2.1 of Chap. 13 as noted above. So the first term is the noise, σ^2.
- The second term $E\left(\tilde{f}(x_{new}) - E\left(\tilde{f}(x_{new})\right)^2\right)$ is the variance (see Eq. 14.12).
- The third term is the bias squared (see Eq. 14.11).

Hence, we have seen in the above example (Example 14.6) that when we assess a trained model on a test set, the error over the test set, also called the generalisation error, can be decomposed into three parts. To minimise the generalisation error, we want to reduce both the bias squared and the variance, since we cannot change the irreducible error (the noise in Example 14.6). However, as illustrated in Fig. 14.5, bias squared decreases and variance increases as the model complexity increases.

14.2 Model Selection

Fig. 14.5 Illustration of the bias-variance trade-off

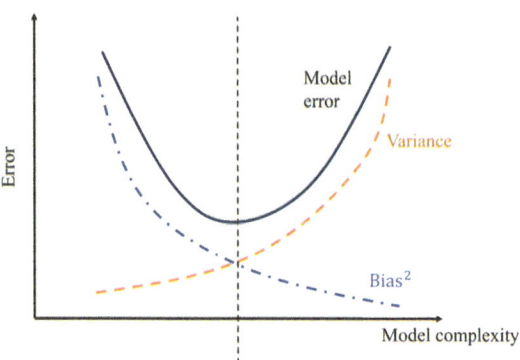

It indicates a trade-off between the bias squared and variance. Where the bias and variance are both relatively small, we get the minimum generalisation error. More details about error decomposition can be read in [21].

In practice, we usually start with several models widely used in many different real-world applications. These models are not interpretable due to their complexity, but they may produce better results with high probability. Then we can look into simpler models that are interpretable. The aim is to consider using a model that is as simple as possible but provides a similar performance to that of the complex models.

14.2.4 Underfitting and Overfitting

Underfitting occurs when models make little to no attempt to fit the data. Models that are high bias and low variance are prone to underfitting. A badly underfitted model is really unable to do the job of either fitting the training data or providing a useful estimation tool in the case of the test data, or any unseen data. It is really not a useful model at all. We need to do something about it such as adding more features and/or using a more complicated model to overcome this underfitting.

Overfitting is the result of the model trying too hard to exactly fit into the training set, resulting in a lower bias but a much higher variance. Since the model fits the training set so well it often does not perform well on the test data or on any new unseen data, that is, its generalisation ability can be poor. To overcome this overfitting, we may use fewer features and/or use more training data. Another approach is to use a regularisation technique to stop the model from being only suitable for the data it has been trained on.

Regularisation is, therefore, a technique to prevent overfitting or to help optimisation. Usually, it is done by adding additional terms in the objective, or cost, function.

In the next two sections, we will introduce two widely used regularisation methods: ridge regularisation and early stopping.

14.3 Ridge Regression

14.3.1 The Closed-Form Solution

In Chap. 8, we have shown that the coefficients of a linear regression model with multiple variables can be computed using the normal equation (see Eq. (8.17) in Sect. 8.3 of Chap. 8), that is,

$$\mathbf{a} = (\mathbf{X}^T\mathbf{X})^{-1}\mathbf{X}^T\mathbf{y},$$

where $\mathbf{X}$ is the input matrix including a column vector of $1's$, and $\mathbf{y}$ is the dependent variable. This result was found by minimising the sum of the squares of the errors between the data and the linear approximation.

To control the model complexity, the ridge regression method proposed in [22] involves adding a penalty term, also called regularisation term, to the least-squares error function. It may, at first sight, seem strange to add another term to the objective function since minimising the error function must produce the best linear approximation to the data. However, that is what overfitting is all about. Using just the sum of the squares of the errors means the training data is fully satisfied, but it must be remembered that the real aim is to make the linear approximation best for the test set or any unseen data. In the case of ridge regression, $\frac{1}{2}\lambda \sum_j^d a_j^2$ is added into the objective function shown in Eq. (8.15) in Chap. 8, where the regularisation parameter is λ and $\lambda \in [0, \infty)$. That is,

$$Ridge\_Q = \sum_{i=1}^{N}(y_i - \sum_{j=0}^{d} a_j x_{ij})^2 + \frac{1}{2}\lambda \sum_{j=1}^{d} a_j^2. \qquad (14.18)$$

Remark 14.3 We can compare Eq. (14.18) with the previous least-squares error formula given in Eq. (8.15) of Chap. 8. Looking at Eq. (14.18), if $\lambda = 0$, we have $Ridge\_Q = \sum_{i=1}^{N}(y_i - \sum_{j=0}^{d} a_j x_{ij})^2$, which is the same as the calculation of the least-squares error, namely, Eq. (8.15), exactly as we would have expected. When $\lambda = \infty$, we consider two cases: (1) if any of the estimated $a_j \neq 0$, we have $Ridge\_Q = \infty$; (2) if all the $a_j = 0$, we have $Ridge\_Q = \sum_{i=1}^{N}(y_i)^2$.

♦

To minimise $Ridge\_Q$, both terms in Eq. (14.18) should be as small as possible. When $\lambda > 0$, minimising $\frac{1}{2}\lambda \sum_{j=1}^{d} a_j^2$ means forcing the a_j to be as small as possible. That is how the ridge regression method controls the model complexity.

Remark 14.4 The complexity of the ridge model is lower than the complexity of its corresponding ordinary linear regression model.

♦

Note that j starts with 1 in Eq. (14.18). That means ridge regression does not penalise the intercept.

14.3 Ridge Regression

We can write Eq. (14.18) in its matrix form as follows:

$$Ridge\_Q = (\mathbf{y} - \mathbf{Xa})^T(\mathbf{y} - \mathbf{Xa}) + \lambda \mathbf{a}_*^T \mathbf{a}_*. \quad (14.19)$$

The difference between $\mathbf{a}$ and $\mathbf{a}_*$ is that $\mathbf{a}_*$ does not include the intercept a_0. To remove the awkward-looking $\mathbf{a}_*$ we can multiply the second term by $\mathbf{I}'$, where $\mathbf{I}'$ is an identity matrix, with a size of $(d+1) \times (d+1)$, except with a zero in the top-left cell, corresponding to the a_0 term. Therefore, the objective function of the ridge regression is given by

$$Ridge\_Q = (\mathbf{y} - \mathbf{Xa})^T(\mathbf{y} - \mathbf{Xa}) + \lambda \mathbf{a}^T \mathbf{I}' \mathbf{a}. \quad (14.20)$$

Example 14.7 Suppose $\mathbf{a}_*^T = [a_1, a_2]$ and $\mathbf{a}^T = [a_0, a_1, a_2]$. Let $\mathbf{I}' = \begin{bmatrix} 0 & 0 & 0 \\ 0 & 1 & 0 \\ 0 & 0 & 1 \end{bmatrix}$. Compute $\mathbf{a}_*^T \mathbf{a}_*$ and $\mathbf{a}^T \mathbf{I}' \mathbf{a}$.

Solution $\mathbf{a}_*^T \mathbf{a}_* = [a_1, a_2] \begin{bmatrix} a_1 \\ a_2 \end{bmatrix} = a_1^2 + a_2^2$.

$\mathbf{a}^T \mathbf{I}' \mathbf{a} = [a_0, a_1, a_2] \begin{bmatrix} 0 & 0 & 0 \\ 0 & 1 & 0 \\ 0 & 0 & 1 \end{bmatrix} \begin{bmatrix} a_0 \\ a_1 \\ a_2 \end{bmatrix} = [0, a_1, a_2] \begin{bmatrix} a_0 \\ a_1 \\ a_2 \end{bmatrix} = a_1^2 + a_2^2$.

Therefore, $\mathbf{a}_*^T \mathbf{a}_* = \mathbf{a}^T \mathbf{I}' \mathbf{a}$.

To obtain a formula for $\mathbf{a}$, we need to find the partial derivative of $Ridge\_Q$ with respect to $\mathbf{a}$. The working of the derivative of the first term in Eq. (14.20) is the same as the one shown in Eq. (8.16) in Chap. 8. The derivative of the second term is $\frac{\lambda \partial \mathbf{a}^T \mathbf{I}' \mathbf{a}}{\partial \mathbf{a}}$. To calculate it, we can apply Eq. (7.4) of Chap. 7 and obtain $2\lambda \mathbf{I}' \mathbf{a}$. Hence, we can obtain the derivative of $Ridge\_Q$:

$$\frac{\partial Ridge\_Q}{\partial \mathbf{a}} = \frac{\partial [(\mathbf{y} - \mathbf{Xa})^T(\mathbf{y} - \mathbf{Xa}) + \lambda \mathbf{a}^T \mathbf{I}' \mathbf{a}]}{\partial \mathbf{a}}$$

$$= -2\mathbf{X}^T(\mathbf{y} - \mathbf{Xa}) + 2\lambda \mathbf{I}' \mathbf{a}. \quad (14.21)$$

Therefore, by setting Eq. (14.21) equal to zero and rearranging the formula, exactly like we did when we set Eq. (8.16) equal to zero in Chap. 8, we obtain the closed-form solution of ridge regression as follows:

$$\mathbf{a} = (\mathbf{X}^T \mathbf{X} + \lambda \mathbf{I}')^{-1} \mathbf{X}^T \mathbf{y}.$$

As we can see, $\mathbf{X}^T\mathbf{X} + \lambda \mathbf{I}'$ replaces the $\mathbf{X}^T\mathbf{X}$ in Eq. (8.17) to give the ridge regression solution. To distinguish this from the solution **a** for the ordinary linear regression, we denote the solution for the ridge regression as $\mathbf{a}_R$. That is,

$$\mathbf{a}_R = (\mathbf{X}^T\mathbf{X} + \lambda \mathbf{I}')^{-1}\mathbf{X}^T\mathbf{y}. \tag{14.22}$$

If $\lambda = 0$, Eq. (14.22) gives the same solution as we obtained from the least-squares technique in Chap. 8.

14.3.2 Bias and Variance of Ridge Regression Coefficients

14.3.2.1 Bias

Substituting $\mathbf{y} = \mathbf{X}\mathbf{a} + \boldsymbol{\epsilon}$ to Eq. (14.22), we have

$$\begin{aligned}\mathbf{a}_R &= (\mathbf{X}^T\mathbf{X} + \lambda \mathbf{I}')^{-1}\mathbf{X}^T(\mathbf{X}\mathbf{a} + \boldsymbol{\epsilon}) \\ &= (\mathbf{X}^T\mathbf{X} + \lambda \mathbf{I}')^{-1}\mathbf{X}^T\mathbf{X}\mathbf{a} + (\mathbf{X}^T\mathbf{X} + \lambda \mathbf{I}')^{-1}\mathbf{X}^T\boldsymbol{\epsilon}.\end{aligned} \tag{14.23}$$

Therefore, the expected estimator of $\mathbf{a}_R$ is given by

$$\begin{aligned}E[\tilde{\mathbf{a}}_R] &= (\mathbf{X}^T\mathbf{X} + \lambda \mathbf{I}')^{-1}\mathbf{X}^T\mathbf{X}\mathbf{a} + (\mathbf{X}^T\mathbf{X} + \lambda \mathbf{I}')^{-1}\mathbf{X}^T\mathbf{X}E(\boldsymbol{\epsilon}) \\ &= (\mathbf{X}^T\mathbf{X} + \lambda \mathbf{I}')^{-1}\mathbf{X}^T\mathbf{X}\mathbf{a},\end{aligned} \tag{14.24}$$

where we have used the assumption that $E(\boldsymbol{\epsilon}) = 0$ (see Sect. 13.2.1 in Chap. 13). Therefore, the ridge estimator is biased, since $E[\tilde{\mathbf{a}}_R] \neq \mathbf{a}$. Substituting Eq. (14.24) to Eq. (14.11), where we have $\tilde{\mathbf{a}}_R$ as the point estimator of **a**, gives

$$\begin{aligned}B(\tilde{\mathbf{a}}_R) &= (\mathbf{X}^T\mathbf{X} + \lambda \mathbf{I}')^{-1}\mathbf{X}^T\mathbf{X}\mathbf{a} - \mathbf{a} \\ &= (\mathbf{X}^T\mathbf{X} + \lambda \mathbf{I}')^{-1}\mathbf{X}^T\mathbf{X}\mathbf{a} - (\mathbf{X}^T\mathbf{X})^{-1}(\mathbf{X}^T\mathbf{X})\mathbf{a} \\ &= \left((\mathbf{X}^T\mathbf{X} + \lambda \mathbf{I}')^{-1} - (\mathbf{X}^T\mathbf{X})^{-1}\right)\mathbf{X}^T\mathbf{X}\mathbf{a}.\end{aligned} \tag{14.25}$$

The middle line in Eq. (14.25) is obtained by multiplying **a** by $(\mathbf{X}^T\mathbf{X})^{-1}(\mathbf{X}^T\mathbf{X})$ which is a matrix and its inverse and so is just the identity matrix. We obtain $E[\tilde{\mathbf{a}}_R] = \mathbf{a}$ only if $\lambda = 0$, which is indeed the linear regression without the ridge regularisation term.

Equation (14.25) shows the ridge estimator is biased if $\lambda \neq 0$. The lower the λ value, the lower the bias.

14.3 Ridge Regression

14.3.2.2 Variance

Since $\mathbf{X}^T\mathbf{X}(\mathbf{X}^T\mathbf{X})^{-1}$ gives an identity matrix, we can multiply it to Eq. (14.22) and obtain the following:

$$\mathbf{a}_R = (\mathbf{X}^T\mathbf{X} + \lambda \mathbf{I}')^{-1}\mathbf{X}^T\mathbf{y}$$
$$= (\mathbf{X}^T\mathbf{X} + \lambda \mathbf{I}')^{-1}\mathbf{X}^T\mathbf{X}(\mathbf{X}^T\mathbf{X})^{-1}\mathbf{X}^T\mathbf{y}.$$

Applying Eq. (8.17) to the second line of the above equation, we obtain

$$\mathbf{a}_R = (\mathbf{X}^T\mathbf{X} + \lambda \mathbf{I}')^{-1}\mathbf{X}^T\mathbf{X}\tilde{\mathbf{a}}. \qquad (14.26)$$

The variance of the estimated value of $\mathbf{a}_R$, $\tilde{\mathbf{a}}_R$, for the ridge regression is

$$Var(\tilde{\mathbf{a}}_R) = Var((\mathbf{X}^T\mathbf{X} + \lambda \mathbf{I}')^{-1}\mathbf{X}^T\mathbf{X}\tilde{\mathbf{a}})$$
$$= (\mathbf{X}^T\mathbf{X} + \lambda \mathbf{I}')^{-1}\mathbf{X}^T\mathbf{X}Var(\tilde{\mathbf{a}})\mathbf{X}^T\mathbf{X}\Big((\mathbf{X}^T\mathbf{X} + \lambda \mathbf{I}')^{-1}\Big)^T$$
$$= (\mathbf{X}^T\mathbf{X} + \lambda \mathbf{I}')^{-1}\mathbf{X}^T\mathbf{X}\sigma^2(\mathbf{X}^T\mathbf{X})^{-1}\mathbf{X}^T\mathbf{X}\Big((\mathbf{X}^T\mathbf{X} + \lambda \mathbf{I}')^{-1}\Big)^T$$
$$= \sigma^2(\mathbf{X}^T\mathbf{X} + \lambda \mathbf{I}')^{-1}\mathbf{X}^T\mathbf{X}\Big((\mathbf{X}^T\mathbf{X} + \lambda \mathbf{I}')^{-1}\Big)^T, \qquad (14.27)$$

where we have used the second property in Sect. 13.2.2.1 of Chap. 13, namely, Eq. (13.15), to simplify the variance, the fourth property in Sect. 3.3.11.1 of Chap. 3 to evaluate the transpose of multiple matrices and $Var(\tilde{\mathbf{a}}) = \sigma^2(\mathbf{X}^T\mathbf{X})^{-1}$ (see Eq. (13.18)) to replace $Var(\tilde{\mathbf{a}})$, where σ^2 is the variance of the error term ϵ.

Example 14.8 Consider an example with $d = 1$, which is just one independent variable. Also let us have 5 points, so $N = 5$. Given the 5 points as $(1, 2)$, $(2, 4)$ and $(3, 3)$, $(4, 4)$ and $(5, 5)$. Perform ridge regression and find the estimated value of $\mathbf{a}_R$, that is, $\tilde{\mathbf{a}}_R$, for (a) λ equal to zero (so we actually get $\tilde{\mathbf{a}}$ and not $\tilde{\mathbf{a}}_R$, since this would be ordinary linear regression), (b) $\lambda = 1$, and (c) $\lambda = 10$. For each value of λ calculate the value of the bias and the variance.

Solution We have $\mathbf{X} = \begin{bmatrix} 1 & 1 \\ 1 & 2 \\ 1 & 3 \\ 1 & 4 \\ 1 & 5 \end{bmatrix}$ and $\mathbf{y} = \begin{bmatrix} 2 \\ 4 \\ 3 \\ 4 \\ 5 \end{bmatrix}$.

(a) For $\lambda = 0$, we wish to find $\mathbf{a} = \begin{bmatrix} a_0 \\ a_1 \end{bmatrix}$.

(continued)

Example 14.8 (continued)

So $\mathbf{X}^T\mathbf{X} = \begin{bmatrix} 5 & 15 \\ 15 & 55 \end{bmatrix}$ and $(\mathbf{X}^T\mathbf{X})^{-1} = \frac{1}{50}\begin{bmatrix} 55 & -15 \\ -15 & 5 \end{bmatrix} = \begin{bmatrix} 1.1 & -0.3 \\ -0.3 & 0.1 \end{bmatrix}$.

Also $\mathbf{X}^T\mathbf{y} = \begin{bmatrix} 18 \\ 60 \end{bmatrix}$. Since $\lambda = 0$ we use Eq. (8.17) from Chap. 8. So

$$\tilde{\mathbf{a}} = (\mathbf{X}^T\mathbf{X})^{-1}\mathbf{X}^T\mathbf{y} = \begin{bmatrix} 1.8 \\ 0.6 \end{bmatrix}.$$

Now for the bias, we use Eq. (14.25) with $\lambda = 0$ and get that the bias is $B(\tilde{\mathbf{a}}) = \begin{bmatrix} 0 \\ 0 \end{bmatrix}$, as expected.

For the variance, we use Eq. (14.27) with $\lambda = 0$, giving

$$Var(\tilde{\mathbf{a}}) = \sigma^2(\mathbf{X}^T\mathbf{X})^{-1} = \sigma^2\begin{bmatrix} 1.1 & -0.3 \\ -0.3 & 0.1 \end{bmatrix}.$$

(b) For $\lambda = 1$, we need matrix $\lambda \mathbf{I}'$ which is

$$\begin{bmatrix} 0 & 0 \\ 0 & 1 \end{bmatrix}.$$

We now need to use $\mathbf{X}^T\mathbf{X} + \lambda \mathbf{I}'$ a lot. For $\lambda = 1$,

$$\mathbf{X}^T\mathbf{X} + \lambda \mathbf{I}' = \begin{bmatrix} 5 & 15 \\ 15 & 56 \end{bmatrix}.$$

Using Eq. (14.22), we get

$$\tilde{\mathbf{a}}_R = \begin{bmatrix} 1.96 \\ 0.55 \end{bmatrix}.$$

For bias, we use Eq. (14.25) with $\lambda = 1$ and get that the bias is,

$$B(\tilde{\mathbf{a}}_R) = \begin{bmatrix} 0 & 0.27 \\ 0 & -0.09 \end{bmatrix}\mathbf{a}.$$

(continued)

14.3 Ridge Regression

Example 14.8 (continued)

If we assume that $\mathbf{a} = \tilde{\mathbf{a}} = \begin{bmatrix} 1.8 \\ 0.6 \end{bmatrix}$, then

$$B(\tilde{\mathbf{a}}_R) = \begin{bmatrix} 0.16 \\ -0.05 \end{bmatrix}.$$

For variance, we use Eq. (14.27) with $\lambda = 1$, giving

$$Var(\tilde{\mathbf{a}}_R) = \sigma^2 \begin{bmatrix} 0.94 & -0.25 \\ -0.25 & 0.08 \end{bmatrix}.$$

c) For $\lambda = 10$, we need matrix $\lambda \mathbf{I}'$ which is

$$\begin{bmatrix} 0 & 0 \\ 0 & 10 \end{bmatrix}$$

So

$$\mathbf{X}^T \mathbf{X} + \lambda \mathbf{I}' = \begin{bmatrix} 5 & 15 \\ 15 & 65 \end{bmatrix}.$$

Using Eq. (14.22), we get

$$\tilde{\mathbf{a}}_R = \begin{bmatrix} 2.7 \\ 0.3 \end{bmatrix}$$

For bias, we use Eq. (14.25) with $\lambda = 10$ and get that the bias is,

$$B(\tilde{\mathbf{a}}_R) = \begin{bmatrix} 0 & 1.5 \\ 0 & -0.5 \end{bmatrix} \mathbf{a}.$$

If we again assume that $\mathbf{a} = \tilde{\mathbf{a}} = \begin{bmatrix} 1.8 \\ 0.6 \end{bmatrix}$, then

$$B(\tilde{\mathbf{a}}_R) = \begin{bmatrix} 0.9 \\ -0.3 \end{bmatrix}.$$

(continued)

Example 14.8 (continued)
For variance, we use Eq. (14.27) with $\lambda = 10$, giving

$$Var(\tilde{\mathbf{a}}_R) = \sigma^2 \begin{bmatrix} 0.425 & -0.075 \\ -0.075 & 0.025 \end{bmatrix}.$$

Notice that we get three different lines for the three different values of λ. These are shown in Fig. 14.6. If you look at the three values of bias we have calculated, you will see that the absolute value increases as λ increases. Also, if you look at the three matrices of values for the variance, you will see that the variance gets smaller as λ increases. So, we have demonstrated that as λ increases, the bias gets larger and the variance gets smaller.

Remark 14.5 As already remarked, the aim of ridge regression is to find a better result on the test set than an overfitted linear approximation that favours the training data too much. This is where the validation set comes in. We could use different values for λ and then test each on the validation data. The value of λ that gives the best result on the validation data would be the one used for the final test on the real test data.

♦

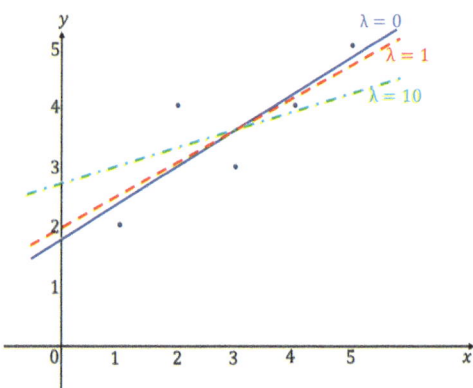

Fig. 14.6 Regression lines for Example 14.8 with $\lambda = 0$ (solid line), $\lambda = 1$ (dashed line), and $\lambda = 10$ (dash-dotted line)

> **Exercise**
>
> **14.6** Perform ridge regression on the $d = 1$, $N = 4$ example with the four points as $(1, 3)$, $(2, 3)$ and $(3, 2)$ and $(4, 1)$. Find the value of $\tilde{\mathbf{a}}_R$ or $\tilde{\mathbf{a}}$, as appropriate, the bias and the variance for (a) $\lambda = 0$ and (b) $\lambda = 10$, respectively. For $\lambda = 10$ assume that $\mathbf{a} = \tilde{\mathbf{a}}$ to get a value for the bias.

14.4 Early Stopping

When training a neural network or linear regression model using the gradient descent method, people usually need to pre-set the number of iterations before the training. After the weights are updated at each iteration, the error on the training and validation sets can be calculated. We can then produce learning curves by plotting these errors against the number of training iterations, as shown in Fig. 14.7. As we can see, as the number of iterations increases, the validation error decreases first and then increases while the training error keeps falling to convergence. Since we want the model to perform well on the validation set and then later on the test set, the training should stop when the training error reaches a reasonably small value, and the validation error reaches a minimum before going up. The number of training iterations that gives the minimum error on the validation set is the optimal number of iterations, presented as the dashed line shown in Fig. 14.7. Training that is stopped at the optimal number of iterations is called early stopping, another technique to prevent overfitting.

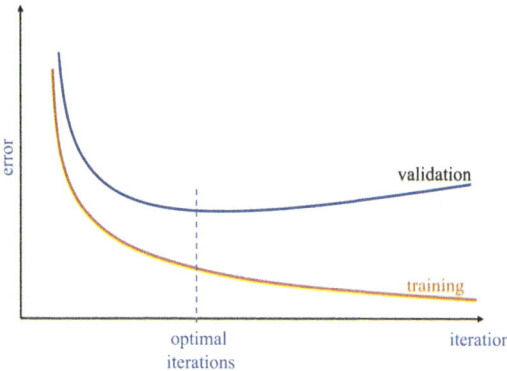

Fig. 14.7 Illustration of learning curves

Algorithm 1 displays the pseudocode for implementing the early-stopping method. First, we initialise a model and pre-set the number of training iterations and the minimum validation error as a huge value. As the training continues, we check whether the validation error is smaller than the previous one. If not, the training stops.

Algorithm 1 Pseudocode for the Early Stopping Algorithm

Initialising a model
Pre-set a number of iterations: Items
best_iteration = None
best_model = None
minimum_validation_error = infinite
for <iteration in 1:Items> **do**
   <model: update the weights>
   <calculate the training error>
   <calculate the validation error>
   if validation error $\leq$ minimum validation error **then**
     minimum_validation_error = validation error
     best_iteration = iteration
     best_model = model
   else
     iteration = Items
   end if
end for

Solutions

Problems of Chap. 1

1.1 d.

1.2

(1) Unstructured.
(2) Unstructured.
(3) Unstructured.
(4) Structured.

1.3

(1) Qualitative.
(2) Qualitative.
(3) Quantitative.
(4) Qualitative.

1.4

(1) Nominal.
(2) Nominal.
(3) Ordinal.
(4) Nominal.
(5) Ordinal.

1.5

(1) Ordinal
(2) Nominal.
(3) Interval.

(4) Ratio.
(5) Interval.
(6) Ratio.

Problems of Chap. 2

2.1

(1) True.
(2) True.
(3) False.
(4) False.
(5) True.
(6) False.
(7) True.

2.2

(1) 6.
(2) 0.
(3) 1.

2.3

(1) $\{\{\}, \{-1\}, \{1\}, \{-1, 1\}\}$. Cardinality is $2^2 = 4$.
(2) $\{\{\}, \{0\}, \{1\}, \{2\}, \{3\}, \{0, 1\}, \{0, 2\}, \{0, 3\}, \{1, 2\}, \{1, 3\}, \{2, 3\},$
$\{0, 1, 2\}, \{0, 1, 3\}, \{0, 2, 3\}, \{1, 2, 3\}, \{0, 1, 2, 3\}\}$. Cardinality is $2^4 = 16$.
(3) $2^8 = 256$.

2.4

(1) $A \cup B = \{b, c, e, f, g, h, i, k\}$.
(2) $C \cap B = \{g, h, i\}$.
(3) $\overline{A \cup B} = \{a, d, j, l, m\}$.
(4) $A \backslash (B \cap C) = \{b, c, e, f\}$.
(5) $A \cup B \cup C = \{b, c, e, f, g, h, i, k, l, m\}$.
(6) $A \cap B \cap C = \{g, h\}$.
(7) $(A \cup B) \backslash C = \{b, c, e, k\}$.
(8) $\overline{(A \cup B) \backslash C} = \{a, d, f, g, h, i, j, l, m\}$.

2.5

(1) $A = \{0, 1, 2, 3, 4, 5, 6, 7, 8, 9, 10\}$.
(2) $B = \{6, 7, 8, 9, 10, 11\}$.
(3) $C = \{0, 1, 2, 3, 4\}$.
(4) $D = \{0, 1, 2, 3, 4, 5, 6, 7\}$.
(5) $E = \{4, 5, 6, 7, 8, 9, 10, 11\}$.

2.6

(1) False.
(2) True.
(3) False.
(4) True.

2.7

(1) $\{(2, 0), (2, 1), (3, 0), (3, 1), (5, 0), (5, 1)\}$.
(2) $\emptyset$.

2.8

(1) True.
(2) False.
(3) False.
(4) True.

2.9

(1) Odd.
(2) Neither odd nor even.
(3) Neither odd nor even.
(4) Neither odd nor even.
(5) Odd.
(6) Even.
(7) Neither odd nor even.
(8) Even.
(9) Odd.

2.10

(1) $f^{-1}(x) = \sqrt[3]{x - 10}$.
(2) $f^{-1}(x) = \arcsin \frac{x}{3}$.
(3) $f^{-1}(x) = e^{(x-4)} - 1$.
(4) $f^{-1}(x) = \log_3 \frac{x}{1-x}$.

2.11

(1) $g \circ f(x) = 25x^2 + 20x + 4$.
(2) $g \circ f(x) = \sin(2x)$.
(3) $g \circ f(x) = e^{2x}$.
(4) $g \circ f(x) = x$.
(5) $g \circ f(x) = \cos^3 x$.

Problems of Chap. 3

3.1

(1) $\begin{bmatrix} 5 \\ 7 \\ 6 \end{bmatrix}$, and $\begin{bmatrix} -1 \\ -5 \\ 4 \end{bmatrix}$.

(2) $\begin{bmatrix} 4 \\ 2 \\ 10 \end{bmatrix}$, and $\begin{bmatrix} -4 \\ -2 \\ -10 \end{bmatrix}$.

3.2

(1) -5.
(2) -11.
(3) 0.
(4) 0.

3.3

(1) $d(\mathbf{w}, \mathbf{z}) = 6\sqrt{2}$,
(2) $d(\mathbf{a}, \mathbf{b}) = 10$.

3.4

(1) The direction of $\mathbf{u}$ is $\theta \approx -0.9828$ radians (or $\theta \approx 2.1588$ radians) and the direction of $\mathbf{v}$ is $\theta \approx 0.6747$ radians.
(2) $\|\mathbf{u}\| = \sqrt{13}$ and $\|\mathbf{v}\| = \sqrt{41}$.
(3) $d(\mathbf{u}, \mathbf{v}) = \sqrt{58}$.

3.5

(1) $\mathbf{u} \cdot \mathbf{v} = 12$.
(2) $\mathbf{u} \cdot \mathbf{w} = 0$.
(3) $\mathbf{u} \cdot \mathbf{s} = -8$.
(4) $\mathbf{u} \cdot \mathbf{t} = 10$.

3.6

(1) $\hat{\mathbf{w}} = (\frac{2}{\sqrt{5}}, \frac{1}{\sqrt{5}})$.
(2) $\hat{\mathbf{s}} = (\frac{3}{\sqrt{10}}, \frac{1}{\sqrt{10}})$.
(3) $\hat{\mathbf{t}} = (\frac{3}{\sqrt{11}}, \frac{1}{\sqrt{11}}, \frac{-1}{\sqrt{11}})$.
(4) $\hat{\mathbf{v}} = (\frac{-1}{\sqrt{22}}, \frac{2}{\sqrt{22}}, \frac{4}{\sqrt{22}}, \frac{1}{\sqrt{22}})$.

3.7

(1) $\mathbf{U} + \mathbf{V} = \begin{bmatrix} -4 & 12 \\ 10 & 0.6 \\ 1 & -4 \end{bmatrix}$.

Solutions

(2) $2\mathbf{U} - 4\mathbf{V} = \begin{bmatrix} -2 & 12 \\ 14 & 1.2 \\ 2 & -14 \end{bmatrix}$.

(3) $-3\mathbf{U} + 2\mathbf{V} = \begin{bmatrix} 7 & -26 \\ -25 & -1.8 \\ -3 & 17 \end{bmatrix}$.

3.8

(1) $\begin{bmatrix} 20 & 10 & 2 \\ 17 & 8 & 14 \end{bmatrix}$.

(2) $\begin{bmatrix} 9 & 6 \\ 0 & 3 \\ 15 & -3 \end{bmatrix}$.

(3) $\begin{bmatrix} 16 \\ 10 \end{bmatrix}$.

(4) $\begin{bmatrix} 27 & 54 \\ 14 & 43 \end{bmatrix}$.

3.9

(1) 8.
(2) 14.

3.10

(1) 30.
(2) -3.
(3) 0.

3.11

(1) 80.6.
(2) -42.3.

3.12

(1) $\mathbf{A}^{-1} = -\frac{1}{18} \begin{bmatrix} 6 & 3 \\ -2 & -4 \end{bmatrix}$.

(2) The inverse matrix does not exist.

(3) $\mathbf{C}^{-1} = \frac{1}{30} \begin{bmatrix} 2 & -1 \\ 4 & 13 \end{bmatrix}$.

(4) $\mathbf{I}^{-1} = \mathbf{I}$.

3.13

(1) $\mathbf{A}^T = \begin{bmatrix} 1 & 4 & 7 \\ 2 & 5 & 0 \\ 10 & -1 & -3 \end{bmatrix}$.

(2) $\mathbf{B}^T = \begin{bmatrix} -1 & 0 \\ 4 & 5 \\ -13 & 8 \end{bmatrix}$.

(3) $\mathbf{C}^T = \begin{bmatrix} 10, & -2, & 23, & -1 \end{bmatrix}$.

(4) $\mathbf{D}^T = \begin{bmatrix} 1 \\ 0 \\ -0.7 \\ 10 \end{bmatrix}$.

3.14

Yes, these three vectors are orthogonal to each other.

3.15

(1) a. $\mathbf{Q}^T = \begin{bmatrix} \frac{1}{\sqrt{10}} & -\frac{3}{\sqrt{10}} \\ \frac{3}{\sqrt{10}} & \frac{1}{\sqrt{10}} \end{bmatrix}$.

b. $\mathbf{Q}^{-1} = \begin{bmatrix} \frac{1}{\sqrt{10}} & -\frac{3}{\sqrt{10}} \\ \frac{3}{\sqrt{10}} & \frac{1}{\sqrt{10}} \end{bmatrix}$.

c. $\mathbf{Q}$ is an orthogonal matrix since $\mathbf{Q}^T = \mathbf{Q}^{-1}$.

(2) a. $\mathbf{Q}^T = \begin{bmatrix} \frac{4}{\sqrt{5}} & -\frac{3}{\sqrt{5}} \\ \frac{3}{\sqrt{5}} & \frac{4}{\sqrt{5}} \end{bmatrix}$.

b. $\mathbf{Q}^{-1} = \frac{1}{5} \begin{bmatrix} \frac{4}{\sqrt{5}} & -\frac{3}{\sqrt{5}} \\ \frac{3}{\sqrt{5}} & \frac{4}{\sqrt{5}} \end{bmatrix}$.

c. $\mathbf{Q}$ is not an orthogonal matrix since $\mathbf{Q}^T \neq \mathbf{Q}^{-1}$. In fact, the columns of $\mathbf{Q}$ are orthogonal but not of unit norm.

3.16

(1) Yes, S_1 is a subspace of R^3.
(2) Yes, S_2 is a subspace of R^3.
(3) Yes, S_3 is a subspace of R^2.

3.17

(1) Linearly independent.
(2) Linearly independent.
(3) Linearly dependent.

3.18

(1) $\det(\mathbf{A}) = 0$.
(2) The columns are linearly dependent.
(3) No.
(4) The rank of $\mathbf{A} = 2$.

Problems of Chap. 4

4.1
Possible solutions are:

(1) $\lambda_1 = 6, \lambda_2 = 2, \mathbf{u}_1^T = [\frac{1}{\sqrt{10}}, \frac{3}{\sqrt{10}}]$, and $\mathbf{u}_2^T = [\frac{1}{\sqrt{2}}, -\frac{1}{\sqrt{2}}]$.
(2) $\lambda_1 = 6, \lambda_2 = 1, \mathbf{u}_1^T = [\frac{1}{\sqrt{17}}, \frac{4}{\sqrt{17}}]$, and $\mathbf{u}_2^T = [\frac{1}{\sqrt{2}}, -\frac{1}{\sqrt{2}}]$.
(3) $\lambda_1 = 2, \lambda_2 = -3, \mathbf{u}_1^T = [\frac{1}{\sqrt{5}}, -\frac{2}{\sqrt{5}}]$, and $\mathbf{u}_2^T = [-\frac{3}{\sqrt{10}}, \frac{1}{\sqrt{10}}]$.
(4) $\lambda_1 = 4, \lambda_2 = 3, \mathbf{u}_1^T = [-\frac{1}{\sqrt{2}}, \frac{1}{\sqrt{2}}]$ and $\mathbf{u}_2^T = [-\frac{3}{\sqrt{13}}, \frac{2}{\sqrt{13}}]$.
(5) $\lambda_1 = -6, \lambda_2 = -3, \lambda_3 = 4, \mathbf{u}_1^T = [\frac{2}{\sqrt{41}}, \frac{1}{\sqrt{41}}, \frac{-6}{\sqrt{41}}], \mathbf{u}_2^T = [\frac{1}{\sqrt{2}}, 0, -\frac{1}{\sqrt{2}}]$, and $\mathbf{u}_3^T = [\frac{4}{\sqrt{74}}, \frac{7}{\sqrt{74}}, \frac{3}{\sqrt{74}}]$.

4.2
We have complex eigenvalues. (Complex eigenvalues of a matrix with non-zero eigenvectors are beyond the scope of this book.)

4.3

(1) $\mathbf{U} = \begin{bmatrix} \frac{1}{\sqrt{10}} & \frac{1}{\sqrt{2}} \\ \frac{3}{\sqrt{10}} & \frac{-1}{\sqrt{2}} \end{bmatrix}$. So

$$\mathbf{D} = \begin{bmatrix} \frac{\sqrt{5}}{2\sqrt{2}} & \frac{\sqrt{5}}{2\sqrt{2}} \\ \frac{3}{2\sqrt{2}} & \frac{-1}{2\sqrt{2}} \end{bmatrix} \begin{bmatrix} 3 & 1 \\ 3 & 5 \end{bmatrix} \begin{bmatrix} \frac{1}{\sqrt{10}} & \frac{1}{\sqrt{2}} \\ \frac{3}{\sqrt{10}} & \frac{-1}{\sqrt{2}} \end{bmatrix} = \begin{bmatrix} 6 & 0 \\ 0 & 2 \end{bmatrix}.$$

(2) $\mathbf{U} = \begin{bmatrix} \frac{1}{\sqrt{5}} & \frac{-2}{\sqrt{5}} \\ \frac{2}{\sqrt{5}} & \frac{1}{\sqrt{5}} \end{bmatrix}$. So

$$\mathbf{D} = \begin{bmatrix} \frac{1}{\sqrt{5}} & \frac{2}{\sqrt{5}} \\ \frac{-2}{\sqrt{5}} & \frac{1}{\sqrt{5}} \end{bmatrix} \begin{bmatrix} 3 & 2 \\ 2 & 6 \end{bmatrix} \begin{bmatrix} \frac{1}{\sqrt{5}} & \frac{-2}{\sqrt{5}} \\ \frac{2}{\sqrt{5}} & \frac{1}{\sqrt{5}} \end{bmatrix} = \begin{bmatrix} 7 & 0 \\ 0 & 2 \end{bmatrix}.$$

The columns of $\mathbf{U}$ are orthogonal, since $\mathbf{u_1} \cdot \mathbf{u_2} = 0$, and both are of unit length so $\mathbf{U}$ is an orthogonal matrix.

4.4

(1) The mean of each variable is 0.
(2) The standard deviation of each variable is $\sqrt{5}$.
(3) The covariance between two variables is 4.
(4)

$$\begin{bmatrix} 5 & 4 \\ 4 & 5 \end{bmatrix}.$$

(5) $\lambda_1 = 9, \lambda_2 = 1, \mathbf{u}_1^T = [\frac{1}{\sqrt{2}}, \frac{1}{\sqrt{2}}]$, and $\mathbf{u}_2^T = [\frac{1}{\sqrt{2}}, \frac{-1}{\sqrt{2}}]$.
(6) The first principal component captures 90% of the total variation, and the second captures 10% of the total variation.
(7) $[3\sqrt{2}, 0]$.

4.5

One possible solution is shown as follows:

$$\mathbf{Y} = \begin{bmatrix} 3 & 3 \\ 0 & 0 \\ -3 & -3 \\ -1 & 1 \\ 1 & -1 \end{bmatrix} = \begin{bmatrix} \frac{1}{\sqrt{2}} & 0 \\ 0 & 0 \\ -\frac{1}{\sqrt{2}} & 0 \\ 0 & -\frac{1}{\sqrt{2}} \\ 0 & \frac{1}{\sqrt{2}} \end{bmatrix} \begin{bmatrix} 6 & 0 \\ 0 & 2 \end{bmatrix} \begin{bmatrix} \frac{1}{\sqrt{2}} & \frac{1}{\sqrt{2}} \\ \frac{1}{\sqrt{2}} & -\frac{1}{\sqrt{2}} \end{bmatrix}^T.$$

4.6

16.30.

4.7

$\lambda_1 = \frac{6^2}{5-1} = 9$ and $\lambda_2 = \frac{2^2}{5-1} = 1$.

Problems of Chap. 5

5.1

(1) 1.
(2) 3.
(3) 0.
(4) 6.
(5) $\frac{1}{2}$.

5.2

(1) 0, $2x$.
(2) 1, $2x + 1$.

5.3

(1) 1.
(2) $6x^5$.
(3) $10e^x$.
(4) $\frac{5}{x}$.
(5) $\frac{1}{x}\sin x + \cos x \ln x$.
(6) $\frac{e^x \cos x + e^x \sin x}{(\cos x)^2}$.
(7) $10e^{(10x+1)}$.

(8) $8e^{2x}$.
(9) $5e^{3x}\cos 5x + 3e^{3x}\sin 5x$.
(10) $\frac{1}{xe^{(x^2)}} - \frac{2x\ln(8x)}{e^{(x^2)}}$.

5.4

(1) $6x\ln x + 5x$.
(2) $a^2 y$.
(3) $a^2 y$.

5.5

(1) $f(x)$ has the maximum value of 31 at $x = -2$ and the minimum value of -77 at $x = 4$.
(2) $f(x)$ has the maximum value of 1 at $x = 0$ and the minimum value of $\frac{3}{4}$ at $x = \frac{1}{2}$.
(3) $f(x)$ has the maximum value of 32 at $x = 2$ and the minimum value of -32 at $x = -2$.
(4) $f(x)$ has the minimum value of 3 at $x = -\frac{1}{2}$.
(5) $f(x)$ has the maximum value of 11 at $x = 1$ and the minimum value of -5 at $x = -1$ and $x = 3$, respectively.
(6) $f(x)$ has the maximum value of $e^{\pi/2} \approx 4.81$ at $x = \frac{\pi}{2}$ and the minimum value of $-e^{\frac{3\pi}{2}} \approx -111.32$ at $x = \frac{3\pi}{2}$.
(7) $f(x)$ has the maximum value of e^{-1} at $x = 1$.
(8) $f(x)$ has the maximum value of $4e^{-2} \approx 0.54$ at $x = 2$ and the minimum value of 0 at $x = 0$.

5.6

(1) $\frac{e^{2x}}{2} + C$.
(2) $-2x^4 + C$.
(3) $6\ln|x| + C$.
(4) $-3e^{-x} + C$.
(5) $-5\cos x + C$.

5.7

(1) $-4\cos x + e^x + C$.
(2) $\frac{\pi^2}{18} + \frac{\sqrt{3}}{2}$.
(3) $\frac{e^x - e^{-x}}{2} + C$.

5.8

(1) Substituting $u = 2 - x^2$ gives $-\frac{(2-x^2)^{\frac{3}{2}}}{3} + C$.
(2) Substituting $u = x - 1$ gives $-\frac{1}{x-1} + C$.
(3) Substituting $u = 4x$ gives $\frac{1}{4}$.
(4) Substituting $u = 1 + x^3$ gives $-\frac{1}{3(1+x^3)} + C$.

(5) Substituting $u = 1 + x$ gives $\frac{1}{8}$.
(6) Substituting $u = \sin x$ gives $\frac{3}{2}$.
(7) Substituting $u = 1 + x^2$ gives $\frac{\ln|1+x^2|}{2} + C$.
(8) Substituting $u = x^2$ gives $\frac{1}{2}e^{x^2} + C$.

5.9

(1) $(x-1)e^x + C$.
(2) $-x^2 \cos x + 2x \sin x + 2 \cos x + C$.
(3) $\frac{x^4}{4}(\ln x - \frac{1}{4}) + C$.
(4) $2\ln 2 - 1$.
(5) $-\frac{x \cos 4x}{4} + \frac{\sin 4x}{16} + C$.
(6) $x^2 \ln 5x - \frac{x^2}{2} + C$.
(7) $\frac{x - \sin x \cos x}{2} + C$.

Problems of Chap. 6

6.1

(1) $\frac{\partial f}{\partial x} = 3x^2 y + 10xy^2 + 2y^3$,
$\frac{\partial f}{\partial y} = x^3 + 10x^2 y + 6xy^2$,
$\left.\frac{\partial f}{\partial x}\right|_{\substack{x=3 \\ y=-1}} = 1$,
$\left.\frac{\partial f}{\partial y}\right|_{\substack{x=3 \\ y=-1}} = -45$.

(2) $\frac{\partial f}{\partial x} = 2x \sin y - 3 \cos y$,
$\frac{\partial f}{\partial y} = x^2 \cos y + 3x \sin y$,
$\left.\frac{\partial f}{\partial x}\right|_{\substack{x=1 \\ y=\frac{\pi}{2}}} = 2$,
$\left.\frac{\partial f}{\partial y}\right|_{\substack{x=1 \\ y=\frac{\pi}{2}}} = 3$.

(3) $\frac{\partial f}{\partial x} = 2y^3 e^{2x} + 3y^2 e^{3x} + 4ye^{4x}$,
$\frac{\partial f}{\partial y} = 3y^2 e^{2x} + 2ye^{3x} + e^{4x}$,
$\left.\frac{\partial f}{\partial x}\right|_{\substack{x=0 \\ y=2}} = 36$,
$\left.\frac{\partial f}{\partial y}\right|_{\substack{x=0 \\ y=2}} = 17$.

6.2

The maximum error in the area is 0.075 cm^2, and this represents 0.3% error.

6.3

(1) $\frac{\partial f}{\partial x} = 12x^3y + 18x^2y^2 - 8xy^3 + y^4$,
$\frac{\partial^2 f}{\partial x^2} = 4y(9x^2 + 9xy - 2y^2)$,
$\frac{\partial f}{\partial y} = 3x^4 + 12x^3y - 12x^2y^2 + 4xy^3$,
$\frac{\partial^2 f}{\partial y^2} = 12x(x^2 - 2xy + y^2)$,
$\frac{\partial^2 f}{\partial x \partial y} = 12x^3 + 36x^2y - 24xy^2 + 4y^3$.

(2) $\frac{\partial f}{\partial x} = 2x\sin y + 18x^2 \cos y$,
$\frac{\partial^2 f}{\partial x^2} = 2\sin y + 36x \cos y$,
$\frac{\partial f}{\partial y} = x^2 \cos y - 6x^3 \sin y$,
$\frac{\partial^2 f}{\partial y^2} = -x^2(\sin y + 6x \cos y)$,
$\frac{\partial^2 f}{\partial x \partial y} = 2x \cos y - 18x^2 \sin y$.

(3) $\frac{\partial f}{\partial x} = 2xe^{(x^2+y^2)}$,
$\frac{\partial^2 f}{\partial x^2} = 2(2x^2 + 1)e^{(x^2+y^2)}$,
$\frac{\partial f}{\partial y} = 2ye^{(x^2+y^2)}$,
$\frac{\partial^2 f}{\partial y^2} = 2(2y^2 + 1)e^{(x^2+y^2)}$,
$\frac{\partial^2 f}{\partial x \partial y} = 4xye^{(x^2+y^2)}$.

(4) $\frac{\partial f}{\partial x} = 3x^2 \ln(x^3 + y^3) + 3x^2$,
$\frac{\partial^2 f}{\partial x^2} = 6x(\ln(y^3 + x^3) + 1) + \frac{9x^4}{y^3+x^3}$,
$\frac{\partial f}{\partial y} = 3y^2 \ln(x^3 + y^3) + 3y^2$,
$\frac{\partial^2 f}{\partial y^2} = 6y(\ln(y^3 + x^3) + 1) + \frac{9y^4}{y^3+x^3}$,
$\frac{\partial^2 f}{\partial x \partial y} = \frac{9x^2y^2}{x^3+y^3}$.

(5) $\frac{\partial f}{\partial x} = 2x - \frac{3y}{x^2}$,
$\frac{\partial^2 f}{\partial x^2} = 2 + \frac{6y}{x^3}$,
$\frac{\partial f}{\partial y} = \frac{3}{x}$,
$\frac{\partial^2 f}{\partial y^2} = 0$,
$\frac{\partial^2 f}{\partial x \partial y} = -\frac{3}{x^2}$.

6.4

(1) $\frac{\partial z}{\partial t} = e^{\sin t} \ln(\cos t) \cos t - \frac{e^{\sin t} \sin t}{\cos t}$.

(2) $\frac{\partial z}{\partial s} = 4s(s^2 + t^2) \sin(st^2) + t^2((s^2 + t^2)^2 + 1)\cos(st^2)$,
$\frac{\partial z}{\partial t} = 2st((s^2 + t^2)^2 + 1)\cos(st^2) + 4t(s^2 + t^2)\sin(st^2)$.

6.5

(1) The gradient vector at the point (1, 1) is [3, 3],
   The gradient vector at the point (2, 1) is [12, 24],
   The gradient vector at the point (1, 2) is [24, 12].
(2) The gradient vector at the point (0, 1) is [0, 2],
   The gradient vector at the point (1, 0) is [0, 1],
   The gradient vector at the point $(\frac{\pi}{2}, \frac{\pi}{2})$ is [0.674, 0],
   The gradient vector at the point $(\frac{\pi}{4}, \frac{\pi}{4})$ is [0.675, 1.547].

6.6

(1) $J = \begin{bmatrix} 2x \sin y & x^2 \cos y \\ y^3 \cos x & 3y^2 \sin x \end{bmatrix}$.

(2) $J = \begin{bmatrix} \cos \theta & -r \sin \theta \\ \sin \theta & r \cos \theta \end{bmatrix}$.

(3) $J = \begin{bmatrix} \sin \theta \cos \phi & r \cos \phi \cos \theta & -r \sin \phi \sin \theta \\ \sin \phi \sin \theta & r \sin \phi \cos \theta & r \sin \theta \cos \phi \\ \cos \theta & -r \sin \theta & 0 \end{bmatrix}$.

6.7

(1) $\mathcal{H} = \begin{bmatrix} y^2 e^x + 2e^y & 2xe^y + 2ye^x \\ 2xe^y + 2ye^x & x^2 e^y + 2e^x \end{bmatrix}$.

(2) $\mathcal{H} = \begin{bmatrix} 6xy^2 z & 6x^2 yz - 2z^3 & 3x^2 y^2 - 6yz^2 \\ 6x^2 yz - 2z^3 & 2x^3 z & 2x^3 y - 6xz^2 \\ 3x^2 y^2 - 6yz^2 & 2x^3 y - 6xz^2 & -12xyz \end{bmatrix}$.

6.8

(2).

6.9

(1) The critical points are at $(-1, -2)$, $(-1, 1)$, $(2, -2)$, $(2, 1)$. The local maximum value at $(-1, -2)$ is 27. The local minimum value at $(2,1)$ is -27.
(2) The critical points are at $(\frac{1}{3}, -\frac{1}{3})$, $(1, -1)$. The local maximum value at $(1, -1)$ is 6.
(3) The critical point is at $(-\frac{1}{3}, -\frac{1}{3})$. The local minimum value at this point is $\frac{1}{3}$.
(4) The critical points are at $(0, -3)$, $(0, 2)$, $(3, -3)$, $(3, 2)$. The local maximum value at $(0, -3)$ is 85. The local minimum value at $(3,2)$ is -67.

6.10

(1)

$$\begin{cases} x = -\frac{1}{2} \\ y = \frac{1}{2} \\ \lambda = 1. \end{cases}$$

The relative extreme of the function is 1.5 obtained at $(-\frac{1}{2}, \frac{1}{2})$ subject to the given constraint.

(2)
$$\begin{cases} x = \frac{1}{2} \\ y = \frac{1}{2} \\ \lambda = -\frac{3}{4}. \end{cases}$$

The relative extreme of the function is 0.25 obtained at $(\frac{1}{2}, \frac{1}{2})$ subject to the given constraint.

6.11

(2).

6.12

(1) 1,
(2) $45\frac{1}{3}$,
(3) 2.

6.13

(1) $\frac{1}{8}$,
(2) $\frac{5}{6}$,
(3) 74.4,
(4) $\frac{16}{315}$.

6.14

(1) $\pi(e^4 - 1) \approx 168.384$,
(2) 10,
(3) $\pi(1 - \cos 1) \approx 1.44$.

Problems of Chap. 7

7.4 Possible answers:

(1) With and without normalisation:

$$\lambda_1 = 1.25, \lambda_2 = 0.75.$$

$$\mathbf{u}_1 = \begin{bmatrix} \frac{1}{\sqrt{2}} \\ \frac{1}{\sqrt{2}} \end{bmatrix}, \mathbf{u}_2 = \begin{bmatrix} -\frac{1}{\sqrt{2}} \\ \frac{1}{\sqrt{2}} \end{bmatrix}.$$

(2) Without normalisation:

$$\lambda_1 = 2, \lambda_2 = 0.4.$$

$$\mathbf{u}_1 = \begin{bmatrix} \frac{1}{\sqrt{2}} \\ \frac{1}{\sqrt{2}} \end{bmatrix}, \mathbf{u}_2 = \begin{bmatrix} -\frac{1}{\sqrt{2}} \\ \frac{1}{\sqrt{2}} \end{bmatrix}.$$

With normalisation:

$$\lambda_1 = \frac{5}{3}, \lambda_2 = \frac{1}{3}.$$

$$\mathbf{u}_1 = \begin{bmatrix} \frac{1}{\sqrt{2}} \\ \frac{1}{\sqrt{2}} \end{bmatrix}, \mathbf{u}_2 = \begin{bmatrix} -\frac{1}{\sqrt{2}} \\ \frac{1}{\sqrt{2}} \end{bmatrix}.$$

(3) Without normalisation:

$$\lambda_1 = 1.8, \lambda_2 = 1.6, \lambda_3 = 1.4.$$

$$\mathbf{u}_1 = \begin{bmatrix} \frac{1}{\sqrt{2}} \\ 0 \\ -\frac{1}{\sqrt{2}} \end{bmatrix}, \mathbf{u}_2 = \begin{bmatrix} 0 \\ 1 \\ 0 \end{bmatrix}, \mathbf{u}_3 = \begin{bmatrix} \frac{1}{\sqrt{2}} \\ 0 \\ \frac{1}{\sqrt{2}} \end{bmatrix}.$$

With normalisation:

$$\lambda_1 = 1.125, \lambda_2 = 1, \lambda_3 = 0.875.$$

$$\mathbf{u}_1 = \begin{bmatrix} \frac{1}{\sqrt{2}} \\ 0 \\ -\frac{1}{\sqrt{2}} \end{bmatrix}, \mathbf{u}_2 = \begin{bmatrix} 0 \\ 1 \\ 0 \end{bmatrix}, \mathbf{u}_3 = \begin{bmatrix} \frac{1}{\sqrt{2}} \\ 0 \\ \frac{1}{\sqrt{2}} \end{bmatrix}.$$

Problems of Chap. 8

8.1

(1) $\hat{a}_0 = -1.0$ and $\hat{a}_1 = 2.0$.
(2) $\hat{a}_0 = 2.0$ and $\hat{a}_1 = 0.5$.
(3) $\hat{a}_0 = 5.0$ and $\hat{a}_1 = -1.25$.
(4) $\hat{a}_0 = 0.75$ and $\hat{a}_1 = 0.75$.

8.3

After the first iteration, $a_0 = 0.97$ and $a_1 = 0.89$. The total error is 1.4428, and the total partial derivatives are $\frac{\partial \text{error}}{\partial a_0} = 1.92$ and $\frac{\partial \text{error}}{\partial a_1} = 6.88$, keeping two decimal places, respectively.

8.4

(1) The sum of the residuals is 0. The sum of the target values is 7.5. The sum of the estimates is 7.5. $R^2 \approx 0.89$.
(2) The sum of the residuals is 0. The sum of the target values is 9. The sum of the estimates is 9. $R^2 = 0.75$.

Problems of Chap. 9

9.1

Initially: $y_1 = 0.5$, $y_2 = 0.25$ and $E = 0.0625$.
After the first iteration: $w_{11} = 0.475$, $w_{12} = 0.5$, $w_{21} = 0.275$, and $w_{22} = 0.25$.
Now: $y_1 = 0.475$, $y_2 = 0.275$ and $E = 0.051$.

9.2

Initially: $y_1 = 0.622$, $y_2 = 0.562$ and $E = 0.0711$.
After the first iteration: $w_{11} = 0.491$, $w_{12} = 0.5$, $w_{21} = 0.248$, and $w_{22} = 0.25$.
Now: $y_1 = 0.620$, $y_2 = 0.562$ and $E = 0.0703$.

9.3

Initially: $y_1 = 0.188$, $y_2 = 0.375$ and $E = 0.0097$.
After the first iteration:
$w_{11}^{(1)} = 0.508$, $w_{12}^{(1)} = 0.50$, $w_{21}^{(1)} = 0.258$, and $w_{22}^{(1)} = 0.25$.
$w_{11}^{(2)} = 0.253$, $w_{12}^{(2)} = 0.252$, $w_{21}^{(2)} = 0.506$, and $w_{22}^{(2)} = 0.503$.
Now: $y_1 = 0.194$ and $y_2 = 0.387$ and $E = 0.0080$.

Problems of Chap. 10

10.1

(a) The number of different "words" from Wales is $5! = 120$.
(b) The number of different "words" from Scotland is $8! = 40320$.

10.2 The number of different photos they can take is $4! = 24$.

10.3

The number of passwords that can be made is $\frac{10!}{6!} = 5040$.

10.4
The number of five digits that can be made is $P(6,5) - P(5,4) = 600$.

10.5
The number of fruit salads that can be made is $\frac{7!}{4!3!} = 35$.

10.6
The number of triangles that can be made is $\frac{12!}{3!9!} = 220$.

10.7
The probability of getting two heads is 0.375.

10.8
The probability of getting four different numbers is $\frac{6 \times 5 \times 4 \times 3}{6^4}$ or 0.2778.
The probability of getting five different numbers is $\frac{6 \times 5 \times 4 \times 3 \times 2}{6^5}$ or 0.0926.
The probability of getting six different numbers is $\frac{6 \times 5 \times 4 \times 3 \times 2 \times 1}{6^6}$ or 0.0154.

10.9
The probability that all six digits are different is $\frac{10 \times 9 \times 8 \times 7 \times 6 \times 5}{10^6}$ or 0.1512.

10.10
See Table S.1.

10.11
See Table S.2.

10.12
$a = \frac{3}{4}$.
$F_X(x) = \frac{3}{4}(x^2 - \frac{x^3}{3})$ for $0 \leq x < 2$.
Below $x = 0$ it is zero, above $x = 2$ it is 1.
$P(X \leq 1) = \frac{1}{2}$; $P(X \leq \frac{1}{2}) = \frac{5}{32}$; $P(X > \frac{1}{2}) = 1 - \frac{5}{32} = \frac{27}{32}$.

10.13
$m = \sqrt[8]{0.5}$ and $n = \sqrt[8]{0.95}$.

Table S.1 Answer to Exercise 10.10: the probability distribution of X, where X is the total value after throwing two fair dice

x	1	2	3	4	5	6	7	8	9	10	11	12	13
$f_X(x_k) = P(X = x_k)$	0	$\frac{1}{36}$	$\frac{2}{36}$	$\frac{3}{36}$	$\frac{4}{36}$	$\frac{5}{36}$	$\frac{6}{36}$	$\frac{5}{36}$	$\frac{4}{36}$	$\frac{3}{36}$	$\frac{2}{36}$	$\frac{1}{36}$	0
$F_X(x) = \sum_{x_k \leq x} f_X(x_k)$	0	$\frac{1}{36}$	$\frac{3}{36}$	$\frac{6}{36}$	$\frac{10}{36}$	$\frac{15}{36}$	$\frac{21}{36}$	$\frac{26}{36}$	$\frac{30}{36}$	$\frac{33}{36}$	$\frac{35}{36}$	1	1

Table S.2 Answer to Exercise 10.11: the probability distribution of X, where X is the number of heads when tossing four fair coins

x	-1	0	1	2	3	4	5
$f_X(x_k) = P(X = x_k)$	0	$\frac{1}{16}$	$\frac{4}{16}$	$\frac{6}{16}$	$\frac{4}{16}$	$\frac{1}{16}$	0
$F_X(x) = \sum_{x_k \leq x} f_X(x_k)$	0	$\frac{1}{16}$	$\frac{5}{16}$	$\frac{11}{16}$	$\frac{15}{16}$	1	1

10.14

(1) $E(Z) = \frac{3}{8}$,
(2) $E(-Z+2) = \frac{13}{8} = 1.625$,
(3) $E(Z^2) = \frac{49}{48}$.

10.15

(1) $E(Z + X_1 + X_3) = 3.875$,
(2) $E(ZX_4) = 1.125$,
(3) $E(3Z - 5) = -3.875$,
(4) $E(X_1 Z) = 0.625$,
(5) $E(X_2 + X_4 + Z) = 4.875$.

10.16

Let X be the random variable that represents the absolute difference between the two numbers. $E(X) = \frac{5}{3}$.

10.17

(1) $E(X) = \frac{8}{9}$.
(2) $E(X^2) = \frac{4}{5}$.

10.18

(1) $E(X^2) = \frac{1}{12}; E(X^4) = \frac{1}{80}$,
(2) $Var(2X^2) = \frac{1}{45}$,
(3) $Var(2X^2 + 5) = \frac{1}{45}$.

10.19

$\mu_X = 4.5$.

10.20

The probability that:

(1) exactly seven students out of 10 pass the module is about 0.13.
(2) exactly eight students out of 10 pass the module is about 0.28.
(3) exactly nine students out of 10 pass the module is about 0.35.
(4) exactly ten students out of 10 pass the module is about 0.20.

10.21

(1) The probability that the doctor will see five patients is about 0.16.
(2) The probability that the doctor will see six patients is about 0.16.
(3) The probability that the doctor will see seven patients is about 0.14.
(4) The probability that the doctor will see eight patients is about 0.10.

10.22

(1) The probability that the number of calls is exactly eight in one minute is about 0.0298.

(2) The probability that the number of calls is more than five per minute is about 0.215.

10.23
The probability that the student waits less than three minutes is 0.36.

10.24
About 16%.

10.25
About 160 students.

10.26

(1) $P(X < 8) = 0.9773$, $P(X < 0.5) = 0.3085$, and $P(0.5 < X < 8) = 0.6688$;
(2) $P(-1 < X < 5) = 68.27\%$;
(3) $C = \mu = 2$.

10.27
$\sigma = 3.125$.

Problems of Chap. 11

11.1

(2) 0.3264.
(3) 0.1841

11.2.
0.0228.

11.3
See Table S.3

11.4
See Table S.4.

11.5
See Table S.5.

Table S.3 The answer to Exercise 11.3

	X=0	X=1	$P(Y = y_i)$
Y=0	$\frac{9}{26}$	$\frac{9}{26}$	$\frac{9}{13}$
Y=1	$\frac{2}{13}$	$\frac{2}{13}$	$\frac{4}{13}$
$P(X = x_i)$	$\frac{13}{26}$	$\frac{13}{26}$	–

Solutions

Table S.4 The answer to Exercise 11.4

	X=0	X=1	$P(Y = y_i)$
Y=0	$\frac{1}{10}$	$\frac{3}{10}$	$\frac{2}{5}$
Y=1	$\frac{3}{10}$	$\frac{3}{10}$	$\frac{3}{5}$
$P(X = x_i)$	$\frac{2}{5}$	$\frac{3}{5}$	–

Table S.5 The answer to Exercise 11.5

	X=0	X=1	X=2	$P(Y = y_i)$
Y=0	$\frac{2}{50}$	$\frac{3}{50}$	$\frac{1}{10}$	$\frac{1}{5}$
Y=1	$\frac{3}{50}$	$\frac{9}{100}$	$\frac{3}{20}$	$\frac{3}{10}$
Y=2	$\frac{1}{10}$	$\frac{3}{20}$	$\frac{1}{4}$	$\frac{1}{2}$
$P(X = x_i)$	$\frac{1}{5}$	$\frac{3}{10}$	$\frac{1}{2}$	–

11.6

(1) $A = \frac{1}{64}$,

(2)
$$F_{XY}(x,y) = \begin{cases} \frac{1}{128}x^2y + \frac{1}{128}xy^2, & 0 < x < 4,\ 0 < y < 4 \\ 0, & \text{otherwise}. \end{cases}$$

(3) $P(0 \leq X < 2, 0 \leq Y < 2) = \frac{1}{8}$,
(4) $P(X + Y < 4) = \frac{1}{3}$.

11.7

$f_X(x) = e^{-x}$.

11.8

(1) $A = \frac{1}{2}$.
(2) $F_{XY}(x,y) = \frac{1}{2}\sin x \sin y + \frac{1}{2}\cos x \cos y - \frac{1}{2}\cos x - \frac{1}{2}\cos y + \frac{1}{2}$ for $0 < x < \frac{\pi}{2}$, $0 < y < \frac{\pi}{2}$; otherwise, $F_{XY}(x,y) = 0$.
(3) $P(0 \leq X < \frac{\pi}{4}, 0 \leq Y < \frac{\pi}{4}) = 1 - \frac{\sqrt{2}}{2}$.
(4) $f_X(x) = \frac{1}{2}\cos x + \frac{1}{2}\sin x$.
(5) $f_Y(y) = \frac{1}{2}\cos y + \frac{1}{2}\sin y$.

11.9

$E(X) = \frac{1}{2}$; $E(Y) = \frac{1}{3}$; $E(X, Y) = \frac{1}{6}$.
$Cov(X, Y) = 0$.

11.10

$E(X) = \frac{7}{3}$; $E(Y) = \frac{7}{3}$; $E(X, Y) = \frac{16}{3}$.
$Cov(X, Y) = -\frac{1}{9}$.

11.11
$$P = \frac{6!}{2!1!0!3!} \left(\frac{2}{10}\right)^2 \left(\frac{3}{10}\right)^1 \left(\frac{4}{10}\right)^0 \left(\frac{1}{10}\right)^3 = 0.00072.$$

11.12
The probability that a given woman will have an episode of depression by the age of 65 in America is 33.34%.
The probability that a given man will have an episode of depression by the age of 65 in America is 20%.

11.13
P(science|female) = 44.4%.
P(female|science) = 46.15%.
P(science|male) = 60.87%.

11.14
P(two heads-up | first comes heads up) = 50%.

11.15
P(son | daughter) = $\frac{2}{3}$.

11.16
$f_{Y|X}(y|x) = \frac{1}{x}$, $y \leq x < 2, 0 < x < 2$.

11.17
$A = 1$, $f_{X|Y}(x|y) = \sin x$, $f_{Y|X}(y|x) = \cos y$.

11.18
$E(Y|x) = \frac{3}{4}$; $E(Y^2|x) = \frac{3}{5}$.
$Var(Y|x) = \frac{3}{80}$.

11.19
A and B are independent, A and C are not independent, and B and C are not independent.

11.20
$B = \{1, 2, 3, 4\}$, $B = \{1, 2\}$, $B = \{1, 4\}$, $B = \{2, 3\}$, and $B = \{3, 4\}$.

11.21
$P = \frac{71}{120}$.
If all balls in one bag, $P = \frac{35}{57}$.

11.22
$P = \frac{2}{5}$.

11.23
$\frac{57}{277} \approx 20.6\%$.

11.24
$\frac{135}{276} \approx 48.9\%$.

Solutions

11.25

$\frac{285}{2279} \approx 12.51\%$.

11.26

$\frac{29}{92} \approx 31.52\%$.

Problems of Chap. 12

12.1

See Table S.6.

12.2

(1) 2.82
(2) 3.76
(3) 2.47
(4) 0.19
(5) 3.16
(6) 0.3

12.3

(1) $cov(x_1, x_2) = -0.125$ and $r(x_1, x_2) = -0.20$.
(2) $cov(x_1, x_3) = -0.01125$ and $r(x_1, x_3) = -0.19$.
(3) $cov(x_2, x_3) = -0.225$ and $r(x_2, x_3) = -0.24$.
(4) $cov(x_1, x_2) = 0.6$, $cov(x_1, x_3) = 0.055$, $cov(x_2, x_3) = 0.9375$, and $r(x_1, x_2) = 0.98, r(x_1, x_3) = 0.95, r(x_2, x_3) = 0.99$.

12.4

Class A and Class B have the same relative dispersion: 20%.

12.5

For the first dataset:

1. Mode = 87.
2. Median = 86.5.
3. IQR = 13.
4. 50 and 110 are outliers.

Table S.6 Answers to Exercise 12.1

Question number	Mean	Median	Mode
(1)	6.25	6	6
(2)	$\frac{23}{6}$	4	0 and 5
(3)	5.1	5.5	5.5
(4)	5.9	5.9	5.8 and 6.0
(5)	8	9	10

For the second dataset:
1. Mode = 55.
2. Median = 54.
3. IQR = 10.
4. 30, 72, and 80 are outliers.

12.6
23.58%.

12.7
49.72%.

12.8
30.85%.

12.9
92.36%.

12.10
(1) 58.32%.
(2) 85.08%.

12.11
(1) $t = 1.753$.
(2) $t = 2.602$.

12.12
(1) $\chi^2 = 21.03$.
(2) $\chi^2 = 26.22$.

12.13
The estimated μ is 1486.4; the estimated σ is 98.53.

12.14
The 95% confidence interval is [161.432, 164.568].
The 99% confidence interval is [160.939, 165.061].

12.15
The 95% confidence interval is [0.522, 0.658].
The 99% confidence interval is [0.500, 0.680].

12.16
The 90% confidence interval of σ is [5.196, 8.425].
The 95% confidence interval of σ is [4.997, 8.903].

12.17
The 95% confidence interval is [0.809, 0.831].

Solutions

12.18
The 90% confidence interval of σ is [0.056, 0.096].

12.19
The 99% confidence interval is [22.7, 23.1].

12.20
$t = -3.2$; reject H_0 at 0.05 significance level.

12.21
$t = -1.69$; do not reject H_0 at 0.05 significance level.

12.22
$t = -1.535$; do not reject H_0 at 0.1 significance level.

12.23
$t = 0.83$; do not reject H_0 at 0.01 significance level.

12.24
$t = 2.48$; reject H_0 at 0.05 significance level.

12.25
$t = 1.88$; do not reject H_0 at 0.01 significance level.

12.26
See Table S.7.
$\chi^2 = 84.75$; reject H_0 at 0.05 significance level.

12.27
$\chi^2 = 7.53$; do not reject H_0 at 0.05 significance level.

12.28
$\chi^2 = 9.62$; do not reject H_0 at 0.05 significance level.

12.29
$\chi^2 = 3.58$; do not reject H_0 at 0.05 significance level.

12.30
$\chi^2 = 3.91$; Reject H_0 at 0.05 signif

12.31
$\chi^2 = 3.8$; do not reject H_0 at 0.05 significance level.

12.32
$\chi^2 = 2.29$; do not reject H_0 at 0.05 significance level.

Table S.7 The distribution of Z of 360 throws

z_i	1	2	3	4	5	6
Observed frequency	90	145	35	50	15	25
Expected frequency	110	90	70	50	30	10

Problems of Chap. 13

13.1
The maximum likelihood estimate of λ is $\hat{\lambda} = \frac{n}{\sum_{i=1}^{n} x_i}$.

(1) $p(x = 1|\lambda = 1) \approx 0.37$, $p(x = 2|\lambda = 1) \approx 0.14$,
$p(x = 1, x = 2|\lambda = 1) = 0.0518$.
(2) $p(x = 1|\lambda = 2) \approx 0.27$, $p(x = 2|\lambda = 2) \approx 0.037$,
$p(x = 1, x = 2|\lambda = 2) = 0.01$.
(3) $\hat{\lambda} = \frac{2}{3}$, $p(x = 1|\lambda = \frac{2}{3}) \approx 0.34$, $p(x = 2|\lambda = \frac{2}{3}) \approx 0.18$,
$p(x = 1, x = 2|\lambda = \frac{2}{3}) = 0.0612$.

13.2

(1) $E(y|x = 1) = -1$.
(2) $\text{std}(y|x = 1) = 0.1$.

13.3
For the first dataset (the left one in Table 13.4), the fitted linear regression is

$$\tilde{y} = 9.14 - 1.93x + \epsilon,$$

where $\epsilon \sim \mathcal{N}(\mu = 0, \tilde{\sigma}^2 = 0.64)$.

For the second dataset (the right one in Table 13.4), the fitted linear regression is

$$\tilde{y} = 9.71 - 2.14x + \epsilon,$$

where $\epsilon \sim \mathcal{N}(\mu = 0, \tilde{\sigma}^2 = 4.57)$.

13.4
The fitted linear regression is

$$\tilde{y} = 0.1 + 0.48x + \epsilon,$$

where $\epsilon \sim \mathcal{N}(\mu = 0, \tilde{\sigma}^2 = 0.004)$.

13.5
The fitted linear regression is

$$\tilde{y} = 1.86 - 1.03x + \epsilon,$$

where $\epsilon \sim \mathcal{N}(\mu = 0, \tilde{\sigma}^2 = 0.16)$.
The 95% confidence interval for a_0 is $[-1.92, 5.64]$ and for a_1 is $[-3.89, 1.83]$. 95% confidence interval for σ^2 is $[0.032, 160]$.
y_{new} is 0.315. Its 95% confidence interval is $[-3.18, 3.81]$.

13.6
The 95% confidence interval for a_0 is $[-0.23, 0.43]$ and for a_1 is $[0.36, 0.60]$.
The 95% confidence interval for σ^2 is $[0.001, 0.16]$.
y_{new} is 1.3. Its 95% confidence interval is $[1.16, 1.44]$.

13.7
$a_0^{new} = 0.456$ and $a_1^{new} = 0.657$.

$$P(y_1 = 1|x_1) \approx 0.37,$$

$$P(y_2 = 1|x_2) \approx 0.45,$$

$$P(y_3 = 1|x_3) \approx 0.61,$$

$$P(y_4 = 1|x_4) \approx 0.66.$$

13.8
$a_0^{new} = 0.293$ and $a_1^{new} = 0.309$.

$$P(y_1 = 1|x_1) \approx 0.354,$$

$$P(y_2 = 1|x_2) \approx 0.427,$$

$$P(y_3 = 1|x_3) \approx 0.504,$$

$$P(y_4 = 1|x_4) \approx 0.653.$$

Problems of Chap. 14

14.1 :

For **X1**:
  Eigenvalues are 0.873 and 0.127.
  Eigenvectors are $[0.526, 0.850]$ and $[-0.850, 0.526]$.
For **X2**:
  Eigenvalues are 0.397 and 0.103.
  Eigenvectors are $[0.615, 0.788]$ and $[-0.788, 0.615]$.

14.2 :
  $f2$ and $f5$ have the largest correlation.
  Average absolute correlation for $f2$ is 0.473
  Average absolute correlation for $f5$ is 0.373
  Therefore, remove $f2$.

14.3:
For the first part:

$$\text{Accuracy rate} = \frac{984}{1000} = 0.984.$$

$$\text{Recall} = \frac{2}{10} = 0.2,$$

$$\text{Precision} = \frac{2}{10} = 0.2,$$

$$\text{F-score} = \frac{2 \times 0.2 \times 0.2}{0.2 + 0.2} = 0.2,$$

$$\text{FP rate} = \frac{8}{990} = 0.008.$$

$$\text{True-negative rate} = \frac{982}{982 + 8} = 0.992.$$

For the second part:

$$\text{Accuracy rate} = \frac{992}{1000} = 0.992.$$

$$\text{Recall} = \frac{10}{10} = 1,$$

$$\text{Precision} = \frac{10}{18} \approx 0.556 \approx 0.56,$$

$$\text{F-score} = \frac{2 \times 1 \times 0.556}{1 + 0.556} \approx 0.71,$$

$$\text{FP rate} = \frac{8}{990} = 0.008.$$

$$\text{True-negative rate} = \frac{982}{982 + 8} = 0.992.$$

14.4:

$$\text{Accuracy rate} = \frac{975}{1000} = 0.975.$$

Solutions

$$\text{Recall} = \frac{500}{510} \approx 0.980 \approx 0.98,$$

$$\text{Precision} = \frac{500}{515} \approx 0.971 \approx 0.97,$$

$$\text{F-score} = \frac{2 \times 0.980 \times 0.971}{0.980 + 0.971} \approx 0.98,$$

$$\text{FP rate} = \frac{15}{490} \approx 0.031.$$

$$\text{True-negative rate} = \frac{475}{490} \approx 0.969.$$

14.5 :
First part:

$$\text{Accuracy rate} = \frac{875}{1000} = 0.875.$$

$$\text{Recall} = \frac{20}{100} = 0.2,$$

$$\text{Precision} = \frac{20}{65} \approx 0.308 \approx 0.31,$$

$$\text{F-score} = \frac{2 \times 0.2 \times 0.308}{0.2 + 0.308} \approx 0.24,$$

$$\text{FP rate} = \frac{45}{900} = 0.05.$$

$$\text{True-negative rate} = \frac{855}{900} = 0.95.$$

Second part:

$$\text{Accuracy rate} = \frac{1000}{1000} = 1.$$

$$\text{Recall} = \frac{100}{100} = 1,$$

$$\text{Precision} = \frac{100}{100} = 1,$$

$$\text{F-score} = \frac{2 \times 1 \times 1}{1+1} = 1,$$

$$\text{FP rate} = \frac{0}{900} = 0.$$

$$\text{True-negative rate} = \frac{900}{900} = 1.$$

14.6:

(a) $\lambda = 0$:

$$\tilde{\mathbf{a}} = \begin{bmatrix} 4 \\ -0.7 \end{bmatrix}$$

$$B(\tilde{\mathbf{a}}) = \begin{bmatrix} 0 \\ 0 \end{bmatrix}.$$

$$Var(\tilde{\mathbf{a}}) = \sigma^2 \begin{bmatrix} 1.5 & -0.5 \\ -0.5 & 0.2 \end{bmatrix}.$$

(b) $\lambda = 10$:

$$\tilde{\mathbf{a}}_R = \begin{bmatrix} 2.83 \\ -0.23 \end{bmatrix}$$

$$B(\tilde{\mathbf{a}}_R) = \begin{bmatrix} 0 & 1.67 \\ 0 & -0.67 \end{bmatrix} \mathbf{a}.$$

If $\mathbf{a} = \tilde{\mathbf{a}}$, then

$$B(\tilde{\mathbf{a}}_R) = \begin{bmatrix} -1.2 \\ 0.5 \end{bmatrix}.$$

$$Var(\tilde{\mathbf{a}}_R) = \sigma^2 \begin{bmatrix} 0.39 & -0.055 \\ -0.055 & 0.022 \end{bmatrix}.$$

References

1. Statista Research Department, *Amount of data created, consumed, and stored 2010–2023, with forecasts to 2028*, accessed Nov 21, 2024. Available at: https://www.statista.com/statistics/871513/worldwide-data-created/
2. Ozdemir, S.: *Principles of Data Science*, Packt, Birmingham - Mumbai (2016).
3. Moss, G.P., Shah, A.J., Adams, R.G., Davey, N., Wilkinson, S.C., Pugh, W.J., Sun, Y.: The application of discriminant analysis and Machine Learning methods as tools to identify and classify compounds with potential as transdermal enhancers. *European Journal of Pharmaceutical Sciences*, **45**, 116–127 (2012).
4. Fisher, R.A.: The use of multiple measurements in taxonomic problems. *Annual Eugenics*, 7, Part II, 179–188 (1936); also in *Contributions to Mathematical Statistics* John Wiley, NY (1950).
5. Bishop, C.M.: *Neural Networks for Pattern Recognition*, Clarendon Press, Oxford (1995).
6. Courant, R., John, F.: *Introduction to Calculus and Analysis*, Volume One, Springer (1998).
7. Kenneth, H.R.,:*Discrete Mathematics and its Applications*, 7th edition, McGraw-Hill (2012)
8. Gilbert, S.: *Linear Algebra and Its Applications, Edition* Fourth, Brooks Cole (2006).
9. Manly, B.F.J., Navarro Alberto, J.A.,:*Multivariate Statistical Methods: A primer, Edition* Fourth, Chapman and Hall/CRC (2017).
10. Ross, S.: *A First Course in PROBABILITY*, 8th Edition, Prentice Hall (2010).
11. Miklavcic, S.J.:*An Illustrative Guide to Multivariable and Vector Calculus*, Springer (2020).
12. Spiegelhalter, D.:*The Art of Statistics: Learning from Data*, Pelican (2020).
13. Lancaster, H.O.:*The Chi-squared Distribution*, Wiley (1969).
14. Cox, D.R., Hinkley, D.V.:*Theoretical Statistics*, Chapman And Hall, London (1979).
15. Wooldridge,J.M.:*Introductory Econometrics - A Modern Approach*, Cengage (2018).
16. Ribeiro, R.P., Moniz, N.: Imbalanced regression and extreme value prediction. In *Machine Learning*. (2020), 109(9): 1803–1835.
17. Haibo H., Garcia, E.A.: Learning from Imbalanced Data. In *IEEE Transactions on Knowledge and Data Engineering*. (2009), 21(9): 1263–1284.
18. Chawla, N.V., Herrera, F., Garcia, S., Fernandez, A.: SMOTE for Learning from Imbalanced Data: Progress and Challenges, Marking the 15-year Anniversary. In *Journal of Artificial Intelligence Research*. (2018), 61: 863–905.
19. Tlamelo E., Thabiso M., Dimane M., Thabo S., Banyatsang M., Oteng, T.: A survey on missing data in machine learning. In *Journal of Big Data*. (2021), 8:140.

20. Kuhn, M., Johnson, K.: *Applied Predictive Modeling*, Springer (2013).
21. Domingos, P.: A Unified Bias-Variance Decomposition and its Applications In *Proceedings of the Seventeenth International Conference on Machine Learning*, (June 2000), 231–238.
22. Hoerl, A., Kennard, R.: Ridge Regression: Biased Estimation for Nonorthogonal Problems. In *Technometrics*, (1970), 12, 55–67.

Index

A
Activation function, 228
Antiderivative, 143
Arithmetic mean, 336
Associative property of multiplication, 60

B
Back-propagation, 230, 239
Basis, 87
Bayes' theorem, 330
Bernoulli distribution, 280
Bias, 428
Bias and variance of ridge regression coefficients, 436
Bias in neural network, 248
Binary relation, 33
Binomial coefficients, 252
Binomial distribution, 281
Bounded function, 41
Boxplot, 343

C
Cardinality, 24
Cartesian Product, 32
Central limit theorem, 296
Chain rule, 132
Characteristic polynomial, 92
Chi-square (χ^2) Distribution, 352
Chi-square test, 369
Classification, 3
Cochran's theorem, 399
Coefficient of determination, 225, 423
Coefficient of variation, 341

Combination, 254
Composite function, 45
Conditional mean, 324
Conditional probability, 319, 320, 323
Conditional variance, 324
Confidence intervals for means, 356
Confidence intervals for proportions, 360
Confidence Intervals for χ^2, 361
Confusion matrix, 424
Continuity, 127
Continuous random variables, 263
Continuous uniform distribution, 286
Covariance, 100, 310, 341
Covariance matrix, 101
Cumulative distribution function, 299, 305

D
Data normalisation, 194, 224
Data pre-processing, 415
Data Science, 1
Data visualisation, 10
Definite integral, 144
Degrees of freedom, 349
Delta rule, 247
Dependent variable, 36, 207
Derivative, 129
Determinant, 68
Diagonalisation, 97
Diagonal matrix, 68
Diagonal of a matrix, 67
Differential, 129
Differentiation of composite Function, 159
Discrete random variable, 260
Discrete uniform distribution, 279

Distributive properties, 61
Dot product, 50, 53
Double integral, 176

E
Early stopping, 441
Eigendecomposition, 91
Eigenvalue, 92
Eigenvector, 92
Even function, 42
Expected random vector of random variables, 395
Expected value, 268
Exponential function, 38

F
Factorial, 252
Feature selection, 420
Feed-forward propagation, 237
The first partial derivative, 155
FP rate, 425
F-score, 425
Function, 34, 35

G
Gaussian distribution, 288
Generalisation error, 429
Gradient, 161
Gradient descent algorithm, 173
Graph of a Function, 36

H
Hat matrix, 399
Hessian matrix, 164
Hyperparameter, 422

I
Idempotent, 398
Identity matrix, 71
Image compression ratio, 118
Imbalanced dataset, 416
Inconsistent data, 416
Indefinite integral, 144
Independent event, 327
Independent variable, 36, 207
Infinite sets, 24
Integral, 142
Integration by parts, 152
Integration by substitution, 148

Integration of double integrals using Polar coordinates, 181
Interquartile range, 342
Interval, 24
Interval estimation, 355
Interval level, 20
Inverse function, 43
Inverse matrix, 71

J
Jacobian matrix, 163, 182
Joint probability distribution, 299, 305
Joint probability mass function, 299, 305

L
Law of large numbers, 294
Law of total probability, 328
Least-squares estimation, 208
Left-singular vectors, 109
Likelihood function, 381
Limit, 121
Linear combination, 80
Linear function, 37
Linearly dependent, 83
Linearly independent, 83
Linear regression, 207, 216
Linear regression with maximum likelihood estimation, 384
Linear transformation, 62
Local maxima, 139, 166
Local minima, 139, 166
Logarithmic function, 39
Logic, 31
Logistic regression, 406
Logistic sigmoid activation function, 232
Lower quartile, 342

M
Marginal probability distribution, 300, 308
Matrix, 55
Matrix addition, 56
Matrix decomposition, 91
Matrix multiplication, 58
Matrix transposition, 72
Matrix transposition properties, 73
Maximum likelihood estimation, 379
Mean, 268
Mean squared errors, 423
Median, 336
Method of Lagrange multipliers, 169, 191
Missing data, 417

Index 475

Mode, 337
Monotonic function, 41
Multinomial distribution, 314
Multivariate normal distribution, 316
Mutual exclusivity, 326

N
Neural network, 14, 227
Nominal level, 19
Norm, 51
Normal distribution, 288
Normal equation, 218

O
Odd function, 42
Ordinal level, 20
Organised data, 17
Orthogonal matrix, 76
Outlier, 343
Overfitting, 433

P
Pearson correlation coefficient, 100, 341
Period of function, 43
Permutation, 253
Point estimation, 354
Poisson distribution, 284
Polynomial function, 37
Power set, 25
Precision, 425
Predictor, 207
Principal component, 101
Principal Component Analysis, 10, 99, 185
Probability, 256
Probability multiplication rule, 327
Procedure for conducting a hypothesis test, 365
The procedure of Data Science, 2
Proper subset, 25

Q
Qualitative, 18
Quantitative, 18

R
Range, 342
Rank, 88
Ratio level, 20
Reading a normal distribution table, 291

Recall, 425
Rectified linear activation unit, 250
Regression, 3
Regressor, 207
Regularisation term, 434
Relation, 33
Repeated data, 417
Representations of simultaneous equations, 64
Residual, 224
Ridge regression, 434
Right-singular vectors, 109
Root mean squared errors, 423

S
Sample mean, 99
Sampling distribution of estimators $\tilde{a}$, 399
Sampling distribution of the linear regression estimators, 394
Sampling distribution of variance $\tilde{\sigma}^2$, 402
Sampling distributions of means, 344
Sampling distributions of proportions, 347
Scalar multiplication, 57
Scalar-vector multiplication, 49
Scatter plots of residual against predictions, 423
The second derivative, 137
The second partial derivative, 158
Sensitivity, 425
Set, 23
Set complement, 29
Set intersection, 27
Set membership, 24
Set subtraction, 28
Sets written in comprehension, 30
Set union, 26
Sigmoid, 41
Sigmoid function, 134, 407
Singular matrix, 88
Singular value, 109
Singular value decomposition, 109
Span, 82
Spanning set, 83
Specificity, 425
Square matrix, 66
Standard deviation, 100, 339
Standard errors, 349
Standard normal distribution, 289
Student's t-distribution, 350
Subset, 25
Subspace, 79
Supervised learning, 3

T

Test set, 422
Trace, 67
Trade-off between the bias squared and variance, 433
Training set, 422
Trigonometric function, 39
True negative rate, 425
T-test, 366
Type I error, 364
Type II error, 364
Types of relation, 33

U

Underfitting, 433
Unit vector, 55
Unstructured data, 18
Unsupervised learning, 5
Upper quartile, 342

V

Validation set, 422
Variance, 100, 186, 275, 339, 428
Vector, 47
Vector addition, 49
Vector direction, 52
Vector magnitude, 52
Vector space, 78
Venn diagram, 26

W

Weights, 228

Z

Z-score, 290

The manufacturer's authorised representative in the EU is Springer Nature Customer Service Centre GmbH, Europaplatz 3, 69115 Heidelberg, Germany. If you have any concerns regarding our products, please contact ProductSafety@springernature.com

Printed and bound by CPI Group (UK) Ltd, Croydon, CR0 4YY
18/02/2026
02055594-0010